THE ECONOMIC AND SOCIAL EFFECTS OF THE SPREAD OF MOTOR VEHICLES

The Economic and Social Effects of the Spread of Motor Vehicles

An International Centenary Tribute

Edited by
Theo Barker
Professor Emeritus of Economic History
University of London

First published 1987

Published by
THE MACMILLAN PRESS LTD
Houndmills, Basingstoke, Hampshire RG21 2XS
and London
Companies and representatives
throughout the world

Printed in Hong Kong

British Library Cataloguing in Publication Data
The Economic and social effects of the spread of motor vehicles.
1. Transportation, Automotive
2. Transportation, Automotive — Social aspects
I. Barker, Theo
388.3'4   HE5611
ISBN 0–333–41299–0

Contents

Notes on the Editor and Contributors   vii

Editor's Preface   ix

1   A German Centenary in 1986, a French in 1995 or the Real Beginnings about 1905?   1
Theo Barker

2   The Beginnings of the Automobile in Germany   55
Otto Nübel

3   The Motor Vehicle and the Revolution in Road Transport: The American Experience   67
John B. Rae

4   The Early Growth of Long-distance Bus Transport in the United States   81
Margaret Walsh

5   Diesel Trucks and Buses: Their Gradual Spread in the United States   97
James M. Laux

6   The Automobile and the City in the American South   115
David R. Goldfield and *Blaine A. Brownell*

7   Some Economic and Social Effects of Motor Vehicles in France since 1890   130
Patrick Fridenson

8   Why did the Pioneer Fall Behind? Motorisation in Germany Between the Wars   148
Fritz Blaich

9   Motorisation on the New Frontier: The Case of Saskatchewan, Canada, 1906–34   165
G. T. Bloomfield

10   The Internal Combustion Engine and the Revolution in Transport: The Case of Czechoslovakia with some European Comparisons   194
Jaroslav Purš

11 Japan: the Late Starter Who Outpaced All Her Rivals 214
  Koichi Shimokawa

12 Motor Transport in a Developing Area: (i) Zaïre,
  1903–59 236
  Epanya Sh. Tshund'olela

13 Motor Transport in a Developing Area: (ii) Soviet
  Central Asia 256
  M. A. Akhunova, B. A. Tulepbaev and J. S. Borisov

14 Death on the Roads: Changing National Responses to
  Motor Accidents 264
  James Foreman-Peck

15 Advances in Road Construction Technology in France 291
  Dominique Barjot

Index 313

Notes on the Editor and Contributors

M. A. Akhunova is Professor and Director of Research at the Institute of History, Tashkent, and Corresponding Member of the Soviet Academy of Sciences.

Theo Barker is Professor Emeritus of Economic History in the University of London, still teaching on a part-time basis at the London School of Economics.

Dominique Barjot is Attaché de Recherche Agrégé, CNRS, Institute of Modern and Contemporary History, Caen.

Fritz Blaich is Professor of Economic History at the University of Regensburg.

Gerald T. Bloomfield is Professor of Geography at the University of Guelph, Canada.

Yu. S. Borisov is Professor and Head of Section in the Institute of History of the Soviet Union at the Academy of Sciences, Moscow.

Blaine A. Brownell is Professor of History and Urban Studies and Dean of Social and Behavioral Sciences, University of Alabama at Birmingham.

James Foreman-Peck is Lecturer in Economics at the University of Newcastle Upon Tyne.

Patrick Fridenson is Directeur d'Etudes at the Ecole des Hautes Etudes en Sciences Sociales, Paris.

David R. Goldfield is Robert Lee Bailey Professor of History, University of North Carolina at Charlotte.

James M. Laux is Professor of History at the University of Cincinnati.

Otto Nübel is Archivist of Daimler-Benz, Stuttgart.

Jaroslav Purš is Professor of History at Charles University, Prague, and Director of the Institute of Czechoslovak and World History at the Czechoslovak Academy of Sciences.

John B. Rae is Professor Emeritus of the History of Technology at Harvey Mudd College, Claremont, California.

Koichi Shimokawa is Professor of Business Administration at Hosei University, Tokyo, and Director of the Japan Business History Society

Epanya Sh. Tshun'olela is Associate Professor of History at the University of Lubumbashi, Zaïre.

B. A. Tulepbaev is Professor and Director of Research at the Institute of History, Moscow, and Member of the Soviet Academy of Sciences.

Margaret Walsh is Senior Lecturer in Economic and Social History at the University of Birmingham.

Editor's Preface

Professional historians still love to write about railways, the heyday of which lies so far in the past that only the elderly can remember it. Few interest themselves in the much more recent, rapidly growing and widespread development of motor transport, and those who do, concentrate upon motor vehicle (and almost exclusively motor car) manufacture. Their books are primarily business histories, studies in the spread of mass production and the new problems of management and industrial relations to which this gave rise. Very little indeed has been written in any organised way about the effects of motor vehicles of all sorts when they left the factory and took to the roads. It is as if the railway historians had confined their attention to the building of engines and rolling stock.

This is surprising, for it is not hard to demonstrate, as these pages will attempt to do, that the internal combustion engine in its locomotive form has had an increasing effect upon people's lives during the past century so that, during the past forty years, certainly in the wealthier parts of the world and probably in developing countries too, it has had a greater influence than railways ever had. Indeed, by the 1980s it is no longer unthinkable to argue that, taking the world as a whole, motorisation has had a greater effect than any other interconnected series of inventions in the history of mankind. We all appreciate the significance of the wheel. We have still to appreciate the greater significance of the mechanised wheel, running not on a fixed track but, much more freely, flexibly and economically, on the world's vastly improved roads.

Motor vehicles were first developed to replace horse-drawn transport over short distances as the railways had replaced it over medium and longer journeys. And, even on longer journeys, the railway had not by any means completely replaced the horse. Only in the case of goods traffic between large customers who owned their own sidings did railways provide a door-to-door service. All other loads had to be carried overland at the beginning and/or end of their journey; and all passengers had to travel by road, either on foot or in horse-drawn vehicles, to and from railway stations. Because of railway stimulation of short-distance traffic and also because of population growth, especially in towns, in only the very largest of which did people use trains to get about, horse-drawn goods and passenger transport by road vastly increased during the Railway Age.

It was this rich prize which the mechanisation of road transport was intended to win. Direct motor competition with the railways for their medium- and long-distance traffic was never originally envisaged. But, as roads were improved (and especially during the past forty years or so with purpose-built motor roads and larger, faster, more powerful and efficient vehicles to run upon them), the new technology gained more and more railway traffic, too. This book aims to show how and when these momentous developments took place in various parts of the world.

We are not, of course, dealing solely with private passenger transport, the motor car and the motor cycle, but also with the motor bus and the motor lorry. These heavier vehicles are now between one-fifth and one-sixth as numerous as the former in mature economies and a far higher proportion, often 1:1, in developing countries. They affect the lives of vast numbers of people all over the world who do not possess their own private vehicles, as well as the millions more who do.

It is hard to think of any major technological development which has not had some disadvantage, even if it was that of increasing the social division between those who could afford it and those who could not. The slaughter on the road, and the noise and air pollution associated with motor vehicles are obvious disadvantages; and in so far as private cars discriminate in favour of the 'haves' against the 'have-nots' by, for instance, causing a deterioration in public transport, motorisation has been socially divisive. The young and the old, the poor and the handicapped are being discriminated against. So are all those who wish to get about easily on foot or by bus in the centres of towns or suburbs, the streets of which are choked by private cars or narrowed by parking.

These disadvantages, as well as the gains, of motorisation are considered in these chapters and concentrated upon in one of them. Few of those who read this book, however, will fail to appreciate the enormous benefits which motorisation has bestowed. The motor vehicles now running on the world's roads have given increased freedom of movement to vast numbers of people, together with their families and friends. The private carriage is no longer the flaunted symbol of wealth, but the often treasured possession of a multitude of ordinary folk. For the growing number of fortunate, the ability to climb into a car or jump on to a motor bike to visit relatives or friends, or to go on some errand quickly and without delay, has given a new dimension to life. Moreover, the private vehicle makes it easier and cheaper to do this as a family or a group. The luggage or the child's push-chair are easily stowed in the car. No time-tables need to be looked up. There are no walks to the station or bus stop, or waits there, perhaps in the wind and the rain.

Or, if there are other goods to be carried, the motor vehicle can also oblige. Shopping habits have been greatly changed. Motors have made possible the supermarket in its most modern form and the hypermarket. They are supplied by motor lorry and customers carry away their purchases by car which can be accommodated in the spacious car parks adjacent. Those women, in particular, who no longer have to carry heavy baskets and bulky parcels, benefit particularly. Life in the countryside, too, has been greatly changed by closer links by car and bus with near-by towns. They were beginning to break down rural isolation even before the coming of radio and electricity.

In the wealthier parts of the world the benefits of motorisation will spread farther down the social scale as economies grow and as these countries realise the need to cater for the remaining 'have-nots'. Buses, for instance, can be adapted to meet modern needs by providing smaller, as well as larger, vehicles and equipping them with radio (more like taxis) which can bring to those who do not own their own private carriages more of the advantages of those who do. And the vast improvement in the speed, cost and comfort of long-distance coaches running on specially built motorways offer obvious advantages to all non-car owners. In less wealthy lands, where the private car is a much rarer sight, motorised public transport is still a necessity for all who can afford to ride; and this number will grow with the inhabitants' real earnings.

All these basic themes are central to the chapters which follow, whether they deal with the spread of diesel engines (which greatly improved and cheapened heavy, long-distance transport), with the effect of motor vehicles on cities in the American South (which deals with traffic congestion in the centre and the relocation of shops) or with the problems of traffic accidents. Such a wide geographical coverage of so many aspects of motorisation inevitably means that we are involved essentially in a ground-clearing exercise. It is the first, not the last, word. Our aim is to indicate the pace of change as the new technology spread and to see to what extent its economic and social results were similar, and to what extent different, in various countries depending upon differences in income per head, state activity, geography or other factors. It is written with the intention of arousing interest in a crucial topic which, despite a massive detailed literature, has up to now hardly been approached in the widely sweeping way which we have tried to do. We hope that others will adopt our approach for other countries which we have not covered, or draw attention to general benefits or disadvantages of motorisation which we may have overlooked or exaggerated.

A book on this scale could not have been written in such a short time by any one person. No doubt this is why it has never been attempted before. My first debt, therefore, is to the International Historical Congress, the meetings of which in Stuttgart in August 1985 gave me the opportunity to assemble a group of specialists in precisely that part of the world where our story began almost exactly a hundred years before. I must thank all the contributors for rallying round so nobly and letting me have their chapters on time.* In doing my best to translate the writing of those whose native tongue is not English into more idiomatic English, I have taken greater liberties with their texts than an editor would normally do. I hope they approve of the results even if they are sometimes rather shorter than the original copy.

Writers on this subject in different countries use the same words to mean different things. Automobile, for instance, is often used to mean more than *motor car*; we have tried to use the latter in the interests of precision. Most readers of English understand that *truck* and *lorry* are synonymous, though the former is more common in America and in those parts of the world that have taken to American English. The word *motorisation*, another American word, is a most convenient piece of shorthand; but we have set our face firmly against the monstrous *transportation*, preferring, whenever possible, the more economical and unambiguous *transport*. *Transportation* was the punishment meted out upon many British people convicted of crime in the eighteenth century which often soon gave them a chance in life they would not otherwise have enjoyed.

Individual acknowledgements will be found in the notes to the various chapters; but I should like here to thank my secretary, Jenny Law, for help in preparing some of the final typescript and Tess

---

*The papers on that occasion were summarised and appear in full here for the first time. A full recording of the Stuttgart session is available on three sound tapes from the Audio-Visual Aids Centre, University of Kent, Canterbury, UK.

Truman for typing and circulating the series of newsletters with which I kept in touch with the contributors while the chapters were being prepared. Other personal acknowledgments will be found at the end of Chapter 1.

THEO BARKER

London School of Economics and Political Science

1 A German Centenary in 1986, a French in 1995 or the Real Beginnings About 1905?

Theo Barker[1]

THE OVERVIEW

There are now over 470 million motor vehicles in use throughout the world. This huge and daunting total means that there is one vehicle for every ten people on earth: men, women and children. Although private cars and taxis preponderate (there are over 320 million of them), and motor cycles add over 60 million more, this leaves over 90 million commercial vehicles of various sorts: vans, lorries and buses.[2] Farm tractors have not been included, though they are sometimes referred to incidentally in subsequent chapters. Important as they are to agricultural development, they are not primarily vehicles which run upon public roads; they would add only a little over 21 million vehicles to the total.

This remarkable outpouring of the motor factories on to the roads of the world has not occurred at a steady pace over the past century. Most of it has taken place since the 1940s. Registration statistics collected at the beginning of 1940, which included China and the USSR, suggest that there were then only about 47½ million vehicles all told, including over 8 million lorries and buses and over 3 million motor cycles.[3] That is to say, there are now ten times as many motor vehicles now as there were then. Motorised two-wheelers now form a larger part of the whole and lorries and buses a somewhat smaller part. When the United Nations began its statistical series after the war, the grand total had already risen to nearly 55 million (12½ million of which were lorries and buses); and these figures excluded not only motor cycles but also all returns from the USSR, mainland China and the countries of Eastern Europe.[4] The true total must then (1948) have been appreciably more than 55 million.

1

The extent of motorisation since the 1940s has been most impressive; even more so when we bear in mind that, in comparing the totals
of the 1980s with those of the 1940s, we are not comparing like with
like. The vehicles of today are very different from those of forty years
ago; and so are most of the roads that they run upon. Present-day
cars are faster and more sophisticated. Lorries and buses are faster,
heavier, more powerful and capable of doing much more work than
their earlier counterparts. If it is an exaggeration to argue that motor
vehicles have had a major impact only since the 1940s, the great
waves of vastly improved mobile machines which have poured on to
the also vastly improved (and often completely new) roads since then
put the first sixty years into perspective.

This certainly *is* an exaggeration in so far as it concerns the United
States and, to a smaller extent, Canada, Autralia and New Zealand.
Motor-vehicle penetration near, or in excess of, 100 per 1 000 population in the 1920s, was already having a considerable economic and
social impact in those countries, as is shown by John Rae in his
chapter on America (Chapter 3) and by Gerald Bloomfield in his on
'Motorisation on the New Frontier' (Chapter 9). Professor Bloomfield has provided a graph (Figure 1.1) which shows this very clearly.
The United States was far and away ahead of everywhere else. If we
go back to 1921/2 when there were only about 15½ million motor
vehicles in the world, 12½ million of these were running upon the
roads of the United States.[5] Already, by 1927, 55 per cent of American families owned a car,[6] a proportion that was not reached in other
advanced countries, such as Britain, until the 1960s. By 1940, despite
greater progress in other parts of the world, the United States still
had more motor vehicles than the whole of the rest of the world put
together: over 30 million of the 47½ million.[7] The process of catching
up on America continued; but it was not until about 1960 that the
total of all the motor vehicles outside that rich country equalled
America's.[8] Those who wish to explore the effects of motorisation,
therefore, will need to examine the pattern first revealed in the
United States and then to see to what extent it was repeated when
motorisation reached similar levels elsewhere.

America's huge lead is explained very largely by the large number
of motor *cars* in the U.S. total. This, in its turn, is explained by the
far greater social depth of demand in America as well as by the
earlier availability of cheaper, mass-produced cars and plentiful
supplies of inexpensive motor spirit. Hence, in 1921/2, over 11
million cars figured in the U.S. total of 12½ million, together with 1.3

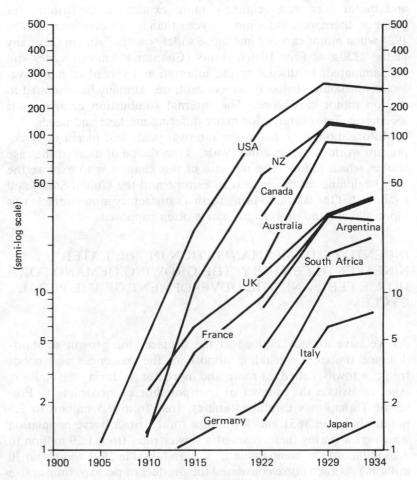

Figure 1.1   Motor vehicle adoption rates in selected countries
(motor vehicles per 1000 population)

Text-figure prepared by G. T. Bloomfield.

million lorries and only 210 000 motor cycles. For 1940 the comparable figures were 26 million, 4.3 million and 110 000. In the developing countries, however, with much lower purchasing power, motor
lorries figured more prominently and sometimes exceeded private
cars in number, as may be seen in Epanya Tshund'olela's particularly
interesting contribution on Zaïre (Chapter 12). Similarly, motor
bicycles, relatively unimportant in the United States where cars could

be afforded, found many more buyers in those countries where purchasing power was less widely spread throughout the community and motor cars were relatively more expensive. In Britain, for instance, there were more motor cycles than motor cars from 1919 to 1924 when motor cars became more widely owned;[9] and in Germany in the 1930s, as Fritz Blaich shows (Chapter 8), motor cycles still predominated. In discussing the international spread of motor vehicles we should obviously not concentrate, as many have tended to do, on motor cars alone. The internal combustion engine in its locomotive form catered for many differing markets and needs.

Motorisation, U.S.-led in the interwar years and increasingly important world-wide since the 1940s, is the theme of most of the case studies which follow. The purpose of this chapter is to explore the very beginnings of the process in Europe and the United States and to show that the labour to bring forth a satisfactory motor vehicle was more difficult and protracted than is often supposed.

INCENTIVES TO MECHANISATION IN THE LATER NINETEENTH CENTURY: THE GROWING DEMAND FOR HORSE FEED AND THE DEVELOPMENT OF THE PEDAL CYCLE

As we have already noticed in the Preface, the growth of short-distance transport, including almost all the passenger and goods traffic in towns, called for more and more horses during the Railway Age. In Britain the number of transport horses, according to Professor Thompson's careful estimates, rose from 0.5 million to 1.5 million between 1851 and 1901, in a total British horse population which grew during these years at a slower rate, from 1.29 million to 3.28 million. (The comparable U.S. grand total in 1901 was about 30 million.) As each horse consumed the product of perhaps four or five acres of farmland per year, all these animals were by then beginning to make very large demands upon the world's agricultural output.[10] Production of oats and hay may have been increased at an even faster rate as new land was opened up in different parts of the world during the last quarter of the nineteenth century; but there was an obvious limit to this process as the requirements of these large creatures competed with those of the world's human population, also increasing at a fast rate. A rich prize obviously awaited those who could discover and develop a mechanical replacement for horse transport over short distances in much the same way as the Stephensons and others had developed the steam railway to replace it over medium and long distances.

The cycle, propelled by human energy, not the carriage drawn by horse, was the important precursor, and indeed prerequisite, of the motor vehicle; and, as with the motor car, the French played a key role at an early stage.[11] In the 1860s a Frenchman, Pierre Michaux, had fitted a crank and pedal to the front wheel of a scooter, thus transforming what had been essentially a child's toy into something which could be developed into the modern pedal bicycle. After visitors to the 1869 International Exhibition in Paris had taken to it, it found purchasers in the rest of Europe and in the United States. This velocipede craze did not last for long, however. In the later 1870s, when the machine was further developed, it was the British who took the lead, especially after the marketing, in 1886, of the safety bicycle. The bicycle had then reached what was basically its present form, with diamond tubular frame, spoked metal wheels of equal diameter, low, centrally positioned pedals and chain drive to the rear wheel. Although Britain now led the new industry, located in Coventry and other parts of the Midlands where motor manufacture was to begin, cycle-making was also growing in the United States and in other parts of Europe. It was often associated, in Britain and elsewhere, with the sewing-machine business or with some branch of metal-working.

Several features of cycle (including not only bicycle but also tricycle and quadricycle) manufacture were soon to be adopted by those who made motor vehicles. On the *technical* side, the light, tubular frame was to be used in many of the earlier, low-powered three- or four-wheelers. Ball bearings, chain drive and, from the end of the 1880s, the pneumatic tyre, were also available. Standardisation of parts was a feature of cycle production, too. On the *marketing* side, the cycle-makers devised methods which were carried over to the selling of motors: regular changes of model at the beginning of each season; hire-purchase buying; the network of dealers and re-pairers; the organisation of shows and races (the first cycle race on a public road was between Paris and Rouen in 1869) and other forms of advertising; the growth of a specialised cycling press to stimulate and maintain interest. (The future Lord Northcliffe gained early experience on *Bicycling News* in Coventry.) Henry Sturmey, who knew the cycle industry intimately, aptly summarised the cycle's contribution to the development of the motor car on the first page of the first issue of Britain's first motoring journal, _The Autocar_:

> The cyclist and the cycle maker have paved the way for the autocar . . . [The cycle-maker] has achieved a mechanical triumph in the combination of great strength to withstand internal strains with great lightness and the successful overcoming of the vibrations and obstacles of the road surface, so that whilst the bicycle rider has accustomed the public mind to the sight of wheeled vehicles without horses . . . the manufacturer has brought the science of road-carriage construction to a point of perfection which enables the power developed by a motor to be utilised to its fullest and best advantage.[12]

Until the later 1890s new bicycles were expensive (about £20 each) and therefore affordable only by the better-off. In those days these swells were for ever forming, or joining, clubs. The Bicyclists' (later the Cyclists') Touring Club (CTC), formed in Britain in 1878, was soon approving places at which to stay or to eat, putting up road signs to warn of steep hills and generally acting as a pressure group. The need to improve the roads was a constant cause of concern, and a Road Improvement Association was formed in 1886 specifically for this purpose. The National Cyclists' Union co-ordinated the many little local clubs which often kitted out their members in smart caps and blazers and arranged regular outings. These privileged people, riding their machines round the countryside, often at speed, brought quiet lanes to life again; but in the process they also frightened the horses, got involved in accidents, raised clouds of dust in dry weather from the untarred macadam roads and generally aroused popular hostility which was to foreshadow the widespread opposition to the early motorists.

The close links between the pedal cycle and the motor car are underlined by many of the names in the cycle trade which were carried over into motor manufacture: in France Armand Peugeot, who had branched out into cycles in the later 1880s from the family business of making steel wire, springs, saw blades and other items of hardware; Gustav-Adolphe Clément, the largest of the French cycle-makers; and Alexandre Darracq, the second largest. In Germany, Heinrich Kleyer of Frankfurt and (going back to sewing-machine manufacture) Adam Opel at Russelsheim; Dürkopp at Bielefeld; Stiwer at Stettin. In Austria, Job Puch at Graz. In the United States

Colonel Albert A. Pope of Hartford, Connecticutt, the leading maker, who marketed the 'Columbia' cycle; Duryea, Winton and Willys. In Britain William R. Morris, the future Lord Nuffield, Riley, Singer, Rover, Humber and Lea-Francis.

THE MECHANICAL OPTIONS

(i) Steam

Those who attempted to mechanise short-distance transport after 1880 had three options. Of these, steam seemed the best bet, for it was tried and trusted. There were many skilled engineers trained in the technology capable not only of servicing steam engines, but also (it was hoped) able to adapt for road transport the heavy locomotives which needed specially built railways to run upon. The problem was to build an engine which could withstand higher steam pressures and so increase the power to weight ratio.

In fact, steam-powered vehicles were being operated commercially upon the roads of Britain and elsewhere in the world from the 1850s. They were, however, still very heavy machines: lumbering traction engines of up to 14 tons weight which travelled at only a few miles per hour. They were used mainly in agricultural districts to pull wagons of farm machinery and also, to a smaller extent, for general haulage purposes. In general haulage they were sometimes known to travel considerable distances. In 1857, for instance, a traction engine pulled a train of wagons 122 ft long, weighing over 30 tons, from Thetford, some 35 miles north-east of Cambridge, to Stratford-at-Bow, on the east side of London, 85 miles away. A few years later, another engine, which had been making short demonstration journeys of eight miles, pulling five wagons of coal (34 tons) between Little Hulton, near Worsley, and Manchester, travelled via Warrington to Liverpool before being exported to Venezuela.[13]

These heavy vehicles were both unpleasant and unpopular, not least with the local authorities responsible for maintaining the roads and bridges which they damaged. Even *The Engineer*, on the whole, as one would expect, a well-wisher to developments mechanical, commented that they were 'unutterably hideous, astoundingly noisy and to the last degree offensive in the matter of smoke'. 'All the moving machinery', we are told on another occasion, 'was visible; there was the barking noise due to the steam blast; they occasionally emitted fire'.[14]

The builders and operators of these snorting monsters sought parliamentary sanction for lower, and uniform, tolls on turnpike roads. Here, after several attempts, they were successful in 1861. A speed limit was set at 10 m.p.h. in open country and 5 m.p.h. in built-up areas; but the engines were limited to 12 tons in weight and 7 ft in width, and their owners were to be responsible for heeding warning notices concerning permissible weights on bridges and for paying for any damage caused. Four years later, however, by which time the English county with most traction engines, Kent, already had 100 running on its roads, the opposition struck back. In what transport historians have come to call the Red Flag Act, these road locomotives were limited in speed to 4 m.p.h. in open country and to 2 m.p.h. in built-up areas and were to be preceded at a distance of 60 yards by a man who was to help horses past and to carry a warning red flag. The 1865 Act underwent some modification in 1878 when local authorities were empowered to enforce their own district regulation. One of them even demanded a man with a red flag should walk behind, as well as in front of, the engine.[15]

This restrictive legislation did not, however, prevent the number of traction engines from increasing. Thomas Aveling, a Kent manufacturer, told a parliamentary committee in 1873 that he had built 800 of them and had even sent a boiler by road from Rochester all the way to Aberdeen. He evidently exported some of his output – there were 'a great many in France working on the highways' – for he estimated that he had sold only 500 to British users. Another maker claimed to have sold 1 000 on the home market.[16] By the 1890s there was in existence a National Association of Traction Engine Owners and Users with more than 100 members. It testified that there were then 8 000 engines at work in England and Wales alone and that they gave work to 30 000 to 40 000 men. There were others in Scotland.[17]

The Red Flag Legislation was very necessary to control the movement of these thousands of heavy, offensive and destructive traction engines. At the same time, Parliament should have taken advantage of the new Bill in 1878 to encourage the development of lighter power-driven vehicles, more suited to road transport work. Instead, legislation covering what were to be called 'light locomotives' had to wait until 1896.

Experiments to produce a lighter steam-driven vehicle were already taking place at the end of the 1860s. John Henry Knight of Weybourne House, Farnham, for instance, in 1868 built a steam carriage weighing 32 cwt. It required two men to work and could carry two passengers at an average speed of 8 miles per hour. Its progress, however, was not without accident and, when driving into the wind, the unfortunate passengers were showered with a sort of black rain. The trials were soon abandoned. Ten years later, in 1878, a much lighter, 180-lb engine and boiler, heated by a liquid fuel burner, was used in a three-wheeler built by A. B. Blackburn of Tunbridge Wells. Three years later still, in 1881, a two-cwt steam tricycle, also using liquid fuel, was built by A. H. Bateman of East Greenwich to the design of Sir Thomas Parkyns of Beckenham. It was exhibited at the fifth annual Stanley Cycle Club Show in London in 1881, 'the first time a self-propelled machine had been displayed there'. 'The problem of steam carriages has been solved', proudly boasted the *Bicycling Times*; but the little machine was deemed to be a locomotive in law and its owner was fined for travelling at 5 m.p.h. in a built-up area, a decision which was upheld on appeal to the High Court. Later in the 1880s, James Sumner suffered a similar fate with a steam tricycle, though he did go on, in due course, to create what was to become Leyland Motors.[18]

Experiments were also taking place in France and without legal discouragement – but also without financial success. Amédée Bollée, a bellfounder of Le Mans, started to produce more sophisticated, but still rather heavy, steam vehicles in the 1870s.[19] Count Albert de Dion, 'playboy aristocrat with a passion for machines and publicity',[20] and soon to be an outstanding leader in the motoring movements in France, employed two Paris craftsmen, George Bouton and Charles-Armand Trépardoux, during the 1880s to build some steam tricycles and one- or two-seater four-wheeled carriages; but they were no more commercially successful than Bollée's. Léon Serpollet, who worked during the 1880s upon the development of a boiler employing pre-heated small-bore pipes which quickly transformed (or 'flashed') water into steam, making it easier to produce sudden increases in power, looked more likely to succeed, especially when he managed to interest Armand Peugeot in his work. One of his steam three-wheelers built by Peugeot in January 1890 made a much-publicised journey from Lyons to Paris, with frequent breakdowns, in two weeks; but then Peugeot lost faith in steam.[21]

Events were to show, however, that steam was still well placed in the mechanisation stakes.

(ii) Electricity

With electric traction the promoters' problem was to transfer electric power from the stationary steam- or water-driven generator to the mobile motor. The problem was solved for tramcars by the use of a trolley fixed to the end of a sprung pole to pick up the current from bare overhead wires (with the metal rails serving as the return) and the series/parallel controller to apply the current to the motors at a controlled rate so that the heavy vehicle could be started and accelerated without burning out the motors' windings. All this was achieved by Frank Julian Sprague in Richmond, Virginia, in 1888. In the United States horse tramways were soon converted and new electrified lines were opened: by 1893 nearly 7 500 miles altogether and by 1895 over 12 500 miles.[22] Outside the United States, where municipal control was often stronger and townspeople took less enthusiastically to large, unsightly, steel poles supporting naked copper wire carrying electric current at about 500 volts pressure not far above the streets, the electric tramway often took a little longer to arrive; but, by replacing tramway horses, it soon provided faster passenger transport at lower cost in more spacious and comfortable vehicles, brightly lit at night. Electric tramways established new settlement beyond existing urban limits and also connected towns together (interurbans).[23]

Electric tramways – or street railways as the Americans more accurately described them – were, however, confined to predetermined routes, as horse tramways had been before them and as steam railways continued to be. Their use was inevitably limited by this lack of flexibility. As soon as manufacturers tried to develop electric vehicles which were truly mobile and, therefore, could not collect current from overhead wires, they ran into great difficulties, for the electricity had to be stored in batteries which had only a limited capacity and had to be frequently recharged. Even worse, being made of lead, they were excessively heavy. From the end of the 1880s, however, electricity was all the fashion and could attract development capital. Electric vehicles had obvious selling points: cleanliness, quietness and lack of smell. They were relatively easy to drive and in towns battery recharging stations could be readily established. Road trials began in England in 1882 when two electrical engineers, William Edward Ayrton and John Perry, professors at the City and Guilds Institute, tried out an electric tricycle. A battery-driven electric cab ran in Brighton in 1885 and an electric carriage used to take its owners for an airing there two years later. In 1888 the Sultan of Turkey ordered a more splendid vehicle, the mechanical hands of which were made by the Acme and Immisch Electric Works in London. He was so pleased with it when it reached Constantinople that he sent for a second. The first electric vehicle in the United States, credited to William Morrison of Des Moines, ran in Chicago in 1892. By 1894 we hear of an electric van carrying parcels around the City of London; it covered 1000 miles that year. Electric taxi cabs, however, had particular attractions. Two thousand of them were manufactured in the United States. Electric cab companies were successfully floated in both London and New York in 1897.[24] At that time many must have thought that manœuvrable, battery-driven electric vehicles were in with a real chance.

(iii) Internal combustion

Although we are now (1986) celebrating the centenary of the motor car, 1886 saw only the first, fitful movement in southern Germany of simple machines driven by crude, experimental petrol engines, mere Mark I mock-ups still requiring much further development. Nobody seemed to have bought any of these new-fangled creations for several years, and even ten years later it was by no means certain that internal combustion would win the mechanisation contest. Steam and electricity still had their fervent devotees.

To this extent the 1986 celebrations are a little premature. It was in France, rather than in Germany that the motor vehicle was developed and seriously exploited commercially in the mid- and later 1890s; and in the United States, with more possible customers, that its full potential was realised more than a decade later still. There will be further opportunities for centenary celebration.

Germany does, however, deserve credit for beginning the story.[25] It was at the Otto & Langen works at Deutz that the stationary gas engine was extensively developed. This handy power source, working on piped town gas, proved increasingly popular for driving small machines in workshops before electric motors became more available and used for this purpose. Gottlieb Daimler (born 1834) had been chief engineer at the Deutz factory from 1872 and Wilhelm Maybach was his chief of design. While there, they had already tried vaporised petrol as an alternative to piped gas. After ten years at Deutz, Daimler (not the easiest of colleagues to get on with) departed, and, taking Maybach with him, began experiments at Cannstatt, now part

of Stuttgart, to adapt the stationary gas engine for transport purposes. Already leading technical experts in this field, their experiments quickly produced results. In 1883/4 they took out engine
patents, as the Archivist of Daimler–Benz explains in the next
chapter. Using a specially devised 'hot tube' ignition system, they
produced a motor capable of 600 revolutions a minute, soon increased to 700 r.p.m., five or six times faster than speeds achieved by
stationary internal combustion engines then in use. They also reduced the weight of the motor to a mere 90 kg. Here was the kind of
power-to-weight ratio which other rivals had been seeking; but, as
with all major inventions, there were still many practical problems to
be overcome before reliable Daimler engines were powering reliable
road vehicles.

Moreover, Daimler was interested only in building *engines*. He did
not want to go into competition with the private-carriage makers but
wished to sell his engines to them or to other business men who
would use them to drive boats, tramcars, fire engines or any other
vehicle. Towards the end of 1885, as Dr Nübel tells us, Maybach 'is
supposed to have made several test runs in an unsprung motor cycle
with iron-rimmed wooden wheels'; and in August 1886 Daimler and
Maybach bought a horse-drawn carriage and converted it for trials
with a more powerful motor. In the following year premises were
acquired in Cannstatt and engine production for the market began;
but it was marine and *stationary* engines which were specifically
mentioned when the first sales information was published-in the
financial year 1890/1.

The activities of the other engineer, Karl Benz (born 1844 and
therefore ten years Daimler's junior), working in Mannheim, about
sixty miles north of Stuttgart, are even more obscure. Like Daimler
and Maybach, he had had a very good training, first as a student of
engineering at the polytechnic school in Karlsruhe and then in a
number of engineering works. He acquired his own machine shop in
1872 and was making stationary gas engines from 1879. In 1882 he
had taken out a patent for a throttle regulator which has been
described as 'one of the basic patents in engine technology'.[26] In the
following year he and two other partners had set up at Mannheim the
Rheinische Gasmotoren Fabrik. All in all, one does not get the
impression that his understanding of the gas engine's possibilities was
up to Daimler's and Maybach's: his petrol motor certainly ran at a
slower speed and had an inferior power-to-weight ratio. It employed
electrical ignition which, in the early years, caused much trouble. But

from the outset he was concerned to market a motor vehicle and not just to sell a mobile petrol motor. He was not a cycle maker and so, presumably, he bought in the tricycle which he specified in his patent dated 29 January 1886. A report of its testing on the Ringstrasse, Mannheim, appeared in the local press on 3 July 1886 and, after further development, a Mark III version attracted much attention when it ran on the Munich streets during the trade and industry fair there just over two years later, all of which Dr Nübel describes in more detail in the next chapter. He also makes the point that, while the Benz vehicles found admirers, they did not attract any customers at that time. Those who could afford the high price asked still preferred their elegant horse-drawn carriages to the mean little motorised rattle-trap which made its driver an object of derision on the frequent occasions when it broke down. In any publicity the Benz was presented as a humdrum commercial machine, not a new and exciting creation which could be used for sport and pleasure; and, as in Britain, the German local authorities imposed all sorts of restrictions and were generally unhelpful.

Neither Daimler nor Benz was a business man. They both lost their business associates at the beginning of the 1890s when their prospects seemed particularly bleak. Benz found others and began to make four-wheeled vehicles: the heavier Viktoria (1893) and (from 1894) the lighter, and lower-priced, Velo, many of which were exported. By then, however, the flair of the French had made their country the centre of world attention.

FRANCE DEVELOPS AND POPULARISES THE MOTOR VEHICLE

The American historian James M. Laux, in his excellent book *In First Gear*, has told us how the French took licences for the Daimler and Benz patents and not only transformed the Germans' primitive efforts into something which began to resemble in basic design the motor car of the future but also publicised it. He explains how Émile Levassor, a product of the École Centrale des Arts et Manufactures, the leading Paris school of engineering, had joined forces with a contemporary there, René Panhard, who had acquired an interest in one of the larger machine shops in Paris specialising in the making of woodworking machines and metal-cutting saws. In 1887 Panhard et Levassor (P. et L.) had been asked by Édouard Sarazin, representative in France for Daimler's high-speed engine patents, to manufacture a few of these motors so as to safeguard the patents under French law. P. et L., who at that time were no more anxious to build motor vehicles than Daimler himself, got into touch with Armand Peugeot, the cycle maker, with whom they already did business and who, as we have seen, had already shown an interest in *voitures sans chevaux*. They asked him to put his firm's expertise to good purpose by providing an appropriate vehicle into which one of these engines could be fitted. Peugeot's cycle works were at Valentigney, just to the south of Montbéliard, half-way between Mulhouse and Besançon in the Jura. Not far from the Swiss frontier, this was a severe and hard-working part of France populated by industrious watchmakers and metal workers. There, at the end of 1888, Daimler himself demonstrated to the satisfaction of Levassor and Peugeot the merits of his engine. (The Daimler vehicle, it is perhaps worthy of note, was carried from Stuttgart to Valentigney by rail, not driven by road.) Very soon afterwards, on 5 February 1889, Louise Sarazin, who had taken over the Daimler agency on the death of her husband, agreed to pay the inventor 12 per cent on all Daimler engines made in France with a guaranteed minimum sum, equivalent to the manufacture of 35 engines a year.[27]

P. et L. also found themselves concerned with Benz tricycles. Émile Roger, an engineer in Paris and owner of the French manufacturing licence for Benz vehicles, arranged with P. et L. to make them until he himself had equipped his own workshop for the purpose. The Benz machine had arrived in four wooden crates in March 1888 and, a little later, Benz himself turned up, too, to supervise its assembly and to get it started, and certainly to demonstrate it on the streets of Paris. Roger advertised the machine without delay but was for some time greeted with same sales resistance that Benz himself had experienced at home in Germany.

The wares of both Daimler and Benz were shown at the International (Eiffel Tower) Exhibition held in Paris in 1889 to celebrate the other famous revolution of a century earlier. A Benz tricycle was on display, but one visit to Paris had apparently been quite enough for its creator; and no attempt was made to run his strange curiosity. The ugly little vehicle, without seats or shaft, 'a tangle of machinery', did not impress the visiting public, especially as it was exhibited in the midst of a row of tall and stately horse cabs.[28] Significantly, P. et L. exhibited, not a road vehicle, but two boats powered by Daimler engines; another Daimler motor was installed in a tram car and a fourth, a stationary one, ran an electric generator. Here was further evidence of Daimler's lack of concern for motor vehicles; but, towards the close of the Exhibition, the most recent and powerful (2-h.p.) Daimler vehicle arrived, two bicycles coupled side by side with the motor in the middle. Maybach drove it for a short distance in the neighbourhood of the Exhibition at 7 m.p.h.

So far neither the Daimler nor the Benz motorised contraptions could be described as a raging success. Their commercial, as distinct from their technical, birth still lay in the future.

In March 1890 the first two Daimler motors built by P. et L. reached the Peugeot works and the man in charge there, Louis Rigoulot, another product of France's good technical education system (he went to the École des Arts et Métiers at Châlons-sur-Marne) began to build a prototype quadricycle, with improved steering and suspension. This reached P. et L. in Paris in the following August. By then Levassor, having thought further about the possibilities of developing saleable vehicles, had, with the agreement of his partner, Panhard, decided to build a prototype vehicle himself. This decision, no doubt, was prompted to some extent by his marriage, in May 1890, to the widow Sarazin, owner of the Daimler patent rights in France who was already committed, as we have seen, to paying Daimler for thirty-five motors a year. The P. et L. prototype, a heavier vehicle (more like a horseless carriage than the Peugeot quadricycle) was completed in September 1890 and by January 1891 it managed to get the twelve miles to St-Cloud and back. Nothing more was heard of the Peugeot prototype, but Rigoulot was still hard at work in remote Valentigney. In September 1891 he and a works foreman drove an improved model, running on solid rubber tyres, to P. et L. in Paris. (It took them four days to cover the 260 miles.) They then joined a bicycle race from Paris to Brest and back. The vehicle did not manage to keep up with the leading pedal cyclists; but it gained some publicity and actually found a customer, back in Mulhouse. It was delivered to him after having covered 1200 miles altogether. (We are not told whether or not the car was sold as new or second-hand.) Peugeot made four cars that year, and Levassor, who also developed an improved model, the first to be made with the engine in front over the wheels, demonstrated this over longer distances. In 1891 he managed to sell one or two more than Peugeot's production. 'You are the father of the modern automobile', Daimler, the engine inventor, wrote to him.

So, five years after the motor vehicle's first fitful appearance, sales had at last begun, and in 1892 P. et L. were emboldened to produce a catalogue, probably the first to be issued. In the next two years, although sales were still minute, they grew quite quickly.

Table 1.1   Petrol vehicle sales in France and Germany, 1892–4

P. et L.		Peugeot	Benz (Roger)	Benz (Mannheim)	Daimler (Cannstatt)
1892	16	29 made	c.12 tricycles	69 made by end 1893,	11 made but 'they
1893	37	24 made	24	24 of them tricycles. Three-fifths of total production sold in France	were unsuccessful'

Source: James M. Laux, *In First Gear*, pp. 18–19.

By the end of 1893 we also have the first clue to the type of customer to whom these vehicles were sold. Levassor, in a letter to Daimler, having noted that P. et L. produced heavier vehicles than Peugeot, whose cycle-type *voitures sans chevaux* ran on ball bearings and could reach higher speeds with the same power but had uncomfortable seats and poor suspension, went on to observe:

> Some people like ours better, others prefer the Peugeot. There is something for every taste. Until now we have sold our cars to quite different types of people, generally those who use them for pleasure, but we have sold them also to people who use them for business purposes, in particular six doctors, five travelling salesmen, and three insurance agents.[29]

So long as motor vehicles remained costly to buy and to run, these categories of motorist were to persist: the well-to-do and leisured classes; and those professional people and business men who were successful enough to afford a car and shrewd enough to perceive that it would enable them to carry out more visits, possibly at a lower cost than would be possible in vehicles drawn by horses, which had to be fed when they were not working. These advantages may have been somewhat illusory in the early years because of breakdowns and rapid obsolescence; but as motor cars became less underpowered and more reliable, professional and business people, as well as the rich and leisured, were to buy them in greater numbers.

Experiments with petrol engines were not confined to France and Germany. In England, for instance, Edward Butler, an engineer, had taken out a patent in 1884, and made a petrol engine capable of 600 r.p.m. using coil and battery ignition. He later ran a petrol-driven tricycle on the roads of Erith, Kent, in February 1889 which 'worked fairly satisfactorily'; but, like Daimler at that time, he decided to concentrate upon stationary and marine engines.[30] During many early morning hours in 1892, James D. Roots held repeated trials of a kerosene engine in Holborn, London but was 'not very successful'.[31] In America, as Charles E. Duryea reported a few years later, active work on petrol-driven cars began in 1891. He and his brother, cycle mechanics, produced an underpowered vehicle in 1892 but rebuilt it and, with a more powerful engine, put it on the road at Springfield, Mass., in September 1893.[32] Elwood Haynes, who had read some chemistry at Johns Hopkins University and then, in due course, had been appointed to direct the field operations of the Indiana Natural Gas Company, started petrol car experiments with Edgar and Elmer Apperson of Kokomo, Indiana. Their petrol car made its first run in July 1894.[33] Hiram Percy Maxim, a Massachusetts Institute of Technology graduate, son of Sir Hiram Maxim, inventor of the machine gun, and himself superintendent of the American Projectile Company of Lynn, Massachusetts, began experimenting there from the spring of 1893. He claims to have built a small engine by the end of 1894.[34] Meanwhile the European machines were reaching America. William Steinway, the piano manufacturer, imported Daimler *engines* for various uses in America from about 1888. Daimler himself attended the 1893 Chicago World Fair to explain their main features. A Benz motor vehicle was also to be found there.[35] Daimler engines were imported into England from 1891 by Frederick R. Simms, but used at first for stationary and marine purposes. Simms had been born in Germany, is said to have seen the Daimler engine there as early as 1886 and to have met Daimler in 1890 at an exhibition in Bremen. In 1893 he became the major shareholder in the Daimler Motor Syndicate which acquired the British patent rights.[36]

THE THREE OPTIONS TESTED IN FRANCE

The pedal cycle had been publicised, as we have seen, by means of endurance tests and races. From the end of the 1880s the advantages of the pneumatic over the solid rubber tyre for both speed and comfort were also being demonstrated in bicycle races, even though these flimsy early pneumatics were still very unreliable and hardly able to withstand for long even the light weight of cycle and rider.[37] By 1894 the time had arrived for the various sorts of motor vehicle to be similarly displayed, not just by individual feats of endurance, but in open competition with one another. Not surprisingly, this was to be in France, where interest was greatest, motor vehicles most numerous, the roads good[38] and the road authorities sympathetic. Here the first *concours des voitures sans chevaux* was organised.

The promoter, Pierre Giffard, an enterprising member of *Le Petit Journal*, a mass-circulation newspaper, had been responsible for the Paris–Brest–Paris cycle race in 1891 which has been mentioned. With a good sense of history, he arranged the first trial for motors along the shorter route from Paris to Rouen (79 miles) where, as we have also seen, the first cycle race had taken place, in 1869. The 1894 competition, however, was intended to be an endurance test, not a race, and no doubt this helped in obtaining approval from the authorities. All the participants had to complete the course within a specified time and points were given for safety, reliability, ease of control and low running costs.[39] One hundred and two vehicles were entered, all but three of them from France; but even after a post-ponement of seven weeks to enable the less prepared to get ready, only twenty-one appeared for the eliminating rounds of the contest, almost all from firms we have already encountered. Five were Peu-geots, four were from P. et L. (Panhards) and seven were steamers. Among the latter was an entry from de Dion-Bouton, which had returned to steam vehicles in 1893 by building a tractor which would haul its owner's carriage. All the petrol vehicles finished the course (though some had to be pushed up the hills) and the first prize was won jointly by P. et L. and Peugeot. The de Dion tractor gained some kudos for steam vehicles by reaching Rouen ahead of the field, but only after its two chauffeurs (stokers) had sweated away for the whole journey amid smoke and sparks. The other steamers failed to arrive within the time limit and four did not arrive at all. It was a great triumph for Daimler engines. Roger had to be content with fifth prize, for design improvement.

The Paris–Rouen trial, modest though it was in aim and scale, began to draw more widespread world attention to recent develop-ments in France. Among those who went to watch it from England was Walter Arnold, a leading traction engine contractor from Kent, accompanied by a tea broker friend, Henry Hewetson. They were obviously impressed by the success of the petrol-driven cars; but instead of buying one with a Daimler engine, Hewetson went to Mannheim and ordered a Benz. It reached England in November 1894, the earliest recorded petrol car import into the country. The police declined to take any action under the Red Flag Acts when it was driven about on the English roads.[40] In France a number of petrol and steam cars were exhibited in December 1894 at the second Paris Bicycle Show;[41] and de Dion-Bouton, who had started in 1893 to develop a car powered by a specially designed high-speed petrol engine,[42] no doubt redoubled their efforts.

The Paris–Rouen event also attracted much press coverage in France; and a leading English technical journal, *The Engineer*, sent a special correspondent not only to cover the road trial, but also to report upon the P. et L. works, the cars made there and those made by Peugeot. He did not conceal the environmental nuisance these early motors caused – the 'very unpleasant' smell and the 'great vibration' – but he concluded that 'the petroleum carriage can go as fast as an ordinary cyclist cares to travel', the point being that petrol vehicles were then capable of travelling as fast over distances as anything else on the road. Horses might have a higher top speed, but they soon tired.[43]

The first little motor trial had been an obvious success. Interested parties, both motor manufacturers (including Levassor, Peugeot, Roger and Serpollet) and well-to-do buyers, who may, or may not, have had a financial stake in motor manufacture (including Baron van Zuylen, Marquis de Chasseloup-Laubat, Henri Menier, the chocolate king, and Count de Dion), got together at de Dion's in November 1894 to organise a far more ambitious event for the following year: an actual race, non-stop over 732 miles, from Paris to Bordeaux and back.[44]

Daimler himself witnessed the start of the race at Versailles on 11 June 1895 in excellent weather. The detailed arrangements had been made in collaboration with the government Department of Roads and Bridges and with the help of the Touring Club de France (then a cycling organisation) and the French Bicycle Association. Among the numerous other observers of the start were no fewer than seven members of the British Traction Engine Owners' Association and three county surveyors from Kent. Eight of the eleven petrol vehicles completed the course, but only one of the six steamers, a heavy fifteen-year-old, ten-seater Bollée omnibus which had been brought back from retirement for the purpose. The one electric vehicle entry, for which charged batteries were located at intervals throughout the course, failed to get far; not, its apologists claimed, because of any deficiencies in the electrical system but because of a warped axle.[45]

A vehicle which did manage to get back to Paris, though not within the stipulated 100 hours' time limit, was to prove worthy of greater notice than it was given at the time. This was a petrol car fitted with pneumatic tyres which, up to that time, had been used only on the much lighter pedal cycles. André Michelin and his younger brother, Edouard, had lately given life to the dying family rubber manufacturing business at Clermont-Ferrand by acquiring rights to a pneumatic tyre patent other than Dunlop's. They publicised their product by entering a Michelin-tyred bicycle in the Paris–Brest–Paris race of 1891 and it came in first. Four years later they took the next logical step: they made heavier pneumatic tyres for a motor vehicle and entered it in the Bordeaux race. All the tyres had to be renewed at least every 90 miles. To quote Laux: 'After experiencing all kinds of troubles, including water in the petrol tank, broken spokes and gears [and] a fire . . ., their car, called the Lightning, not for its speed but because it zigzagged due to the lack of a differential, finally limped back into Paris.'[46] It had demonstrated, however, the possibilities of another vital element in the interrelated series of technical developments which were to prove essential to the ultimate success of the motor vehicle.

The real hero of the event, however, was 53-year-old Émile Levassor, who came in first, having completed the race in a running time of 48 hours and 48 minutes. Steering his P. et L. car, into which a new and improved, 4-h.p. Daimler Phoenix engine had been fitted, and accompanied by a mechanic, 'he supervised his machine constantly, except when ascending an occasional incline when the speed was comparatively slow, and then he entrusted the lever to his mechanic'.[47] For a motor vehicle to have been on the road from 9 a.m. one day until nearly 1 p.m. the next day but one and to have covered over 730 miles at speed which averaged 15 m.p.h. in just under 49 of those 52 hours (the stops were to take on petrol and water, with 15 minutes for cleaning at the Bordeaux end),[48] was an astonishing achievement which caught the world's imagination. It was for motor vehicles what the Rainhill trials had been for railways. As the *Autocar*, writing primarily for English readers, put it: 'it was the means of calling the attention of Englishmen to the fact that the autocar had been brought by French engineers to a state of practicality which removed it at once from the stage of mere experiment'.[49] It will certainly deserve its own centenary celebrations when the time comes.

SALES PROMOTION IN FRANCE

The race gave a great boost to French motor car manufacture. 'We receive visitors [at the P. et L. works] from morning to night, always English, in a never-ending stream', wrote Panhard's son a few days afterwards, 'And every day we receive at least ten letters from England.'[50] A host of other manufacturers (including the big cycle-maker, Clément) entered the business. De Dion–Bouton went into petrol-driven vehicles with a new motor which ran at 1 500 r.p.m., twice that of the Daimler engine; the De Dion–Bouton tricyle, soon to achieve great popularity, began to be marketed later in 1895. In that year Léon Bollée of Le Mans constructed a motor tricycle which was marketed as a two-seater *voiturette*, the passenger sitting behind the driver in tandem.

The first motoring journal, *La Locomotion Automobile* (1894) was joined by others in France and elsewhere, which publicised the new industry. Motor shows began to be organised, following the earlier example of the cycle trade. The committee which had arranged the Bordeaux Race transformed itself into the Automobile Club de France with the influential banker, Van Zuylen, as its President. Menier and de Dion became Vice-Presidents, and a long list of French society leaders formed the committee. The noisy, smelly, dangerous motor car suddenly gained respectability and powerful pressure group influence. There were perhaps some 300 motor vehicles on French roads by the end of 1895.[51]

The atmosphere in which these enthusiasts took to motoring was to be well recaptured by Baron Henri de Rothschild in a talk he gave in London a few years later. The car in which he travelled on the occasion in question was owned by a friend and had its own chauffeur, whose job, as was customary in the early days, was to drive only when the owner did not wish to do so himself. At other times the chauffeur acted as a mechanic whenever the vehicle broke down.

It was a few weeks after my marriage in the month of May of the year 1896, [Rothschild recalled]. One of my friends had invited me to go with him from Paris to Chantilly, where I pass the summer holidays. There were about 25 miles to go over, and at that time proper road maps and signposts did not exist. . . . For us these 25 miles were quite a journey, and for getting over them we relied on a small 6 horse power Peugeot car (one of the first supplied to amateurs) in three or four hours' time, that is to say, at an average speed of six to eight miles an hour.

My friend was driving the car. We had already gone over about six miles from Paris when I asked my guide if it were a difficult thing to drive a motor car. He replied that nothing was easier. He persuaded me to take his place, and indicated the pedals on which it was necessary to place the foot in order to stop or slacken speed as well as the levers to be worked backwards or forwards for changing speed or for working the brake. The steering bar was similar to the steering bar of a bicycle. Travelling in a straight line it was easy to control the car. I commenced therefore by getting along without hindrance, and I was already very proud to find myself the driver of a motor car. We were going up hill when I commenced steering, and we were travelling at about four miles an hour. On arriving at the top of the hill I felt fully confident on seeing the road stretching straight downhill. I therefore let the car go a little quicker, but suddenly, at the bottom of the hill, I saw a cart drawn across the road. An annoying obstacle, I thought, but doubtless easy to avoid. But when I arrived within 100 yards of the unfortunate cart, I abruptly received a sudden and terrifying demonstration of my inexperience in matters of motor cars; I could no longer find either pedals or levers! It was absolutely impossible for me to stop my [sic] car, which, urged on by the speed acquired, was now going at a dizzy pace. In a few seconds more we should inevitably be upon the cart; at all costs this was to be avoided. I gave a violent tug to the left of my steering bar, but I could not tell you what happened afterwards. In any case I picked myself up seven or eight yards away, with clothes torn and hands bleeding. I was also covered with dust, but, relatively, without having suffered great hurt. Behind me I saw the car completely overturned, broken into a thousand pieces, and at first I could see neither my companion nor the *chauffeur*. It was only on raising the broken car that I was able to drag my friend from a very perilous position; he had a rib broken and was very seriously bruised. To make the story short, I was fortunately able to find a carriage, and we arrived home in a disreputable condition.[52]

This rather alarming experience, however, did not put the young enthusiast off motor cars. On the contrary, he promptly ordered one for himself, in which he soon had yet another accident. The remains of the vehicle had to be towed away by horse; but such was the demand for cars at that time that he received nearly twice as much for the severely damaged vehicle as he had paid for it in the first place when brand new.

The popularity of motoring in France owed much to the quality of its roads, a legacy of the eighteenth century which the authorities and engineers had continued for most of the nineteenth, as Dominique Barjot stresses in Chapter 15. Britain's *Autocar* went further. 'The freedom of traffic and the splendid thoroughfares enjoyed by France', it believed, 'are mainly responsible for the great progress that has been made both in the bicycle and in the autocar'.[53] The widespread availability of petrol was another prerequisite. It was said to be already used 'so extensively' in France in 1895 that it could be 'generally purchased in towns of any size'. The firm of Deutsch was in process of establishing petrol depots throughout the country.[54] The estimated total of motor vehicles on French roads, though still small, grew rapidly: from about 300 in 1895 to over 14 000 (11 250 of them cycles) in 1900.[55]

Perhaps 15 000 de Dion–Bouton (frequently improved) motor tricycles had been sold to buyers in France and abroad between 1895 and 1901, and many of their engines were fitted into the machines of other makers.[56] More annual races were held on public roads until the higher-powered engines and greater speeds caused an unacceptable level of slaughter among participants and bystanders. This happened in 1903 when, after 10 fatalities and many more injuries, the race from Paris to Madrid had to be stopped at Bordeaux. (Among the casualties was Marcel Renault, one of the three brothers, sons of a successful textile and clothing manufacturer, who had, in 1898, begun to make cars on family property at Billancourt, then still a largely rural area to the west of Paris.) The Automobile Club de France also arranged its own motor shows from 1898. The French industry boomed as would-be purchasers tumbled over one another to buy a motor car. As they often paid half the purchase price with the order and the rest on delivery, the customers were to a large extent financing the development of a still far from reliable product. The infant motor industry, having suffered great vicissitudes before 1895, began to enjoy great advantages, for the development costs of most new products to the point of reliability is normally borne by their promoter. Unfortunately Émile Levassor lived to see only the beginnings of this prosperity and success. He died suddenly in April 1897, perhaps from the effects of hitting a tree in his car during the annual race of the previous year from Paris to Marseilles; if so, he was another casualty of the relentless need for publicity which only racing seemed to be able to satisfy.

THE UNITED STATES, QUICK OFF THE MARK, LOSES TIME BY BACKING THE WRONG TECHNOLOGIES

As at the outset in France, it was a newspaperman who first took the initiative in the United States. H. H. Kohlsaat, proprietor of the Chicago *Times-Herald*, who had come across the petrol motor at the World Exhibition there in 1893, announced in July 1895, only a few weeks after the Bordeaux Race, that he would offer prizes for a contest which, like the Paris–Rouen trial, would test qualities other than speed. The French held their races in the summer. Kohlsaat's, however, was arranged for 2 November when the weather in that part of North America was much less favourable. In fact, only two machines were ready by then, an American Duryea and a Benz modified by an American, Hieronymous Mueller. The main contest had to be postponed in the hope of attracting more contestants; but the two machines did run on 2 November the 92 miles from Chicago to Waukegan and back. This was rather a non-event as it transpired, for the Duryea was damaged in an accident and the adapted Benz got back $9\frac{1}{2}$ hours after the start.

Of the thirty-one vehicles entered for the postponed contest, six appeared: the earlier two, two other adapted Benz cars and two electric vehicles. The run this time was only from Chicago to Evanston, part of the previous route, a total of only fifty-four miles; but the postponed date was Thanksgiving Day, 28 November, and by then snow had fallen. The electric vehicles were quite unsuccessful; one of the Benz cars dropped out before it could get out of Chicago and another did not return within the time limit. But on this occasion the Duryea came in first (having been repaired by a blacksmith) and Mueller's Benz second. The event lacked the warmth and excitement, not to mention the vehicles, of the French contests; but it nevertheless attracted press attention throughout the world, and commentators were found who were prepared to proclaim that it had advanced knowledge by five years. They showed astonishment that 'a self-propelled vehicle should be driven 54 miles through a sea of slush and mud at a speed which would kill any team of horses'.[57]

If the result of the Chicago competition had been more positive, perhaps progress with petrol-driven vehicles would have been more rapid in the United States. As it was, although the market was so favourable to motorisation because of the greater social depth of demand, Rouen was not to be followed by Bordeaux and, therefore, not by the wave of publicity, pressure-group activity and selling which resulted from that epic contest. Although *Horseless Age* began publication in New York in 1895, *Motor Age* (Chicago) did not follow until 1899 and the other American motoring publications came later still. The Madison Square Gardens motor shows did not start until 1900 (though motor vehicles had been displayed there alongside pedal cycles before that), and, so late as 1901, the American publication *Automobile Topics* complained that, unlike the Automobile Club de France, the Automobile Club of America (formed in 1899), was 'un-American, a follower of foreign fashion instead of a leader in a national environment'. It was intended to co-ordinate the activities of local clubs which were springing up; but, it asked, 'Will our American clubs be able to accomplish anything if they leave out of consideration all regard to our national pride and national prosperity? Can they ever hang together individually on a diet of regular namby-pamby Saturday excursions, a little social pap, a little mutual admiration, now and then an emasculated endurance contest with the speed cut down and no equivalent for speed to give full significance to the results?'[58]

Apart from the lack of publicity and salesmanship, progress in America was slowed down by a failure to concentrate upon petrol vehicles. Although Henry Ford, Ransom Eli Olds, Alexander Winton and others already understood where the future lay, electricity was still new and popular in the United States. Fortunes had recently been made by the building of electric tramways. Much more attention was paid, therefore, to the possibility of battery electric vehicles. Colonel Pope, the leading pedal-cycle maker, believing that 'You can't get people to sit over an explosion', concentrated in the second half of the 1890s upon electric cars and cabs. They were easier to start (no cranking) and to drive, important selling points, especially where women motorists were concerned. The Americans also favoured light, high-pressure steam engines with liquid fuel burners and flash boilers. The Stanley and the White steamers and Locomobiles enjoyed considerable popularity for a time, the latter achieving peak output of about 1 600 vehicles in the year 1900.[59] American manufacturers concentrated more on petrol motors after this, however, despite the tribute exacted from many of them under the Selden Patent;[60] but until then much effort had been mistakenly dissipated by backing steam and electricity.

Another explanation of the slow start in America may be found in the appalling state of its roads outside the bigger cities, as dreadful as those in France were good. Their condition was due partly to the huge mileages involved in that great continent, much of it recently settled, and partly to the antiquated form of road authority. At the beginning of the present century passengers relied upon the railway or the tramway. The few earlier turnpikes had fallen into disuse and highway maintenance outside the larger cities depended upon semi-autonomous road districts within each township, a primitive method which had been introduced in England as long ago as the middle of the sixteenth century but subsequently superseded. Local overseers of the highway notified employers of labour that all males over 21 were expected to attend on a few specified days with the necessary equipment. Not surprisingly, despite the Good Roads Movement already started by the pedal cyclists and now taken up by the early motorists, by 1904 only 7 per cent of U.S. roads could be classified as 'improved'. Or, as one authority has put it, with some self-restraint, 'By comparison with European standards, the American system of road administration was obsolete and the quality of the roads strikingly inferior.'[61]

A recurring theme in this book, however, is that poor roads never arrested the ultimate spread of motor vehicles. In richer countries, the well-to-do and influential who acquired private cars quickly brought pressure to bear upon the authorities to spend public money on road improvement and were prepared, when necessary, to make their own contribution by a tax on their vehicles and/or on the petrol they used. In poorer countries, where the wealthy were fewer in number, the state had to take the initiative, motivated by the need to encourage economic growth. (The later chapter–Chap. 12–on Zaïre gives a good example of this.) In the United States, where there were more potential car buyers than anywhere else in the world, the bad roads merely served to heighten the challenge; and American manufacturers soon came to build vehicles more suited to these conditions. The motorists' remarkable feats helped to spread the appetite for motor-car ownership, which the recently created motoring clubs and motoring press did their utmost to stimulate.

The publicity was of two sorts. Imported machines, especially the latest Mercedes, a new, high-powered *marque* which brought success to Daimler from 1901,[62] were raced along good stretches of road, such as those on Long Island, by their wealthy owners. Already, before the winter of 1901 had set in, W. K. Vanderbilt jr managed to run a measured mile in his new, 36-h.p. Mercedes in $58\frac{2}{5}$ seconds. American-built models also entered speed contests.[63] During that same autumn the Detroit Automobile Association organised what it called a 'Race Meet'. Henry Ford, until recently superintendent of the Detroit Edison Illuminating Company's engineering department, was one of the contestants. He had built his first motor vehicle, a quadricycle, in his spare time during the first half of 1896 and had left Detroit Edison in the middle of 1899 when the Detroit Automobile Company had been formed to develop and market another Ford vehicle. The venture did not succeed and Ford, realising that he must build and drive a racer if he was to catch the public attention, found himself at the Race Meet of 1901 pitted against the champion driver from Cleveland, Alexander Winton, an established motor manufacturer. Ford won.[64]

In the meantime endurance tests of the French type were also being held in America, despite the wretched roads. Winton himself had created a considerable stir by driving one of his cars from Cleveland to New York in just under 79 running hours between 28 July and 7 August 1897.[65] In the late summer of 1901 the recently formed Automobile Club of America arranged a large-scale trial of vehicles from New York to Buffalo. It had to send on ahead an official, described as the chairman of its Guide Board Committee, to put up signs on parts of the road where participants might take a wrong turning. Of the 81 vehicles which left New York on 13 September 1901, 41 reached Rochester four days later, having driven along roads which were in parts a sea of mud. Enthusing journalists, however, pointed out that nobody would ever dream of imposing such 'incredible severity' upon horse-drawn vehicles; and, when the contest was eventually completed, they allowed themselves the comforting reflection that 'certain American automobiles constructed to sell for $1000 or $2000 are found equal to imported French machines costing over $10 000 when confronted with the task of driving over American roads in their worst conditions . . . '.[66]

Other hardy pioneers ventured forth, usually equipped with rope, pulley and tackle with which to drag their cars along particularly bad stretches of road. Farmers often did good business, too, rescuing the stranded. Yet notable feats were regularly reported from California right across the country. In 1903 H. Nelson Jackson, a 22-year-old doctor, and a chauffeur, made the first coast-to-coast crossing. Start-ing from San Francisco in a Winton car on 23 May, they took 54 days (36 on the road) to Chicago and 63 days over the whole journey. They were fêted when they reached New York; and the Winton company was not slow to advertise that the car, an ordinary model, had started out with only one extra tyre, four extra spark plugs, a shovel, an axe, a block and tackle and cooking and camping equip-ment.[67] Encouraged by this example, another driver set off from San Francisco on 6 July 1903 in a little curved-dash Oldsmobile and was in New York 60 days later.[68]

This simple little one-cylinder runabout, weighing only 700 lb and selling for $600–$650 was already pointing the way ahead. 'Rough roads, muddy roads, snow, frost and ice have no terrors for the Oldsmobile', boasted the company in 1902. It then already had 11 agencies throughout the United States. Six months later it had 20 more.[69] By the end of 1904 Oldsmobiles were to be found far and wide: 'out on the Arizona deserts, far north in the Klondyke, way east in Borneo and Sumatra, or with Japanese tea garden settings, or in South African surroundings, in Madagascar or in the Kremlin at Moscow'.[70] No doubt the numbers in these far-flung places were very strictly limited indeed, though Oldsmobile already had agencies in Yokohama, Cape Town and elsewhere.[71] The company's output had grown from 425 in 1901 to 5000 in 1904, plus a further 6500 straight-dash models in the latter year.[72]

Within five years America had become the main motor-car pro-ducing and motor-car using country. Summarising the position at the beginning of the New Year 1905, *The Motor Age*, published in Chicago, claimed with some justice that Detroit already made 'more motor vehicles than any other city in America, or in the whole world for that matter'. The industry, it claimed, had been 'unknown' six years before.

> A few pioneers like Henry Ford had experimented in the new horseless carriage field, but the results of their experiments looked crude, even to themselves, and the appearance in their streets of one of their noisy, and at that time oderiferous, machines was the signal for the gathering of a curious crowd, especially in the case of the too frequent breakdowns.

R. E. Olds had been the first to start 'a real automobile factory' in Detroit, for his Oldsmobile; but by the end of 1904 the city had 17 manufacturers (including the Ford Motor Company, set up in 1903) and 13 accessory makers. In 1904 *The Motor Age* claimed, Detroit made more than 9000 of the 27 000 cars produced in the United States, with Oldsmobile and Cadillac in the lead. Production on a much smaller scale was scattered among several other centres.

Even more impressive than the scale of manufacture was its concomitant, the rapidity with which motor cars had started to penetrate American towns within these few years. Boston was said to have more than 7000 (1000 of them second-hand) and only eight of these imported; Chicago had 2000 and Cleveland more than 1600. At Minneapolis and St Paul a brisk second-hand market was also reported. 'Many of them go to the country; some to the city.' In Tennessee the moneyed people in the cities and larger country towns were competing fiercely to buy the few cars that were available. In southern California there were 2000 cars, though 1904 saw 'practically the first year of real business' in San Francisco. Denver had 500 cars. In St Louis the number of garages had increased from five to fifteen. And so on.[73] According to official sources the number of motor vehicles in the United States grew from about 8000 in 1900 to over 78 000 in 1905.[74] The figures, however, include all types of motor vehicle, steam and battery electric as well as petrol. *Automobile Topics* in the middle of 1902 thought that it was not until 1899 that the number of petrol vehicles throughout the country reached the hundreds and put the total for 1900 at between 1000 and 1500.[75] By 1905, however, with the switch away from steam and battery electric, the overwhelming majority was petrol driven.

Leadership had already passed from the old world to the new. With the market in the Unites States more favourable to mass production, motorisation was to grow at a far faster pace there than was possible anywhere in Europe. It was in America that the advantages and disadvantages of the new technology were first to be fully experienced.

BRITAIN ALSO BACKS LOSERS FOR A TIME

The American motor industry's dynamic growth between 1900 and 1905, between five to ten years after the Bordeaux Race, suggests that there was more to Britain's rather disappointing start than the legal closing of her roads to the new motor vehicles by the Red Flag legislation. The latter was amended in 1896. From 14 November of that year, motor vehicles weighing under three tons unladen – there was no limit to the weight they were allowed to carry – were allowed to travel on British roads provided they did not exceed 14 miles per hour, a speed raised to 20 m.p.h. by another Act in 1903.

In fact, as we have seen, experiments had already taken place before 1896 and an imported Benz was already running on English roads, undisturbed by the police, from November 1894. Very soon after the Bordeaux Race, Evelyn Ellis, sixth son of Lord Howard de Walden and a partner with Frederick Simms in the Daimler Motor Syndicate, brought back to England his P. et L. car which he had previously kept in France. Having imported it via Southampton on 3 July 1895, he ran it ostentatiously and with much sought-for publicity (the Syndicate was looking for orders) to his home in Datchet, near Windsor, and then on to Malvern in Worcestershire. The police looked the other way. In October the wealthy Sir David Salomons, a Cambridge science graduate and keen experimenter who had contacts in Paris, imported a Peugeot. He was then mayor of Tunbridge Wells and, on 15 October, he organised a display of motor vehicles on private land there. His Peugeot, Ellis's P. et L., a de Dion tricycle and a de Dion steam tractor were all displayed to a crowd said to number 5 000.[76] Other vehicles were also imported that year. T. R. B. Eliot, of Clifton Park, Kelso, was first to own a motor car, a P. et L., in Scotland; Walter Arnold imported a Benz; and Joseph Adolphus Koosen of Southsea, who had seen a Lutzmann car in Germany while on a visit, ordered one. It arrived in crates and took some putting together. It was even more difficult to persuade the machine to work, for Koosen did not realise that it ran on petrol. By the end of the year we begin to hear of police prosecutions. Koosen found himself summonsed, and so did John Henry Knight, whom we last encountered nearly thirty years earlier experimenting with a steam carriage. According to a recent writer who has made a close study of the subject, there may already have been twenty-four self-propelled passenger vehicles in Britain by the end of 1895.[77]

The Tunbridge Wells display was soon followed by further activity. *The Autocar* made its appearance on 2 November 1895 and in December, Salomons, a founder member of the Automobile Club de France, took steps to form a British club, intended in the first place to press for amendment of the Red Flag legislation. The inaugural meeting was held in London on 10 December 1895. To it came many of the interested parties, many of them in the trade. It was, however, far from being the impressive group that the French had been able to assemble. Nor was the body then created given a particularly helpful name: The Self-Propelled Traffic Association (S.P.T.A.).[78] It was not an auspicious start. Worse was to follow.

Scenting the impending legislative changes which would open Britain's roads to the new vehicles, unscrupulous company promoters of the worst kind had come upon the scene in search of easy prey. Chief among them was Henry (Harry) John Lawson, the son of a brass-turner who had started in a very modest way in the cycle trade and, having taken out a number of patents, had come to appreciate their appeal. He floated companies to exploit these potentially profitable new ideas, taking for himself and his associates a large slice of the capital so raised in exchange for the patent rights. He had already been involved with the notorious Nottingham speculator Ernest Terah Hooley, in the promotion of the Dunlop tyre and the Swift and Humber cycle companies. In October 1895 he bought the Daimler Motor Syndicate for £35 000, thereby acquiring the Daimler patent rights in Britain. It was he who started the *Autocar* and then set up another company, the British Motor Syndicate, to acquire licences for other motor vehicle patents in an attempt to gain a stranglehold over the whole new industry. Frederick Simms and Evelyn Ellis, partners in the Daimler Motor Syndicate, joined the Lawson camp. Simms understood engineering matters and was concerned, with Robert Bosch, in the development of magneto ignition. That Lawson's knowledge of internal combustion engines was severely limited, is clear from his being taken in by a glib American, Edward Joel Pennington, whose patented motor vehicle usually developed some undisclosed fault just before any contest was to take place.[79]

By mid-January 1896 Lawson and his associates had broken away from Salomons' S.P.T.A. and formed their own pressure group which they named The Motor Car Club. This then included French-born Léon L'Hollier, a perambulator manufacturer in Birmingham who had acquired British rights to the Benz patents, together with other motor manufacturers and carriage builders. The British Daimler Motor Company was also launched in January 1896 (£100 000 in £10 shares and oversubscribed) with Lawson as its chairman, 'to carry on business as electricians, motor manufacturers, cycle makers, mechanical engineers, machinists, fitters and founders'. The de Dion and Bollée rights were acquired and, in May 1896, the Great Horseless Carriage Company was floated. It paid the British Motor Syndicate £500 000 for its patent rights, leaving £250 000 as working capital to start building motor vehicles. But although the new company bought from the Daimler Company part of the remains of the

Coventry Spinning and Weaving Mills, burnt out by fire five years before, it was not until October 1896 that anything is heard of a single British Daimler being produced.[80] Even a year later, the Daimler shareholders were informed that 'very few' vehicles had actually been completed, though 24 were then said to be 'going through the shops' and finished at the rate of four a week.[81] In the meantime, however, Lawson and his friends had reaped rich rewards from their stake in the British Motor Syndicate. In November 1896, the month in which legal restrictions on motor cars were removed, shareholders in the British Motor Syndicate were paid a dividend of 110 per cent in fully paid shares of the Great Horseless Carriage Company, having already received 40 per cent in cash.[82] The British Motor Syndicate then boasted that it was importing more cars from the Continent than any other firm in England, 10 per cent of the price being paid in advance; it threatened any infringer of its patents with immediate prosecution.[83] That enterprise in Britain was being seriously thwarted by the restrictive practices of these predatory company promoters, seems beyond doubt. They had also split the pro-motoring lobby by setting up the Motor Car Club. Salomons, on his side, turned his back on petrol vehicles, over which the syndicate had so much control, and was stumping the country preaching the merits of Serpollet and steam.[84]

The company promoters were nothing if not good publicists. In April 1896 the British Motor Syndicate held what was described, somewhat prematurely, as its first annual meeting. This gave Lawson the opportunity to conjure up large-scale and profitable visions of the industry's future which would create many ancillary jobs. He also indicated the market he saw for the Syndicate's vehicles, which suggested that it was to consist of 'carriage folk':

> There is no doubt that in every great manufacturing town, a considerable industry will spring up in connection with these horseless carriages because there will always be required repairing and alterations, and there will be new employment in regard to the driving of the cars. . . . The motor man will have to take the place of the coachman, and he will require, of course, to be educated for his work. It would be quite out of character for a lady, say, attending an evening party to drive her own motor car; that cannot be expected.[85]

A month later he was on his feet again, this time at a banquet held by the Motor Car Club, by then said to have 300 members, following a demonstration of (imported) motor vehicles in the grounds of the Imperial Institute. On that occasion he countered the opposition which was already beginning to appear from horse lovers. Motor cars, he argued, ought to appeal to such people, for the new vehicles would, at long last, liberate these fine animals whom mankind then so often overworked and ill-treated: 'The friends of the horse blindfold him with blinkers', he said, warming to his theme, 'gag him with iron bits and bearing reins, and sit over him with whips, putting him under the lash as a reward for his hard work.'[86]

Lawson's main publicity coup was the arranging, for the very day that legal restrictions were removed, 14 November 1896, of what he originally called an autocar ride to Brighton[87] but has since been celebrated annually as The Brighton Run. Originally intended as a sedate procession, beginning and ending with an impressive meal and punctuated by a luncheon stop at Reigate, and to be headed, of course, by Lawson, dressed in his uniform as President of the Motor Car Club, riding in the P. et L. car in which Levassor had won the Bordeaux Race, it became, in fact, a rather disorganised scramble to attract publicity. It began, as planned, with a splendid meal (10 shillings a head with wine) attended by 150 people at the Metropole Hotel, Northumberland Avenue, the first car to leave at about 10.30. Of 58 machines, 32 or 33 started; but the crowds on the first stretch, out to Brixton, were dense, the weather was wet and the cars were, as ever, unreliable. As the *Autocar* politely put it: 'By the time Brixton Rise had been surmounted, the procession was considerably scattered.' When, soon after Brixton Rise, the pilot car's engine stopped and it had to be allowed to cool, the followers broke ranks and turned the procession into a race. The cars that reached Reigate arrived at such considerable intervals, reported *The Times*, 'that there was little enthusiasm excited'. It must have been something of an anti-climax for the guests who had reached the luncheon stop there by special train.

A Bollée tricycle, which by-passed Reigate altogether, arrived in Brighton first, at 2.25 p.m., followed by a second half an hour later (they were driven by Léon and Amadée Bollée themselves and must have averaged 16 or 17 miles per hour). Then, at 3.46, came the American, Frank Duryea, in his own car, who later claimed to have covered the distance in the fastest running time, for he had obeyed the rules and stopped in Reigate for just over an hour.[88] The *Autocar*, however, contented itself by noting that the two Duryea riders were 'carrying enough mud on their hats, faces and clothes to start a small estate'. The Duryea car was certainly responsible for the only serious injury: it knocked down a publican's daughter in Crawley and fractured her skull.[89] Other vehicles had mixed fortunes. The battery-electrics, we are told, 'had not started with any intention of going farther than Brixton and these one by one found their way to the railway station and made their way to Brighton by this means'. The correspondent of the *Daily Graphic* took four hours to cover the 21 miles to Reigate. He, too, took the train to Brighton. A breakdown van itself broke down and limped into Brighton at 3 o'clock the following morning.[90] Meanwhile Lawson the Leader had got there at 4.30, behind the Bollées and the Duryea; but he did have the satisfaction of knowing that he held the British patent rights for the Bollée tricycles and for the P. et L. vehicles, of which there were no fewer than nine in the first 13 which had arrived by 6 p.m. Once again, the petrol engine had been shown to be the most successful.

At the dinner at Brighton that evening, Gottlieb Daimler, who had set off from London in car number 2 immediately behind Lawson, was presented with the Motor Car Club's silver trophy, to be given by him to P. et L. Levassor himself was not present, presumably because of his recent accident in the Paris to Marseilles race. Lawson, for his part, lost no time in capitalising on the publicity derived from the Brighton Run. The British Motor Syndicate Ltd offered the public £1 000 000 in £1 shares, then said to be standing at £3 each.[91]

The journal *Engineering* put Lawson's company-promoting optimism into a more realistic perspective:

To the thoughtful engineer' [it commented in November 1896] last week's saturnalia could not have been otherwise than a melancholy spectacle . . . The proceedings only further illustrated what was pretty well known before – that at present the science of designing mechanical road carriages is in a very elementary stage, and much remains to be done before it can be claimed that a motor car has been produced to take its place on the highway as a commercially successful carriage, and one which can be relied upon to do general work in competition with horse-drawn vehicles.[92]

The Bordeaux Race on June 1895 may have removed the new vehicles 'from the stage of mere experiment', as *The Autocar* claimed, but, more than a year later, they were still in an early stage of development and far from reliable. The main works in France did not produce, on average, more than one car per day.[93] The continental manufacturers, that is to say, did not yet have any great lead. Britain had the best 'class' market in Europe which could have been satisfied by its own industry if this had been quickly and carefully developed during 1895–7. Unlike the French, however, who used their patent rights to build and demonstrate their own cars, Lawson and his associates used theirs to raise capital from the public, most of which went into the promoters' pockets, not into the new industry. While the French put their own vehicles to the test in annual contests from 1894, the British licensees used the first opportunity of a showing on the roads to display foreign vehicles which had been imported into the country (duty free) from which they collected royalties and commissions.

There were, however, British engineers, practical men, who were following the continentals' lead and carrying out their own experimental work; but they lacked the capital to which Lawson had access and the encouragement which he could give. Herbert Austin, for instance, manager in Birmingham of the Wolseley Sheep-Shearing Machine Company Ltd, a manufacturing business which assembled bought-in parts and had produced a Bollée-type three-wheeler by the middle of 1896, in collaboration with Mulliner, the carriage-maker. It was little more than a motor 'on wheels'; but he persuaded his principals to advance £2 000 'to take up the manufacture of motor carriages', a grand-sounding aim which, in fact, resulted only in further development work as a sideline while he was building up the company's main business in sheep-shearing equipment. Progress was inevitably slow. The first Wolseley motor tricycle was exhibited at the National Cycle Exhibition in December 1896 and an improved three-wheeler was driven, with two passengers, the 250 miles to Rhyl, on the North Wales coast, and back in June 1898 without a breakdown. By December 1899 Austin had built a four-wheel car. 'Runs splendidly', he reported, 'This is also a business we have a good chance in.' What might Austin have achieved had Lawson used some of his resources to back him from 1895? In the event, it was Vickers, Son & Maxim, the arms manufacturers (Maxim, the father of the United States motor-car experimenter whom we have already encountered), who, spurred on by the Boer War, acquired the machine-tool and motor-car part of the Wolseley business and Austin with it. They formed the £40 000 Wolseley Tool and Motor Car Company to develop and market Wolseley cars. But this was not until 1901. Austin did not set up his own company at Longbridge, just outside Birmingham, until 1906.[94]

Another talented motor engineer, Frederick W. Lanchester, who had seen the Daimler and Benz machines at the Paris Exhibition in

1889, worked in 1895 on what has been described as 'a unique concept that owed nothing to Daimler and Benz with their stationary engine, horse carriage and cycle background'. It was running on the public road by February 1896. But Lanchester lacked capital and business guidance, too. Although he set up his own company in 1900, he soon lost control of it, becoming eventually, early in the First World War, technical consultant to Daimler (by then in good hands soundly based and successful).[95]

The established pedal-cycle makers, such as Singer, Riley, B.S.A., Swift and Sunbeam also went into motor manufacture in, or just after, 1898 and were followed, a few years later, by Lea-Francis and Rover. That they did not diversify earlier, was clearly because of the great boom in the cycle industry between 1895 and the middle of 1897 which kept them all extremely busy and very profitable making pedal cycle. When the boom burst, prices fell and the machinery installed during the years of acute labour shortage was used to cut costs. 'The vast majority [of cycle firms]', commented *The Economist* at the end of 1898, 'have nothing before them but reconstruction or bankruptcy.'[96] With cheaper new cycles, and even cheaper second-hand ones, ordinary people could afford to take to the roads. The swells in their cycling club uniforms felt out of place and began to turn to motor vehicles. Membership of the Cyclists' Touring Club, over 54 000 in 1898, was down to 20 000 two years later.[97]

In the meantime the growing number of motorists joined the Automobile Club of Great Britain and Ireland (later the Royal Automobile Club). This had been founded at the end of 1897 to cater for the few members of the carriage-owning class who were becoming enthusiastic about motoring. Lawson's star was already on the wane – in the absence of motor cars, the Great Horseless Carriage Company's £10 shares were down to 22*s* 6*d* and Daimler £10 shares fetched £4[98] – and his name is not to be found among the Automobile Club's sponsors. It was Simms who guaranteed the Club's original premises, six rooms on the ground floor of 4 Whitehall Court in London, overlooking Embankment Gardens.[99] By 1902, when the membership had grown to over 2000, the Club moved into more appropriate premises at 119 Piccadilly, formerly the home of a marquess. It was quickly converted into 'a spacious and comfortable club house' with 'a fine set of rooms, including morning coffee, card, billiard and bedrooms'. Large livery stables to the rear were turned into a garage.[100]

Among original applicants for membership in 1897, said to total

200, were those associated with motor engineering, such as W. Worby Beaumont, a consultant and speaker for the S.P.T.A., W. H. Preece, engineer-in-chief at the Post Office, and Hiram Maxim; and others concerned with road transport, like the Earl of Shrewsbury and Talbot, the first person to run horse cabs with rubber tyres in London and Paris, and Evelyn Ellis, whom we have already encountered.[101] Also among the founder members was the Hon. Charles S. Rolls, who was to play a major part in the development of motors and motoring before he was killed in a flying accident in 1910 at the age of 32.

The third son of the recently ennobled Lord Llangattock, Charles Rolls had interested himself in the practical aspects of electricity at Eton before going up to Cambridge, where he read Mechanical Engineering and Applied Science from 1895 to 1898 as well as captaining the university bicycle team. He took an early interest in motor cars and imported a Peugeot just before the repeal of the Red Flag legislation. (It took him nearly 12 hours to drive it, illegally and at night, from Victoria Station to Cambridge. On the way a friendly policeman asked for the experience of a ride.) Rolls soon became a well-known motorist and, in partnership with Claude Johnson, the Automobile Club's first secretary, he formed Rolls & Company in 1902 to sell Panhard (P. et L.) and other imported cars and to provide repair facilities. Early in the following year they described theirs as 'the first and only properly-equipped repair shops in the country'.[102]

Another leading figure in the motoring world in Britain and in the Automobile Club had a background in many ways strikingly similar to Rolls's. John Montagu also went to Eton, was a keen bicyclist and most interested in practical engineering – he worked for a time as an ordinary apprentice in a railway workshop. He was also the son of a recently ennobled peer, though heir to the Beaulieu estates which he was to inherit in 1905. He was eleven years older than Rolls, a Member of Parliament from 1895 and a little later to take up motoring. He first rode *on* a friend's Panhard from London to Windsor in 1897. (Motorists were perched high up in the open in these superior vehicles.) His recollection of the thrilling experience, travelling at about 15 m.p.h., was, no doubt, typical of many others:

> The people by the roadside were not only interested but alarmed. Small boys and girls shouted rude remarks at us as we passed, some people rushed into their houses, most horses cocked up their ears and were terrified at our approach and in some cases reared up on their

hind legs and attempted to bolt. Public attention, in fact, was divided between curiosity and cursing.

But it is impossible to describe now the vivid impression of that first journey. I could hardly sleep that night. It was difficult to think of anything for a few days afterwards, and I at once determined to get a car for myself. . . .

I cannot describe the joy which this first Daimler car gave me when it was delivered at my country house at the end of the summer of 1898. I began to realise at once that the mechanical vehicle was going in time to produce a wonderful revolution in our transport methods. Often I waxed eloquent before older men about its possibilities, but, of course, was generally laughed at and even severely snubbed. Later on, when some of my prophecies became true, I became disliked, especially by my 'horsey' friends. . . . The results of this abuse were seen in various directions. Stones were often thrown at passing motorists, and many persons would hardly speak to a well-known motorist like myself. Indeed, I was considered by some of my relations to be a dangerous revolutionary. . . . Hotel keepers generally regarded us as people not to be admitted. . . . One irate proprietor said he was not going to have any of these contraptions near his place, for they might blow up at any time.[103]

Such hostility was not shown, however, by the Prince of Wales, shortly to become Edward VII. He had already become interested in cycling and had encouraged other members of the royal family and their friends to follow his example.[104] Now, unlike the German Kaiser, he supported motoring and gave the signal to others at the top of British society. When, in 1900, he took delivery of his first motor vehicle, he announced: 'I shall make the motor car a necessity for every English gentleman',[105] John Montagu, who was among those who had given the Prince a ride in his car a little earlier and had joined the Automobile Club in January 1899, was soon to be the motorists' spokesman in Parliament, author of many articles on the subject in the press and editor of *The Motor Car Illustrated* which he launched in May 1902. Two years later he was to bring about great changes in the leadership of the Club itself.[106]

David Salomons also became a member of the Club's committee but the name of Harry J. Lawson was absent. He had been sentenced in 1902 to a year's hard labour for making false statements to persuade the gullible to buy shares.[107] Meanwhile the British Daimler

Company, which had experienced great difficulty, received the Royal Warrant in 1902 and was later to be put on a sound financial footing.[108] But many promising years had been squandered.

The Automobile Club, respectable from the outset and increasingly prestigious, followed the example set by its French counterpart, acting as both a national organisation to arrange trials and rallies and as a focus for the various regional clubs which were springing up. Many members were aspiring, rather than actual, motorists; they did not yet own vehicles of their own. The most outstanding and memorable of the early trials was one of over 1 000 miles held in 1900. Sixty-five vehicles, tricycles and quadricycles as well as motor carriages, set out from London on 23 April and proceeded via Bristol, Birmingham, Manchester and Carlisle to Edinburgh; then south again, through Newcastle, Leeds, Sheffield and Nottingham to London. The aim was to display the vehicles on the road, to put them through their paces on hills and to display them for a whole day in the main towns en route. Rolls, Montagu, Austin and Lanchester were among those taking part. Thirty-two vehicles managed to cover the distance under their own power and Rolls won the gold medal.[109] Petrol vehicles were certainly shown off to the public and seen to be a little less unreliable than they had been a few years before; but most of the total of about 1 500 in 1900 had been imported.

BRITAIN BECOMES INTERESTED IN HEAVY GOODS VEHICLES AND MOTOR BICYCLES

Underpowered and undeveloped petrol engines were capable of moving light passenger cars and vans before they could propel heavier goods vehicles. Here, however, Britain, well-advanced in steam engine technology and with 8 000 traction engines on its roads, was well poised to develop a satisfactory steam wagon.

The initiative came from Liverpool, which, in the mid-1890s as in the mid-1820s, was concerned with the difficulty and cost of sending goods to Manchester and east Lancashire. In the 1820s the Liverpool men had resolved their problem by building the world's first intercity railway to compete with the intransigent waterways: in the 1890s they were confronted with a similar challenge. Manchester was building a vast ship canal which would soon establish that city as one of the country's leading ports, thus bypassing Liverpool altogether. The Liverpool men, anxious to retain its traffic to Manchester and east Lancashire, begged the railways to improve their services and lower their charges. When the railways refused, these Liverpool men, like their predecessors, sought an alternative form of carriage. Mechanised road transport provided a possible answer.

The agency whereby this possibility was explored was a branch in Liverpool of the Self-Propelled Traffic Association. The leading spirit was a young man of 21, Edmund Shrapnell-Smith, a Liverpool University College chemistry student employed by the United Alkali Company in Widnes and, more importantly for our purposes, a keen motoring enthusiast who had attended the foundation meeting of the S.P.T.A. in London in December 1895. (Sir) Alfred Jones, senior partners in the Elder Dempster shipping concern, a leading shareholder in the Liverpool Cartage Company and a prominent local figure was also very much involved.[110]

On 9 September, two months before the repeal of the Red Flag legislation, W. Worby Beaumont gave a talk to the Liverpool Chamber of Commerce in which he argued that the steam engine, because of its greater flexibility, was better suited to heavy haulage than the petrol engine. 'Great as the promises are of motor vehicles for light work', he went on, '. . . the complete motor vehicle for heavy traffic has yet to be designed, although the experience of our road locomotive makers will perhaps make that design comparatively easy. . . .'[111] The Liverpool press was quick to relate his remarks to the port's particular problem:

> The competition of Manchester, in regard to the shipping trade of Liverpool has now been sufficiently felt to cause the shipowners and merchants of the latter city to become alive to the necessity to take special measures to retain their trade . . . It is intended to introduce autocars, each of which shall be capable of drawing three wagons carrying 10 tons of produce. The wagons are to be loaded at the ship's side, and the special advantage of their use will be that goods can in this way be conveyed without transfer to other vehicles direct to the warehouse, wherever this may be situated. . . . It is expected that by this means the cost of conveying merchandise to and from Manchester and other towns will be reduced to a point with which the Ship Canal cannot compete, and the further expectation is that the railway companies will be compelled in self-defence to lower their rates. . . .[112]

A month later a group of Liverpool business men, supported by Henry Selby Hele-Shaw, professor of engineering at the University

College there, waited upon Lord Derby, told him that they intended to offer prizes 'in order to get the best inventions and productions to cheapen the means of transit of goods [shades of the Rainhill locomotive trials again] and by lectures and discussions to increase public interest in the matter'; and invited him to become their President.[113] He agreed; and on 27 October David Salomons gave the inaugural address to the Liverpool branch of the S.P.T.A. The steam engine, he claimed, was 'simplicity itself and of a character known to almost every village smith'.[114]

Unfortunately the development of a satisfactory steam wagon was not so easy as that, as the enthusiasts in Liverpool were to discover. Intense feeling against the railways spurred them on. 'The only argument the railways would listen to', declared the Liverpool shipowner, Alfred Holt, 'was that of competition and so far they had punished Lancashire severely'.[115] The first trials were held over five days in May 1898 and included circuits of 36 miles round Liverpool's hinterland with gradients of up to 1 in 20. Four steam vehicles took part, one built by the Liquid Fuel (Lifu) Company which carried a load of two tons, two by Thornycroft of London (a tractor and trailer carrying eight tons and a vehicle carrying two and a half tons) and one by the Lancashire Steam Motor Company of Leyland, Lancashire (four tons) which won first prize (£100). The judges' verdict was that 'they had demonstrated the practicability of moving weights up to five tons at good commercial speeds on vehicles which weigh under three tons', the maximum unladen weight permitted under the 1896 Act.[116]

Two further trials were held in Liverpool, in 1899 and 1901, after which Shrapnell-Smith and some friends formed The Road Carrying Company Ltd there to acquire the 14 vehicles on trial and to use them for haulage between the docks, various parts of the city and as far afield as Preston; but the venture failed after a year. Shrapnell-Smith took a job in London and, in 1905, became editor of a new journal *The Commercial Motor*. Meanwhile a number of the steam wagon makers, such as Leyland Steam Motor, managed by Henry Spurrier, who had gone to America and spent eight years in railway shops there, and Fodens in Cheshire, were making progress. By the end of 1903 the number of steam wagons on British roads was put at about 1000.[117] 'The English concerns seem to have operated far ahead of their Continental rivals so far as the commercial vehicle goes', commented an American journal in 1902.[118] Early in the following year Thornycroft was advertising there, and Coulthard of Preston participated in an American commercial vehicle contest.[119] After March 1905 an increase in the permitted unladen weight in Britain, from three to five tons, helped British vehicle manufacturers. Some of them, like Leyland, diversified into heavier petrol lorries and buses. 'Below net loads of two tons', Shrapnell-Smith was to report a few years later, 'the petrol-propelled vehicle has a virtual monopoly of use in England; above that, it shares the trade with steam.'[120] Steam lorries did, in fact, carry loads of raw cotton from Liverpool to Manchester in, or soon after, 1905. They were to be seen on Britain's roads until the Second World War.

Motor bicycles also sold well in Britain. Costing about £40 new and perhaps £15 second-hand early in the present century,[121] they appealed to those members of the middle classes who could not yet afford a motor car either in its more substantial, Panhard-type, form or as a quadricycle or tricycle. These vehicles sold for upwards of £150. There were technical problems, however, in the development of an engine compact enough to be put into a bicycle frame. A German machine, made by Hildebrand and Wolfmüller, was tried out in the mid-1890s but without much success. Some British attempts were made about that time. One pioneer, A. G. New, rode a mile in just over $2\frac{1}{4}$ minutes on Catford Cycle track, London, and claimed it to be 'the fastest unpedalled mile ever done on a cycle in this country'; but customers were no more enthusiastic to acquire his machine than they had been to buy the original motor cars. It was described as 'somewhat ungainly looking' with 'an excessive amount of vibration from the handles'.[122] Michel and Eugène Werner, Russians who had moved to Paris, produced a more satisfactory model in 1896 and started to sell it in greater quantity from 1900 when, instead of a motor in front of the handlebars driving the front wheel, a larger motor was placed just above the pedal crank and drove the rear wheel. 1200 Werner bicycles were sold in 1900 and by September 1902 the company had sold 6000 bicycles altogether.[123]

By 1902 the motor bicycle was becoming popular in Britain. At the end of March in the following year *The Autocar* thought it worth while to launch a sister journal, *The Motor Cycle*. 'Two brief years ago', one of the new journal's writers recalled in April 1903, 'the motor bicycle was considered and looked upon as an ingenious toy . . . and . . . it was predicted that its rise and fall would take place without creating any serious disturbance in the cycling world.' Instead, the better-known manufacturers were deluged with orders.[124] At first, many bicycle motors were bought in, often from abroad; but by the beginning of 1904 it was reported that Britain no longer imported much 'except certain parts that are to be obtained cheaper abroad . . . and these are very few now'.[125] Many of these early machines appear to have been of poor quality, however, and to have gained a very bad name indeed;[126] but better machines were produced from 1905. Sidecars and forecars were fitted to the more powerful models. Britain forged ahead in motor bicycle production, although it was not until 1908 that exports of motor cycles exceeded imports by value. By 1910 they were three times as large and by 1911 seven times.[127]

THE END OF THE BEGINNING

Great progress obviously occurred in Europe and America during the decade after Levassor's remarkable feat of 1895. Motor vehicles became more substantial, more powerful, more sophisticated and more reliable; motor bicycles were introduced, some progress was made with heavier goods vehicles and motor buses began to appear. Perhaps most significant of all, during the first few years of the present century such strides were made in the United States that the scene was set for the advent of the Model T Ford and for the assembling of complex machinery on a far greater scale than had ever been known before.

In the mid-1890s the world was not only still horse-drawn but was also being urbanised at an unprecedented rate. The streets in all the growing towns were becoming increasingly congested and made more odious by the rising volume of horse droppings which, apart from the unpleasant smell and filth underfoot (despite the activities of crossing-sweepers), were breeding grounds of disease, transmitted by the vast swarms of flies. The earliest motor vehicles, because of primitve methods of oiling, dropped oil on the road and, because of inefficient combustion, poured smoke into the atmosphere. They also created a vast amount of noise. But these forms of pollution grew less with time.

At first, like high-spirited horses, motor vehicles were a challenge and appealed to the sporting instinct. The hardy motorist and passengers, sitting out high up in the open without the protection even of a windscreen, presented a fearsome aspect in goggles, masses of warm clothing, heavy gauntlet gloves and other motoring gear. In the early days human beings as well as horses were frightened by the appearance of the occupants as well as by the vehicles.

There was also the risk – indeed the probability – of breakdown. In the absence of garages the motorist had to take with him not only a set of tools if he ventured far from home, but also a range of spare parts as well. As one motoring writer advised his readers:

> An assortment of nuts and bolts of the various sizes used in the engine should . . . be carried and a complete set of valves and valve springs. . . . A set of brasses and a supply of insulated wire are useful 'spares'. Spanners to fit all the nuts should, of course, form part of the regular equipment of a car, as well as a Stillson's wrench and a monkey wrench. A small table vice is a very good thing to carry and does not take up much room; there should also be files, punches, a hammer, screwdriver, a cold chisel, gas pliers and a soldering outfit. A good roll of copper and steel wire, a little copper piping, asbestos washers and a length of asbestos cord, and a couple of yards of india rubber tubing of the same gauge as the pipes of the water system must also be carried. . . .[129]

These recommendations appeared in a book published in England in 1904; but they obviously referred to earlier years, for by then garages with repair facilities, few and far between before about 1900, were spreading rapidly. The Automobile Club issued lists.

At first petrol was also often hard to find despite an established distribution network which already supplied Russian and American oil for lamps and lubrication. At first, however, volatile spirit, of the sort needed for motor vehicles, was not in much demand, apart from its use as a cleaning agent. In any case it had a high flash point, was dangerous to store in quantity and was subject to strict fire regulation in Britain. The French, as has been noticed earlier, seem to have been better served both in quality and availability. French visitors certainly complained about the standard of the English product at the time of the Brighton Run. In France, they said, *essence de pétrole* was all doubly distilled. 'So careful are they as to the quality of their fuel', reported a correspondent from Paris in mid-December 1896, 'that some of them carry instruments to measure the density of the petroleum spirit they purchase at the stores; but this is being rendered unnecessary now by the enterprise of the leading oil companies who are supplying specially prepared petroleum in tin cans . . . and this can be procured in at least one oil shop in each town throughout the country'.[130] This brought an immediate response before the year was out from Carless Capel & Leonard of Hackney Wick in East London:

There can be no doubt that your warning against using the ordinary lamp oil or paraffin in spirit motors . . . is most timely and that even deodorized benzoline, such as is usually sold in oil shops etc. . . . is not suitable. In fact, no satisfactory results can be depended upon unless a pure doubly-distilled spirit can be used. . . . We think we, as English manufacturers, should be allowed that we can and do produce a spirit from petroleum called petrol which is both purer and more efficient for use in motors than the French naphtha. . . . It is probable that the air supply in their motors is regulated so as to be best adapted to the heavier and less pure spirit, whereas we have little doubt that, if a slight adjustment were made in the air supply, the advantage in using the purer and more volatile oil would be at once apparent. Petrol has been used continuously and almost exclusively for the Daimler launches ever since they were introduced into this country more than five years ago, and quite lately it has been employed largely for the motors of autocars. Indeed, it was adopted recently by the Motor Car Club for the Brighton run as the best spirit for use in petroleum motors.[131]

It is clear that Carless Capel already had a clear understanding of what was required. They even seem to be using the word 'petrol' as a trade name for their double-distilled product. The main distribution difficulty seems to have arisen from the slow growth in motorists' demand. An 1898 list of petrol stores, forty-one in number, were all in the south of England.[132] When the 1000-mile trial was organised in 1900 the Automobile Club advised entrants to write ahead to the various stores on the route to make sure that the petrol was available. A number of the dealers named were cycle shops and chemists who acted as local agents either for Carless Capel or for the Anglo-American Oil Company (Standard Oil's UK subsidiary) and two other major importers.[133] By then Standard Oil was already distributing its petrol in French blue tins. Marcus Samuel, about to market Shell petrol, distilled from oil imported in his tankers from the East, decided that his product should be sold in red tins.[134] (Petrol was not generally sold in Britain from pumps until after the First World War.) The number of suppliers grew much more rapidly after 1900 and by about 1905 the coverage was extensive. UK imports of motor and other spirits were rising fast: from 9 m. gallons in 1903 to near 19 m. in 1905 (and to $33\frac{1}{2}$ m. in 1907). The oil companies altered the specific gravity about this time from 0.680 to 0.715 to obtain more motor spirit from a given volume of crude oil.[135]

The availability of more reliable vehicles and the spread of repair facilities and petrol supplies encouraged more people to become motorists. That more of them were using their vehicles for business purposes is clear evidence of their advantages over the pedal cycle, horse-drawn transport and even, in some cases, of the train. Levassor's very early market survey was confirmed. Doctors were among the first to find motor vehicles very helpful to them and so were travelling salesmen. In 1903 one of the latter, apparently based in the Birmingham area, decided to use his recently purchased motor bicycle to pay an urgent call upon a customer seventeen miles away, despite his customer's urgent plea that he should not risk a breakdown. He was able to set off at once instead of having to wait an hour and a half for the next train, and in fact reached his customer within the hour and was able to make two other calls before the day was out.[136] Farmers, too, found motor vehicles useful for getting quickly into the local town and back again in half a day instead of taking the whole day on the errand. 'While a man would be getting a horse ready, he is often at his journey's end', wrote one of them.[137] This sort of experience was repeated in France, the United States and elsewhere by this time.

The adventurous and the well-to-do had been running motor vehicles in other parts of the world some years before this – in Melbourne, Cape Town and Rio by 1898, for instance[138] – and the pioneers were soon followed by others. British motorists went on tour not only on the Continent but also as far afield as Egypt, China and Japan.[139] In Britain, more of the well-to-do were buying the superior models which were reaching the market by 1905. The 'first simple, silent Rolls-Royce' was advertised in December 1904 and a Rolls-Royce Pullman Limousine was exhibited at Olympia in 1905.[140] The rich were soon changing their splendid horse-drawn carriages for more economical, and equally splendid, motor cars. Stables became garages. The vast population of transport horses at last stopped growing, and so did the city nuisances associated with them. In public transport, horse-drawn trams had been largely replaced by electric traction. Motor buses were introduced on a large scale, somewhat prematurely, in 1905 and motor taxicabs followed soon afterwards. The years of early development were over and the pendulum was at last swinging definitely in favour of petrol-driven motor passenger vehicles in the more advanced countries. For heavier goods transport the transition was much more protracted, thus cushioning horse transport operators, and the industries dependent upon them, against sudden unemployment.[141]

Such statistics as we possess for the United States, France and Britain support the view that another landmark was reached about 1904/5. These figures are difficult to interpret, however, partly because of the problem of disaggregating petrol-driven vehicles from all those which were self-propelled and partly because of the definition of motor cycles, which earlier meant light three- or four-wheelers and later came to mean mainly bicycles.

Table 1.2   Estimates of motor vehicles in use in the USA, France and Britain, 1900 and 1904/5

	USA	France		Britain	
		Cars	Cycles	Cars	Cycles
1900	*c.* 1500	(Dec) 2897	11 252	*c.* 1500	
1904	54 590 + 700 trucks	(Dec 12 984 1903)	19 816	(Apr) 14 339	15 637
		(Dec 17 107 1904)	27 435		
1905	77 400 + 1400 trucks	(Dec 21 523 1905)	29 954	(Sep) 35 145	35 367

Sources: Historical Statistics of the United States (Washington, 1960) p. 462.
Appendix 1 to Dominique Barjot's Chapter 15, later in this volume.
   Return on Motor Cars, British Parliamentary Paper (1904) 292 LXXIX
Royal Commission on Motor Cars, vol. II, 1906 [Cd 3081] XLVIII

Britain had fuller and more reliable returns from 1 January 1904 when official registration (and number plates and a higher, 20 m.p.h., speed limit) were introduced.[142] If the American estimate relates to the end of 1905, it is broadly in line with returns collected for the British Royal Commission on Motor Cars a few months earlier: 63 373 motor vehicles in sixteen states (22 420 of them in New York State and 14 050 in New Jersey).[143]

Between 1900 and 1904/5 both the United States and the United Kingdom caught up upon, and went ahead of, France, which had been far in the lead in 1900, even though the rate of growth in France was quite impressive. Britain and the United States were level pegging by 1905, which reflected well on Britain with half the United States' population. Although many of Britain's motor vehicles were still imported, her industry was growing stronger, particularly in the manufacture of motor bicycles and heavier vehicles (buses and lorries). From 1905 onwards, however, the United States gained from its advantageous market position, for it had more potential buyers of cars not only because of the larger population but also because of the greater social depth of demand. 'The same class of people that find it easy to pay $650 for a car in the United States, find it difficult to lay aside half that sum here', an experienced dealer in U.S. cars in Britain was soon to record.[144] By 1908 motor-vehicle registration in the United States had reached 200 000; in Britain it had reached 154 000 and in France only 65 000. America was poised for a massive take-off which, as we saw at the beginning of this chapter, was to bring motor-vehicle registration there to 12 500 000 by 1921/2 (and the whole of the rest of the world to only 3 000 000, 827 000 of these in Britain and 330 000 in France). America's great lead was due mainly to widespread ownership of private transport. Elsewhere, buses and lorries were relatively more important and to that extent the effect of the spread of motor vehicles was not so geographically unbalanced as might appear at first glance. The new technology was to have a growing influence in all countries, developed or developing, as many of the subsequent chapters in this volume will show.

Notes

1. Some years ago a small Social Science Research Council grant enabled me to employ during a summer vacation four graduate students (Stephanie Diaper, Jeanne Golay, David Hebb and Peter Lyth) to compile a reading list and to work through relevant British and American consular reports of the pre-First World War period. I am glad to acknowledge their help and encouragement at an early stage of this work. Miss Janet Jempson assisted me when I wrote a brief history of her family's road haulage business. To her and to her brother, Jonathan, I am grateful for enabling me to gain first-hand insights into the freight transport aspects of this subject. Similarly, Michael Mason, one of the BBC's most able and gifted radio producers, gave me the opportunity to research into, carry out interviews for and write a programme about the development of motor-vehicle maintenance and garage provision, another rather neglected topic. I should also like to acknowledge Megnad Desai's help in securing some financial support from the LSE Economics Research Division, which enabled me to employ Patience Champion briefly as a part-time research assistant. Another LSE colleague, Robert Estall, saved me much time by introducing me swiftly to

the most helpful reading on America's road development. Members of the Modern Economic History seminar at London University Institute of Historical Research made valuable comments upon a much earlier, and rather half-baked, version of this chapter.

2. Figures for 1980 rounded up to the nearest 10 million (*United Nations Statistical Year Book for 1981* (New York, 1983) table 190. These statistics exclude the USSR; but in the chapter of this present book (Chap. 13) on Soviet Asia, it is stated that there are now 20 million cars in private ownership in the Soviet Union. This figure has been included in the total in the text. The United Nations table, for some extraordinary reason, excludes motor cycles. The Motor Cycle Association of Great Britain Ltd has, however, collected world statistics for 1980, including the 10.3 million cycles in the Soviet Union, but excluding China and all countries in South America apart from Brazil and Venezuela. I am indebted to Charles Smart of the Motor Cycle Association for this information. The U.N. statistics also exclude ambulances, hearses, military vehicles operated by the police or other government security organisations and I have not been able even to hazard a guess at the total number of these. Clearly the figure of 470 million is a very conservative one, partly because of these omissions and partly because of the increase in the number of vehicles since 1980. Tractor totals are to be found in ibid. table 98.

3. *The American Automobile (Overseas Edition)* (Mar 1940).

4. *United Nations Statistical Year Book for 1955* (New York, 1957) table 135.

5. U.S. Department of Commerce, *Commerce Reports*, 12 Feb 1923, pp. 416–17 and correction in ibid. 19 Feb 1923, p. 485.

6. James J. Flink, 'Henry Ford and the Triumph of the Automobile', in Carroll W. Pursall Jr, *Technology in America. A History of Individuals and Ideas* (Washington, D.C., 1979) p. 181.

7. *The American Automobile (Overseas Edition)* (Mar 1940).

8. *United Nations Statistical Year Book for 1970* (New York, 1971) table 148.

9. B. R. Mitchell and Phyllis Deane, *Abstract of British Historical Statistics* (Cambridge, 1962) p. 230.

10. F. M. L. Thompson, 'Nineteenth-Century Horse Sense', *Economic History Review* (Feb 1976).

11. This and the next three paragraphs are derived from Anthony Edward Harrison, 'Growth, Entrepreneurship and Capital Formation in the United Kingdom Cycle and Related Industries, 1870–1914' (Ph.D., University of York, 1977). Dr Harrison's untimely death robbed the world of a most promising transport historian. Some main points of the thesis, posthumously edited by Professor Charles Feinstein, are contained in A. E. Harrison, 'The Origins and Growth of the U.K. Cycle Industry to 1900', *Journal of Transport History* (Mar 1985).

12. *The Autocar* (2 Nov 1895).

13. T. R. Nicholson, *The Birth of the British Motor Car*, vol. 2 (1982) pp. 170 f.

14. Quoted ibid. p. 186.

15. Ibid. pp. 218–19, 279–82.
16. Report from the Select Committee on Locomotives on Roads, 1873 [312] XVI, p. 3, and evidence of Thomas Aveling, questions 172 and 174.
17. Select Committee on Traction Engines on Roads, 1896 [272] XIV, p. iv, and evidence of E. B. Chittenden, questions 224, 238, 241.
18. Nicholson, op. cit. vol. 2, pp. 239f, 274, 284–6, 309, 332. For Knight see *The Autocar* (11 Apr 1896) and for Blackburn, ibid. (11 Jan 1896).
19. Nicholson, vol. 2, p. 283.
20. James M. Laux, *In First Gear. The French Automobile Industry to 1914* (Liverpool, 1976) p. 3.
21. Nicholson, op. cit. vol. 2, pp. 314–16; Laux, op. cit. p. 19.
22. The general spread of electric tramways is discussed in T. C. Barker and Michael Robbins, *A History of London Transport*, vol. 2 (1974) pp. 16–19.
23. Ibid. pp. 21–34; John P. McKay, *Tramways and Trolleys. The Rise of Urban Mass Transport in Europe* (Princeton, 1976); George W. Hilton, *The Electric Interurban in America* (Stanford, 1960).
24. John B. Rae, *American Automobile Manufacturers* (New York, 1959) pp. 7, 68. For the British activities, see Barker and Robbins, op. cit. vol. I, pp. 296, 300, and vol. 2, pp. 120–1 (which also mentions Radcliffe Ward's attempts between 1888 and 1899 to develop a satisfactory electric bus) and Nicholson, op. cit. vol. 2, pp. 305–6; vol. 3, pp. 335–6. For Ayrton see the *Dictionary of National Biography*.
25. The car of the Viennese Siegfried Markus dates only from about 1890 and not from the 1870s as has been claimed (see Laux, op. cit. p. 6 n.). The claim of Édouard Delamare-Deboutteville, member of a family cotton-spinning business at Montgrimont, near Rouen, has been ably discussed in the course of a general survey by Patrick Fridenson in 'Les premiers inventeurs de l'automobile', *L'Histoire*, no. 73, Dec 1984. Delamare-Deboutteville was a dilettante who became involved in a dictionary of Sanskrit, a collection of Norman birds and the culture of mussels before turning his attention to petrol vehicles because the family cotton mill was ill-served by rail and hard to reach from Rouen by road. With the help of Léon Malandrin, a skilled craftsman, in the middle of 1883, just about the time that Daimler and Maybach were starting work, he experimented with a motorised tricycle and then, from September 1883 until April 1884, with a four-wheeled vehicle. He took out a patent in February 1884. The machine ran in the Montgrimont workshop but, according to Malandin, during a trial at only half-speed, the chassis was strained and parts of the mechanism broken. The experimenters then concentrated on stationary engines, one of which won first prize at the Paris Exhibition in 1889.
26. Friedrich Schildenberger, 'Karl Benz', in Eugen Diesel, Gustav Goldbeck and Friedrich Schildenberger (eds), *From Engines to Autos. Five Pioneers in Engine Development and Their Contributions to the Automotive Industry* (Chicago, 1960) pp. 131f. An earlier version of this book was published in German in Stuttgart in 1957.

27. Laux, op. cit. pp. 14–15. This and the next three paragraphs are based upon this source unless otherwise stated. Daimler's tribute to Levassor is quoted in Fridenson's *L'Histoire* article.
28. P. Siebertz, *Karl Benz*, cited by Dr Nübel in Chapter 2, note 35.
29. Jacques Ickx, *Ainsi Naquit L'Automobile* (Lausanne, 2 vols, 1961), vol. II, pp. 241–2, quoted by Laux, op. cit. p. 18.
30. Nicholson, op. cit. vol. 2, pp. 306–8; *Autocar*, 30 Nov 1895.
31. Ibid. p. 308.
32. *Automobile Topics*, 1 Feb 1902; *The Autocar*, 2 Nov 1895; John B. Rae, *The American Automobile* (Chicago, 1965) pp. 5, 9.
33. Ralph D. Gray, *Alloys and Automobiles* (Indianapolis, 1979) pp. 68 ff. Professor Charlotte Erickson drew my attention to this book.
34. Hiram Percy Maxim, *Horseless Carriage Days* (1937) pp. 1–25; Rae, *American Automobile*, pp. 10–11.
35. Friedrich Schildenberger, 'Gottlieb Daimler', in Diesel *et al.*, op. cit. pp. 114, 117; Gray, op. cit. p. 67.
36. Kenneth Richardson, *The British Motor Industry, 1896–1939* (1977) p. 14.
37. For a good technical account, well illustrated, of the pneumatic tyre, see Eric Tompkins, *The History of the Pneumatic Tyre* (1981).
38. See Dominique Barjot's chapter in this book (Chap. 15).
39. T. R. Nicholson, op. cit. vol. 2, pp. 339–41, upon which this paragraph is based.
40. Ibid. vol. 3, pp. 338–9. For an Arnold Vehicle in 1896 see *The Autocar* (19 Sep 1896).
41. Laux, op. cit. p. 22.
42. Ibid. p. 20.
43. Cited in Nicholson, op. cit. vol. 3, pp. 341–2. See also p. 487 n. 5.
44. Marquis de Chasseloup-Laubat, 'A Short History of the Motor Car', in Alfred C. Harmsworth (ed.), *Motors and Motor Driving* (1902) p. 10.
45. Nicholson, op. cit. vol. 3, p. 350, *The Autocar*, (23 Nov 1895).
46. Laux, op. cit. p. 26.
47. Chasseloup-Laubat in Harmsworth, op. cit. p. 12.
48. Ibid. p. 13.
49. *Autocar*, 23 Nov 1895.
50. Laux, op. cit. p. 23.
51. Ibid. pp. 26–32; Appendix 15.A.1. cf. *The Autocar* (2 May 1896).
52. *Automobile Club Journal*, (15 Jan 1903).
53. *The Autocar* (30 Nov 1895).
54. Ibid.
55. Appendix 15. A. 1.
56. Laux, op. cit. p. 29. For the best account of the boom in the industry, see ibid. chaps 4 and 5.
57. *The Autocar* (18 Jan 1896). The rest of this account is based upon the issues of this journal for 2 Nov, 7, 28, Dec 1895, 21 Mar, 11 Apr 1896; Nicholson, op. cit. vol. 3. pp. 351, 353.
58. *Automobile Topics*, 3 Aug 1901. The American Automobile Association was formed in March 1902.

59. Rae, *American Automobile*, pp. 11f.; Jean-Pierre Bardou, Jean-Jacques Chanaron, Patrick Fridenson and James M. Laux, *The Automobile Revolution* (Chapel Hill, 1982) p. 39.
60. For the Selden Patent see Rae, *American Automobile*, pp. 33–8.
61. Charles L. Dearing, *American Highway Policy* (Washington, D.C.) p. 44.
62. For Emil Jellinek, Daimler's agent in Nice, and the origins of the high-powered cars named after his elder daughter, Mercédès, see Laux, op. cit. pp. 76–7.
63. *Automobile Topics* (9 Nov 1901).
64. Ibid. (19 Oct 1901): Allan Nevins, *Ford. The Times, The Man, The Company* (New York, 1954) pp. 152–6, 172–184, 190–1, 204–5.
65. Rae, *American Automobile*, pp. 30–1.
66. *Automobile Topics* (22 June, 21 Sep, 26 Oct 1901).
67. Ibid. 25 July. 1, 8 Aug; Flink, op. cit. p. 22.
68. *Automobile Topics* (22 Aug, 26 Sep 1903).
69. Ibid. (19 Apr, 25 Oct 1902).
70. *The Motor Age* (5 Jan 1905).
71. *Automobile Topics* (18 Apr 1903).
72. Arthur Pound, *The Turning Wheel. The Story of General Motors Through Twenty-Five Years, 1908–1933* (New York, 1933) p. 54.
73. *The Motor Age* (5 Jan 1905).
74. *Historical Abstract of the United States* (Washington, D.C., 1960) p. 462.
75. *Automobile Topics* (23 Aug 1902).
76. T. R. Nicholson, op. cit. vol. 3, pp. 356 ff. upon which this paragraph is mainly based.
77. *The Autocar* (9 Nov, 21 Dec 1895; 4 Jan 1896; Nicholson, op. cit. vol. 3, pp. 366, 379; Harmsworth (ed.) op. cit. pp. 364–5.
78. *The Autocar*, 14 December 1895.
79. For Hooley and Lawson see David J. Jeremy (ed.), *Dictionary of Business Biography*, vol. 3 (1985). The Simms–Bosch connection is dealt with in Eugen Diesel, 'Robert Bosch', in Diesel *et al.*. op. cit. pp. 247 ff. For Pennington see Rae, *American Automobile*, p. 12.
80. Nicholson, op. cit. vol. 3, pp. 361–88, 431; *The Autocar* (4 Apr, 9, 16 May 1896).
81. *The Automotor and Horseless Vehicle Journal* (16 Nov 1897).
82. *The Autocar* (7 Nov 1896).
83. Ibid. (31 Oct 1896).
84. For Salomons, in addition to Nicholson, op. cit. vol. III, and the columns of *The Autocar* and *The Automotor and Horseless Vehicle Journal* (the S.P.T.A.'s periodical) see Albert M. Hyamson, *David Salomons* (1931). *The Victoria County History of Warwickshire*, vol. III (City of Coventry and Borough of Warwick) (ed. W. B. Stephens, Oxford University Press, 1969) contains a useful summary of these events on p. 177.
85. *The Autocar* (25 Apr 1896).
86. Ibid. (9 May 1896).

87. Ibid. (21 Nov 1896), upon which this and the next paragraph are based unless otherwise stated.
88. Ibid. (28 Nov 1896); *Nature* (19 Nov 1896); *Times* (16 November 1896)
89. Nicholson, op. cit. vol. 3, p. 461.
90. Ibid. pp. 460–1.
91. *The Autocar* (28 Nov 1896).
92. Quoted in *Nature* (26 Nov 1896).
93. *The Autocar* (13 June 1896).
94. Z. E. Lambert and R. J. Wyatt, *Lord Austin–The Man* (1968) chaps 2 and 3. For a more comprehensive business history of the Austin Motor Company Ltd see Roy Church, *Herbert Austin: The British Motor Car Industry to 1941* (1979).
95. Nicholson, op. cit. vol. 3, pp. 378–9; P. W. Kingsford, *F. W. Lanchester: A Life of An Engineer* (1960).
96. *The Economist* (10 Dec 1898), quoted in Harrison, op. cit. p. 311.
97. Ibid. pp. 319–20. The rest of this paragraph is based on this source.
98. Richardson, op. cit. p. 19.
99. *The Automotor and Horseless Vehicle Journal* (15 Dec 1897).
100. *Motoring Annual and Motorists Year Book* (1903) pp. 97 ff., which includes a complete list of members.
101. *The Automotor and Horseless Vehicle Journal* (15 Dec 1897).
102. *Automobile Club Journal* (1 Jan 1903). For Claude Johnson see David J. Jeremy (ed.), *Dictionary of Business Biography*, vol. 3 (1985), and for Rolls, ibid. vol. 4 (1985), and Ian Lloyd, *Rolls-Royce: The Growth of a Firm* (1978). The story of the policeman was recalled by Rolls's cousin in a BBC Interview (recording LP 24908) and the journey to Cambridge in *The Autocar* (12 Dec 1903).
103. Lady Troubridge and Archibald Marshall, *John Lord Montagu of Beaulieu: A Memoir* (1930) pp. 81–4.
104. Harrison, op. cit. p. 193
105. Paul Tritton, *John Montagu of Beaulieu, 1866–1929: Motoring Pioneer and Prophet* (1985) p. 47.
106. Ibid. chaps 7 and 8.
107. David J. Jeremy (ed.), *Dictionary of Business Biography*, vol. 3 (1895).
108. Richardson, op. cit. pp. 29–30.
109. Tritton, op. cit. chap. 6; Official Programme of the 1000-mile Trial, issued by the Automobile Club.
110. For Shrapnell-Smith see *Who's Who* and for Jones, P. N. Davies, *Sir Alfred Jones, Shipping Entrepreneur Par Excellence* (1978).
111. *The Autocar* (12 Sep 1896).
112. Ibid. (26 Sep 1896).
113. Ibid. (10 Oct 1896).
114. Ibid. (31 Oct 1896).
115. *The Automotor and Horseless Vehicle Journal* (15 Mar 1898).
116. Ibid. (15 July 1898).
117. For Spurrier see *The Commercial Motor* (30 Mar 1905) and for Edwin Foden, ibid. (4 May 1905). See also *The Autocar* (14, 21, 28 Nov 1903).
118. *Automobile Topics* (27 Sep 1902); Laux, op. cit. p. 96.

119. Ibid. (2, 23 May 1903).
120. Article on 'Heavy Commercial Vehicles', *Encyclopaedia Britannica*, 11th ed. (Cambridge, 1911).
121. *The Motor Cycle* (31 Mar, 8 Apr 1903).
122. *The Autocar* (16 May 1896). Humber were advertising a motor bicycle driven either by light oil motor or storage cells at the end of 1897 (*The Automotor and Horseless Vehicle Journal* (15 Dec 1897)).
123. Laux, op. cit. pp. 96–7.
124. *The Motor Cycle* (15 Apr 1903).
125. Ibid. (12 Jan 1904).
126. Lord Montagu (ed.), '*The Car' Road Book and Guide* (1906 ed.) p. 202.
127. *The Economist* (17 Aug 1912).
128. *The Autocar* (12 Dec 1903).
129. A. B. Filson Young, *The Complete Motorist* (1904) pp. 223–4.
130. *The Autocar* (12 Dec 1896).
131. Ibid. (19 Dec 1896).
132. *Automotor and Horseless Vehicle Journal* (15 July 1898).
133. Official Programme of the 1000-mile Trial, issued by the Automobile Club.
134. Robert Henriques, *Marcus Samuel* (1960), p. 290.
135. *Petroleum Times*, vol. LIII (1949) pp. 434–5. (I am grateful to Dr Geoffrey Jones for this reference and to Dr. R. W. Ferrier, the historian of B.P., for the point that 'the manufacturing, siting and accounting of millions of petrol tins was a complicated and costly business of supply-planning and stock control'. C. S. Rolls mentions the changed specific gravity of petrol in *The Encyclopaedia Britannica*, 11th ed. (Cambridge, 1911) under 'Motor Vehicles'.
136. *The Motor Cycle* (7 Oct 1903).
137. Ibid. (31 Oct 1904).
138. *The Automotor and Horseless Vehicle Journal* (15 Jan, 16 Apr, 16 May 1898).
139. T. C. Barker, 'The Spread of Motor Vehicles Before 1914', in Charles P. Kindleberger and Guido di Tella (eds), *Economics in the Long View* (1982), vol. 2, p. 154.
140. Lloyd, op. cit. pp. 11–12.
141. T. C. Barker, 'The Delayed Decline of the Horse in the Twentieth Century', in F. M. L. Thompson (ed.), *Horses in European Economic History* (Reading, 1983).
142. By an Act passed in 1903. Registration was by county and county borough and full details were printed in British Parliamentary Papers. Many of the original registers are to be found in county record offices. They give information about the owner, the type of vehicle, including its horse-power and weight, and whether it is used for private or trade/professional purposes. Some preliminary work on this valuable source has been published (e.g. R. G. A. Chesterman, *Laughter in the House. Local Taxation and the Motor Car in Cheshire* (Cheshire County Council 1978), but much more research is required.
143. Royal Commission on Motor Cars, vol. I, 1906 [Cd 3080] XLVIII, p. 69.
144. *U.S. Daily Consular and Trade Reports* (21 June 1911) p. 1269.

2 The Beginnings of the Automobile in Germany

Otto Nübel

For centuries European civilisation has been concerned with the idea, inherited from the ancients, of a self-propelled means of transport. Even in the *Iliad* Homer spoke of self-propelled tripods. In early modern times there were attempts to apply mechanical power of various sorts. In the mid-seventeenth century, for instance, a clockwork-driven carriage is supposed to have run through the streets of Nuremberg. Steam was tried also, but the early steam engines were cumbersome and in locomotive form required special iron roads to run upon if they were to operate satisfactorily.

A few steam traction engines, increasing in number, did lumber along German roads and, in the last part of the nineteenth century, more efficient and lighter versions were being produced; but by then means were being found to adapt for road transport the stationary gas engine, a convenient form of lighter power source used in workshops, with the development of which the German Nikolaus August Otto had been particularly connected. In 1872 Gottlieb Daimler became chief engineer at Otto & Langen's factory at Deutz, just across the river from Cologne, and Wilhelm Maybach went there as his chief of design. They experimented for a time in the mid-1870s upon vaporised petrol as a substitute for gas as fuel. When Daimler and Gasmotoren Fabrik Deutz A. G. parted company in 1882, he and Maybach began their own experiments in a garden shed at Cannstatt where Daimler then moved.[1]

Meanwhile Karl Benz, who had been working on gas engines for some time and had patented a throttle regulator in 1882, formed with others Benz & Company, Rheinische Motorenfabrik in Mannheim in the following year. So Daimler with Maybach, and Benz, without knowledge of the other's work, were independently hard on the track of an internal combustion engine which could be used for transport purposes; and all three saw that the main obstacle in their way was the construction of a light, high-speed engine which could drive such a vehicle.

While Benz never lost sight of the concept of the automobile as a whole and actually created such a vehicle with his patent motor car of 1886, Daimler and Maybach concentrated solely on a universally applicable engine which would serve as the motive power for every means of transport imaginable and not just for the automobile. Gottlieb Daimler had no intention of entering into competition with the long-established guild of coach and wagon builders. In his opinion they should continue to be able to build and sell their products, but with the new engine instead of shafts.

GOTTLIEB DAIMLER

As early as 1883 the hoped-for breakthrough occurred in Cannstatt.[2] On 16 December Daimler obtained patent DRP 28022 on the hot tube ignition system; and on 22 December patent DRP 28243 on an ingenious form of engine governor, by means of which the exhaust valve was worked.[3] A small model engine had been made by the Kurtz bell foundry in Stuttgart according to Daimler's instructions in August 1883. Maybach noted in his diary on 5 May 1884 that such a motor had reached 600 revolutions.[4] This indeed showed the amazing technical progress he had made; until then not more than 100 to 150 revolutions had been attained, even at Deutz.

Daimler and Maybach soon left behind the horizontal design of their first test engine. Already in the patent specification of 22 December 1883 there was the diagram of Gottlieb Daimler's vertical vehicle motor which was to pass into automobile history as the first high-speed internal combustion engine to be built. But it took about another year and a half before the engine functioned reliably. 'It was a long road', reported Daimler, 'requiring endless attempts and the incessant dedicated work of the engineer with practical experience in order not to lose heart despite the initial extremely discouraging results'.[5]

It was an epoch-making step forward. Maybach had made the engine as small and as compact as possible so that at a weight of around 90 kg., it appeared exceptionally delicate and almost like a toy·in comparison with the contemporary Deutz motors which, at 10 h.p., weighed around 4 600 kg.[6] This first high-speed internal combustion engine, soon known by Daimler's workers as the 'grandfather clock' because of its appearance, reached about 1 h.p. at 600 revolutions with a bore and stroke of 70 and 120 mm. respectively. It was covered by patent DRP 34926 of 13 April 1885.[7] Engines of this design powered all means of transport motorised by Daimler until 1889.

The brilliance of Daimler's idea lay in increasing the rotational speed of the internal combustion engine by many times what had previously been possible, so that the engine could be built light enough. 'Today's motor vehicle engine originates from these ideas which Wilhelm Maybach put into practice'.[8] Daimler's '1885 type engine is undeniably the direct ancestor of the present-day car engine. Other pioneers who preceded him in building vehicles used adaptations of stationary gas engines which, by their very nature, were virtually incapable of further development; but development of the Daimler type has been continuous and has still further to go.'[9] In 1894 Daimler himself wrote proudly that he had 'created the basis of a whole new industry'.[10] All the interest in the little shed in the garden of the Daimler villa focused from that moment, since the source of power for self-driven vehicles which had been sought in vain for centuries had been found, on the actual, original goal: the motorisation of transport. The 'grandfather clocks' was quickly installed into the most varied forms of transport in order to prove the usefulness of the engine.

First, still in 1885, the first motor cycle in the world was successfully driven by means of one of the 'grandfather clocks'. It reached 1/2 h.p. at a cubic capacity of 264 cm³ and 700 r.p.m.[11] Towards the end of the year Wilhelm Maybach is supposed to have made several test runs on an unsprung motor cycle with iron-rimmed, wooden wheels.

In 1886 came Daimler's first automobile, a converted horse-drawn carriage like an Americain, which he had bought from the Wimpff company in Stuttgart on 18 August. With the help of the Maschinen-fabrik Esslingen which had a branch in Cannstatt, Daimler's engine was built into this carriage, this time a somewhat more powerful version, 462 cm³ and 1.1. h.p. at 600 revolutions. Over a long period of time Daimler and Maybach undertook many test runs. However, it was only in 1888 that there were press reports on them.[12]

Daimler also knew how to promote his engine for power boats and to propel vehicles running on rails. During the Cannstatt Fair of 1887 he ran a small tram, powered by the 'grandfather clock', through the streets. In 1888 there were Daimler self-propelled cars for the Stuttgart horse-drawn tramway and the railway. There was Daimler's motorised fire-engine which was tested at the Deutscher Feuerwehrtag in Hanover, and countless uses of the 'grandfather clock' for industrial purposes, for example in saw and wood-splitting machines, and finally, on 12 August 1888, the successful ascent of the Woelfert airship. In order to keep up with the universal application of his engine Daimler acquired factory premises in Cannstatt in 1887 and started production on a larger scale. In the financial year 1890/91 about 60 marine and stationary engines were produced there.

Daimler concentrated his energies solely on the motorisation of transport generally and not on automobiles in particular. He saw his inventions as having much wider application than the automobile seemed to offer. 'Herr D[aimler] is mainly interested in demonstrating the varied uses of the engine', stated Maybach.[13] As we have seen, Daimler believed that he should offer his engine only to the existing coach-building trade in order to smooth the way for the 'motorised coach'.

With this in mind, the test runs with the motor carriage continued during 1888. 'Herr D[aimler] and I made long runs at slow speed with this simple coach', reported Maybach, who, only through the impression made by the modest mileage achieved by the motor carriage, managed little by little to convince Daimler of the necessity of a new and co-ordinated design, a carriage intended to be an automobile.[14] 'In the further course of events I insisted on a better mechanical construction in the shape of a steel chassis and four wheels, with the steering acting on both front wheels . . . and the transfer of power from the engine by means of . . . gears for changing speed,'[15] which, as Maybach continues elsewhere, 'never pleased Daimler'.[16]

So the wire-wheeled car came into being, a delicate-looking but extremely successful vehicle. It introduced the modern gearbox and was also fitted with the first twin-cylinder V-type engine in the history of the automobile. Protected by patent No. DRP 50839 of 9 June 1889, it achieved 1.5 h.p. at a cubic capacity of 565 cm^3 in its first version and later 2 h.p. at 1026 cm^3 and 620 revolutions.[17] The wire-wheeled car was ready just in time for the Paris International Exhibition in 1889. 'Mr. Levassor in Paris preferred it to the (earlier) belt-driven cars', wrote Maybach, 'and he kept the four-wheeled car presented by us for the first time in Paris in 1889 as a sample', after Panhard and Levassor had obtained the licences.[18]

Herein lie the roots of the French automobile industry. The French were the first to realise 'the opportunity offered by the Daimler engine to begin making small numbers of sturdy cars, and they had no trouble selling them',[19] an event which fitted perfectly with Daimler's plan. Panhard and Levassor, as well as Peugeot, were among the established, prosperous firms in the metal-working industry when, in addition to a wide range of their own products, they turned to the automobile in order to establish it in France within a very few years by building under Daimler's licences.[20] France must be given the historical credit for the economic realisation of the automobile's economic possibilities after the basic technical problems had been solved in Germany. The contribution of both countries was indispensable for the ultimate success of the automobile. There is therefore no need to go into any disputable detail (by referring to the personality of Delamare-Deboutteville for example[21]) in order to underline the decisive importance of France in these early days.

KARL BENZ

The issuing of patent DRP 37435 of 29 January 1886 for his three-wheeled motor car was the first important event for Karl Benz and the most vital in the history of the automobile.[22] In the months which followed, several independent contemporary reports appeared which clearly set out the continued development of this car. According to Karl Benz's own account, the car was already running a year earlier, but there were obviously several improvements to be made because it was not until 4 June 1886 that a report appeared in the *Neue Badische Landeszeitung* for 'friends of the velocipede' which said: 'At present in this firm' [Benz] 'a three-wheeled velocipede is being built which is driven by an engine'.[23]

This tricycle must have been nearing completion at this time, for on 3 July 1886 a second report appeared in the same newspaper, this time about the first public demonstration of the Benz Patent Motor Car on the Ringstrasse in Mannheim: 'A velocipede powered by Ligroin gas which had been developed in the Rheinische Gasmotorenfabrik of Benz & Co. and about which we have already written in this newspaper, was tested early this morning on the Ringstrasse with satisfactory results.'[24] This small press item hidden away among the other news of the day was no less than the birth certificate of today's automobile.

Karl Benz must have known that he had solved the problem of propelling vehicles by a lightweight power source. Even his native Mannheim realised it, although it was naturally unable to grasp the far-reaching significance of the event. In a third newspaper report, this time in the *Generalanzeiger* of Mannheim of 5 September 1886, it was said: 'It was clear to us at the first trial that with his invention Benz had solved the problem of producing a car with primary power.' The report went on to describe the development of the vehicle which had taken place since. 'As was to be expected, many faults came to light in the motor car which had to be rectified by constant tests and improvements. This work, as difficult as the invention itself, can now be said to be completed.'[25] Enough has been written about it 'to show the impossibility of stating that any one person was the "inventor" of the motor car. The idea germinated in many different minds and took many different forms; but if the meaning of "inventor" is narrowed to signify the man who first designed and produced light self-propelled carriages for sale to the public, then Karl Benz of Mannheim has the greatest claim to the honour.'[26]

After overcoming the technical problem it was Karl Benz's intention that production for the market should be started without delay. According to the already quoted newspaper article of 5 September 1886, 'Herr Benz will now start to build such vehicles for practical use.'[27] Benz Patent Motor Cars were, in fact, already being made, 'several new model cars from 1886', as Benz himself reported,[28] first Model II, of whose performance Benz himself was not totally convinced (the problems were mainly to do with the steering, since even then Benz wanted to change to a four-wheeled car.[29] After that came the much improved Model III, several of which were again produced and an example of which can be seen in the Science Museum in London.[30]

Production for the market involved advertising. Benz, unlike Daimler, did not believe in it. He felt that the quality of his work should speak for itself and disliked any hint of self-praise.[31] The French, on the other hand, were, from 1894 onwards to carry out trials and races on public roads which gained world-wide publicity. Benz nevertheless did attempt a public demonstration in 1888 at the beginning of the trade and industry fair in Munich after having persuaded a very doubting police officer to issue him a licence. His drives through the city every day in the Benz-Patent-Motorwagen Model III caused a great sensation.[32] 'The amazement of the passers-by, who did not know what to make of the sight, was as universal as it was great', reported the *Münchener Tageblatt*.[33] No newspaper missed the opportunity to report on the strange machine, constantly surrounded by a swarm of shouting children. To be able to sit next to Benz on the seat of the 'vehicle travelling at a smart pace' was one of the biggest attractions of the whole fair. There were even a few people (Ritter von Paller, the engineer from Nuremberg, for example) who began to realise the full importance of Benz's invention, and this was not simply because the Patent Motor Car won the Gold Medal, the highest award of the Munich jury. More important was the fact that it left such a strong impression on everyone. The *Münchener Neueste Nachrichten* thought that it would 'soon enjoy a larger circle of fans', for it must 'truly be a great pleasure to fly along in this vehicle without having to rely on horses etc.'[34]

Despite such newspaper 'puffs', these demonstrations produced no serious buyers. 'It was one of the most distressing memories of my life', Benz confessed as it dawned upon him that his technical triumph had failed to attract an immediate queue of eager customers. A year later, at the Paris International Exhibition, he made no effort to run his motor vehicle. It was exhibited in a row of horse cabs without any form of description. The scant attention it received was often unfavourable. One expert observer reported, for instance: 'In that hall a high, three-wheeled car of a most displeasing appearance was displayed which was without seats or shaft and looked like a tangle of machinery.'[35]

Benz found possibly his first buyer in Emile Roger, his French agent for stationary engines, who bought a Patent Motor Car in 1887 and sent it, packed in four crates, to Panhard and Levassor in Paris. This tricycle arrived in March 1888, was reassembled and with the help of Karl Benz, who on this occasion made his first and only trip to Paris, was put into motion.[36] In Germany, on the other hand, nothing had happened. The hopes of the Rheinische Gasmotorenfabrik that they would succeed in selling automobiles faded. Benz's two partners, Esslinger and Rose, thought in spring 1890 that they could foresee the collapse of the firm. They withdrew from it with the well-meaning advice to Benz: 'Leave motor cars alone'.[37] Certainly, the few potential buyers whom we hear found their way to Mannheim did not inspire confidence: an allegedly dying man who intended to enjoy his last days on earth, a woman teacher from Hungary who was an automobile fan, and the equally fanatical Bohemian Baron von Liebig, whose first long journeys made automobile history; a postmaster from Württemberg who considered the motion of the car to be too uncomfortable after all, and, lastly, a rather eccentric doorman from a Munich hotel who had come into a fortune.[38] Few, if any, serious customers came to Mannheim in those early days. Benz did, however, gain two new partners, both of them merchants, who brought in some fresh capital and commercial acumen to the Rheinische Gasmotorenfabrik.

From this point onwards affairs at Mannheim started to be looked upon from the business, rather than from the technical, point of view. As a result of this new and more realistic policy, Benz constructed a new type of car, the Velo, his first small and relatively inexpensive motor vehicle. This sold well from 1894 onwards and helped to prepare the ground for motorisation in many parts of the world. It was the foundation of commercial success.

BOTH PIONEERS FIND THE GOING HARD

Progress was slow, particularly as Daimler still refused to concentrate upon motor-vehicle production at Canstatt. His concern remained with engine manufacture and it took much patient persuasion from Maybach to convince him that to put engines into other people's horseless carriages was not enough. Besides that, Daimler had differences with his partners after the creation of the Daimler Motoren Gesellschaft in 1890. This actually caused him to withdraw from the company. It almost collapsed before his return in 1895. As no other German entrepreneur entered the field before 1894, responsibility for the growth of the industry in Germany rested with Benz. He was to become the world's leading manufacturer by 1900.

It was, however, an uphill struggle, for when admiration of the inventors faded, German public opinion began to treat motor vehicles with growing scepticism. There were several reasons for this. Benz at first advertised his products as vehicles for solid commercial work, capable of carrying goods more cheaply, of running on bad roads and of climbing steep hills. Elsewhere, however, and especially in France, motor cars were seen as exciting luxuries, used for pleasure and sport. The German product, even as late as 1900, lacked flair and, compared to the splendid French cars, they seemed heavy, massive and indeed a little clumsy.[39] Nobody, as Benz later explained, 'could credit that one could possibly give up an elegant horse-drawn carriage for such an unreliable, poor, puffing, rattling iron thing'.[40] For the Germans, motor vehicles were merely pieces of machinery lacking the least touch of elegance; 'most ugly vehicles', contemporary newspapers called them.[41] 'The whole vehicle had the habit of shaking like a dog coming out of the water as soon as the engine started running', Benz later confessed.[42] 'Using such a car, one was sure to be admired as Mister Engineer', one of the first Benz owners wrote from Dresden, 'but if the engine failed, there was nothing but malicious grins. Nobody was ever prepared to lend a helping hand.'[43]

The German industry suffered from not organising competitions at an early stage as the French did from the Paris–Rouen contest onwards; and when, in 1899, some were belatedly held in Berlin, a leading magazine dismissed them as of little value, for by then the car with the largest engine was bound to win and, it claimed, driving skill hardly counted at all.[44] The German Emperor certainly refused to treat motor vehicles as suitable for sport. For him they were mainly a 'replacement for railways' and, when he finally decided to buy them, he painted them in the livery of the imperial train.[45] Eighteen years of hesitation and sceptical rejection passed, for most of which time he refused even to sit in a motor car; but in March 1904, he made his first trip in his own vehicle, though even then he did not manage to persuade the Empress to join him. It was not until 1907 that she started to use cars.[46] Such a poor example did not encourage other members of German society or influential aristocrats to support the new industry.

On top of this, officials, as well as governments, generally tended to over-emphasise the danger of explosions, fire and traffic accidents, especially after some German motorists started to drive faster than police and public opinion thought to be safe. Leaving clouds of annoying dust behind them, they were soon chased by angry or frightened bystanders, subjected to police traps, and even stoned now and them. Very few motorists realised how much they terrified people and how intensely they were disliked. Even so late as 1910 it was inadvisable for them to stop after an accident involving animals; the treatment they were likely to receive from the public might be more severe than the damage they had caused. Motor vehicles were generally considered the most dangerous form of transport.[47] They seemed to endanger everybody else on the road and forced the general public to behave differently. They were a little-appreciated newcomer used by the priviledged classes only.

Benz is known to have been strictly against any form of speeding, from the earliest days until 1903 when he retired from active management at Mannheim. Road safety was his main concern. He demanded that his sons and his drivers took special care. Any form of reckless driving annoyed him intensely. Not that there was much scope for speeding in Germany. When, in November 1893, Benz first managed to get permission to run cars on the roads of the Grand Duchy of Baden, it was for the year 1894 only and speed limits were imposed of 6 km. per hour in towns and 12 km. per hour on the open road. Speeds had to be reduced to a minimum whenever horses were encountered. Government attitudes to the motor vehicle, Benz found, 'had the attractiveness of a barbed wire fence'.[48] His unceasing efforts to get these unfavourable regulations changed were all in vain. In December 1895 he had to be content with an extension of the existing Grand Duchy regulations indefinitely – but with the threat of revocation at any time.

The position elsewhere in Germany was equally uncertain. Even the smallest territory took advantage of its political powers to issue its own set of rules; and their enforcement varied considerably from one place to another. As traffic signs were unknown in the early days, it was hazardous indeed for a driver to enter unfamiliar territory as he rarely knew what regulations were in force. Many local authorities looked to fines from the unwary passing motorist to fill the municipal coffers. All this discouraged the purchase of motor vehicles. One could never be certain whether in the future their use would be more strictly limited or even prohibited altogether.

Last and by no means least, motor vehicles were also very expensive to buy in terms of German purchasing power. Benz felt positive embarrassment when people asked him the price of his early cars, for he had to talk in terms of several thousands of marks.[49] Nobody would believe him and few were prepared to spend such a large amount. In 1890 a workman in German industry could earn only on average 711 marks a year; by 1900 this average had risen to 843 marks and in 1905 to 928 marks.[50] In those three years the equivalent for professional men, such as lawyers and doctors, was 1978, 2091 and 2163 marks respectively.[51] The Benz Velo already cost 2000 marks in 1893 and the Benz Viktoria, far from the most expensive model in the range, 3800 marks. By 1906 car prices had risen very considerably. According to a contemporary newspaper, a small 9-h.p. car for use in towns cost about 6000 marks and a further 6000 marks a year to maintain it if a chauffeur was employed. (1200 marks for the chauffeur, 600 marks for tyres and 300 marks each for insurance and garaging).[52] When German purchasing power and motoring costs are taken into account it seems remarkable that there were already by 1904 more than 1500 motorists with their own vehicles in Berlin alone.[53]

In Germany a motor vehicle was plainly a possession enjoyed by the wealthy alone, and even they had to spend a year's income or more to buy one. In the Reichstag debate of May 1906 on the taxation of motors one of the main arguments in favour of such a tax was that it would have to be paid by the well-to-do. As one member of parliament put it, 'a poorer person will never be able to afford a motor-vehicle anyway'.[54] This seemed to be true for most members of parliament, too. When the Reichstag had to decide on compulsory insurance for motor vehicles in the same year, 1906, its committee gladly accepted the automobile industry's invitations for a test drive in and around Berlin. The gentlemen enjoyed the ride; with few exceptions none of the twenty members of the committee had ever before been in a motor vehicle.[55]

Notes

1. This information about events before the 1880s is taken from Eugen Diesel, Gustav Goldbeck and Friedrich Schildberger, *From Engines to Autos. Five Pioneers in Engine Development and their Contribution to the Automotive Industry* (Chicago, 1960). The book was published first in Stuttgart in German in 1957.
2. P. Siebertz, *Gottlieb Daimler*, pp. 103 ff.
3. For reproduction of both patents see H. Ch. Seherr-Thoss, *Zwei Männer – ein Stern* (Düsseldorf, 1984), vol. I, pp. 18, 25 ff.
4. F. Sass, *Geschichte des deutschen Verbrennungsmotorenbaues* (Berlin, 1962) p. 86.
5. Seherr-Thoss, op. cit. vol. I., pp. 39, 55 ff.
6. Sass, op. cit., p. 79 ff.
7. For reproduction see Seherr-Thoss, op. cit. vol. I, pp. 31 ff.
8. Sass, op. cit., p. 79 f.
9. A. Bird, *The Motor Car, 1765–1914* (London, 1960), p. 27.
10. Siebertz, op. cit., p. 116.
11. Patent DRP 36 423 of 29 Aug 1885, in Seherr-Thoss, *op. cit.* vol. I, pp. 75 ff.
12. Seherr-Thoss, op. cit. vol. I, p. 135; Siebertz, op. cit. pp. 126 ff.
13. Sass, op. cit. p. 175.
14. Ibid., pp. 172 ff.
15. Ibid.
16. Ibid.
17. Ibid. pp. 168 ff.; Seherr-Thoss, op. cit. vol. I pp. 119 ff.
18. Sass, op. cit. p. 175.
19. J. M. Laux, *In First Gear: The French Automobile Industry to 1914* (Liverpool, 1976) p. 20.

20. The beginnings of the French automobile industry are convincingly set out by Laux, *passim*.
21. On Delamare-Deboutteville see p. 50 note 25.
22. For a reproduction see Seherr-Thoss, op. cit. vol. II, pp. 49 ff.
23. For a reproduction ibid. vol. II, p. 108.
24. For a reproduction ibid. vol. II, p. 109.
25. For a reproduction ibid. vol. II, p. 109.
26. Bird, op. cit. p. 30.
27. Seherr-Thoss, op. cit. vol. II, p. 109.
28. Carl Benz, *Lebensfahrt eines deutschen Erfinders* (Leipzig, 1943), pp. 79 ff.
29. P. Siebertz, *Karl Benz*, (München–Berlin, 1943) p. 125.
30. Benz's letter of 18 Aug 1914; for a reproduction see Seherr-Thoss, op. cit. vol. II, pp. 105 ff.
31. Siebertz, *Benz*, p. 152; Benz, op. cit. p. 192.
32. For a reproduction of the Munich advertising leaflet with a good illustration of the Benz-Patent-Motorwagen see Seherr-Thoss, op. cit. vol. II, p. 114.
33. 18 Sep 1888, Seherr-Thoss, op. cit. vol. II, p. 110.
34. Seherr-Thoss, op. cit. vol. II, p. 111.
35. Siebertz, *Benz* p. 113.
36. Benz, op. cit. p. 94; Laux, op. cit. p. 11; Siebertz, *Benz*, p. 111 note 1.
37. Benz, op. cit. p. 115.
38. Ibid. p. 107.
39. *Allgemeine Automobil Zeitung* (AAZ) (Wien, Nr. 4, 1903) pp. 7 ff.
40. *AAZ* (Berlin, 1913) Nr. 1, p. 19.
41. *AAZ* (Wien, 1903) Nr. 4, p. 8.
42. J. Neren, *Automobilens Historia*, (Stockholm, 1937), p. 103.
43. *Motor*, Juli/August 1928; see Benz, op. cit. p. 106.
44. *Motorwagen*, in *Zeitschrift des Vereins deutscher Ingenieure*, Bd XXXXIV (1900) pp. 1 ff.
45. *AAZ* (Berlin, 19. Januar 1912) p. 17.
46. *AAZ* (Berlin, 2. Februar 1912) pp. 22 ff.
47. *AAZ* (Berlin, 1906) p. 51.
48. Benz, op. cit. p. 87.
49. Seherr-Thoss, op. cit. vol. II, p. 109.
50. W. G. Hoffmann/H. J. Müller, *Das deutsche Volkseinkommen 1851–1957* (1959) tab. 108.
51. Ibid. tab. 118.
52. *Jahrbuch der Automobil- und Motorbootindustrie*, edited by E. Neuberg, vol. III (Berlin, 1906), pp. 48 ff.
53. *Moderne Kunst*. Gordon-Bennett Nummer, 21. Heft (Berlin, 1904).
54. *AAZ* (Berlin, 1906) Nr. 20, p. 34.
55. *AAZ* (Berlin, 1906) Nr. 22, p. 101.

3 The Motor Vehicle and the Revolution in Road Transport: The American Experience

John B. Rae

The history of highway transport in the United States shows clearly that large-scale movement by road for medium or long distances or for heavy loads is a phenomenon of the twentieth-century. Before the coming of the railroad water transport was used by preference where it was available; after that the railroad made highway traffic almost exclusively local.

The roads themselves were not the problem. We have known how to build good roads for a long time. The ancient Persians had their Royal Road, which might be called the world's first super-highway, complete with express mail service. Herodotus said of the couriers who carried the Great King's dispatches, 'nor snow, nor rain, not heat, nor gloom of night stays these messengers from the swift completion of their appointed rounds',[1] now the motto of the United States Postal Service – which I suppose might move some of us to inquire what it is that stays our modern messengers. We all know about the Roman roads. The beginning of modern highway building goes back just about two hundred years, to the period that gave the word 'macadam' to our language, although McAdam himself had less to do with road development than his fellow Scotsman Telford or the French engineer Trésauget.

The difficulty was in the vehicles. Whether horses, oxen, mules or dogs were used, animal-drawn vehicles were slow, inefficient and expensive. They could be improved and were. In the United States Conestoga wagons and Concord coaches were masterpieces of design for their function but they were never effectively competitive with water or rail transport. Before the coming of the railroad it cost as much to haul goods thirty miles inland by road in the United States as it did to carry them across the Atlantic in the first place,[2] and a hundred years ago land more than fifteen miles from a railroad was looked on as worthless, regardless of its quality.[3]

The coming of the motor vehicle produced the highway revolution that is my topic. It is still a recent event; this is the hundredth anniversary of the invention of the gasoline automobile. The rival claims of Benz and Daimler have been adequately assessed by German scholars,[4] and are discussed by Otto Nübel in Chapter 2. The United States was a laggard for a while, although it caught up spectacularly fast. The first American-built gasoline automobile was run in Springfield, Mass., by Charles and Frank Duryea in 1893 (contrary to what many Americans still seem to believe, Henry Ford did not invent the automobile, but what he did with it was far more important), and as late as 1900 the U.S. Census lumped motor vehicles among 'miscellaneous manufactures'.

DREADFUL ROADS GRADUALLY IMPROVED

For the first time in history we had a road vehicle that provided flexible and economic mobility. It took some time for the effects of this change to be felt and at least partially understood and I doubt if they will ever be completely understood. It also took time for an automobile industry to grow, and beyond that this new way to move by road needed better roads to move on. The highway revival of the early nineteenth century shrivelled before the phenomenal growth of railroads. When road traffic became predominantly local (don't be misled by the stage coaches and wagon trains of western movies; they survived only as long as there was no rail service available) it was no longer worth while to spend money on trunk highways. In America responsibility for road construction and maintenance was left to local authorities, with the result that the country's highway system deteriorated. In 1900, outside of large cities, there were fewer than 200 miles of hard-surfaced road in the entire United States.[5] Country roads were usually dust-choked in summer and impassable in winter.

There was agitation for road improvement from farm groups and others who wanted to find an alternative to the railroads, and from the bicyclists who appeared in great numbers in the 1890s, but very little progress was made until the 1920s. The Federal Highway Act of 1921 provided support for a national highway network, the present U.S. routes, and at the same time the various states, beginning with Oregon in 1919, discovered that taxes on gasoline offered an apparently painless method of financing road construction and maintenance. (This first Oregon gasoline tax was one cent a gallon.) By this time the number of motor vehicles in use in the United States had reached the ten million mark and would exceed 25 million by 1930. By 1929 there were enough motor vehicles in use in the United States to carry the country's entire population at one time, and at rush hours and on weekends it must have appeared that this was actually being done. This growing volume of cars created an irresistible pressure on governments at all levels to improve roads and streets; it also made the impact of this new mode of transport highly visible. The fact that the motor vehicle was a major instrumentality of social and economic change was clearly understood, even if the full implications were still in the future and indeed may never be completely evaluated. In the early 1930s a commission appointed by President Hoover to study recent social changes in the United States had this to say about the automobile: .

Between 1913 and 1931 the increase in (motor vehicle) registration in the United States was twentyfold.

It is probable that no invention of such far-reaching importance was ever diffused with such rapidity or so quickly exerted influences that ramified through the national culture, transforming even habits of thought and language. . . .

This phenomenal growth involved a displacement of earlier vehicles, such as the horse carriage and the bicycle. It also involved habituation to the use of the automobile of classes in the population who formerly owned no vehicle of private transportation. Within the space of a few years, for vast numbers motor travel ceased to be a novelty and came to be regarded as a necessity. . . .

In no inconsiderable degree the rapid popular acceptance of the new vehicle centered in the fact that it gave to the owner a control over his movements that the older agencies denied. Close at hand and ready for instant use, it carried its owner from door to destination by routes he himself selected, and on schedules of his own making: baggage inconveniences were minimized and perhaps the most important of all, the automobile made possible the movement of an entire family at costs that were relatively small. Convenience augmented utility and accelerated adoption of the vehicle.[6]

In this chapter it is impossible to go into the whole story of the effect of the motor vehicle on American life. I want to focus on the economic implications as they have been manifested in transportation.

To accommodate this swelling flood of motor vehicles the first recourse, understandably, was to improve existing roads by hard-surfacing them, widening them, reducing curves, etc. Then massive building of new roads got under way. Even with the limitations of these early highway programs, the effects were immediate and striking, because the motor provided a rapid, convenient, and economical way to carry both people and goods by road for the first time in history.

ROAD/RAIL COMPETITION

Rail passenger traffic reached its peak in the United States in 1920 and declined steadily thereafter, except for the interlude of the Second World War.[7] Short runs and branch lines were particularly affected. For the years before the Second World War this effect on rail transport was directly and exclusively due to highway competition. There was some passenger travel by water coastwise and on the Great Lakes, but the total volume was negligible, and air travel was still embryonic. The same thing was happening to urban transit systems and street railways. The idea currently popular in some quarters that this change of travel mode was imposed on the American public against its will or without its knowledge can be dismissed as nonsense. Professor Mark Foster of the University of Colorado at Denver has demonstrated convincingly in his book *From Street Car to Superhighway*[8] that the American people welcomed the freedom of movement that the automobile gave them, as compared with having to meet fixed and frequently inconvenient schedules, ride in coaches that were usually uncomfortable and often dirty, and wait indeterminate lengths of time for connections. In the Los Angeles area there is a large body of opinion that bemoans the passing of the 'Big Red Cars', the interurban service of the Pacific Electric Company, whose last passenger run was made in 1961. I have had a number of students go into the history of the Pacific Electric, and its balance sheets from the 1920s all seem to show that, while people may have loved the Big Red Cars, they increasingly refrained from riding on them as paying passengers.

Railroad freight traffic was also affected. Truck competition cut sharply into short-haul and less-than-carload business as well as long-distance shipment of high-priced, class-rate freight. The highway carriers, both freight and passenger, had the additional advantage of being subject to minimal state regulation and no federal control whatever until the passage of the Federal Motor Carriers Act in 1937, which imposed a very limited degree of regulation on commercial interstate road traffic.

ACCIDENT RATES AND BETTER ROADS

We know the various side effects of this explosive growth of road transport: congestion, accidents, atmospheric pollution, and so forth. None of these were original with the automobile. Cities have always been congested. Ancient Rome banned wheeled vehicles within the city limits in daytime; it also prohibited women from driving chariots on the Appian Way. We do not know how the present incidence of traffic accidents compares with pre-motor vehicle days, because such statistics were not kept. We do know that the ratio of injury accidents to vehicle-miles travelled by motor vehicles in the United States has declined fairly steadily over the last fifty years (Table 3.1). 'Smog' is a

Table 3.1   Traffic fatality rates

Year	Rate per hundred million vehicle miles
1936	15.1
1941	11.5
1946	9.7
1950	7.1
1955	6.1
1960	5.1
1965	5.54
1970	4.88
1975	3.34
1980	3.29
1983	2.64

Sources: Automobile Manufacturers Association, *Automobile Facts and Figures* (1970) p. 19; Motor Vehicle Manufacturers Association, *Motor Vehicle Facts and Figures* (1984) p. 89. Figures are taken from National Safety Council, *Accident Facts*.

term originally coined in nineteenth-century Britain to describe the combination of coal, smoke and fog that periodically smothered London. Perhaps I should include the statistical genius of a century ago who made projections of American urban growth, based of course on the technology of that time, and came to the conclusion that by 1890 all major United States cities would be buried under six feet of horse manure.

One cause of these problems has been that basic highway design followed the conventional patterns of the past and was not adapted to the characteristics of fast-moving, self-propelled vehicles. This problem was recognised quite early. In 1912 T. Coleman du Pont proposed to build a new-type highway through the state of Delaware at his own expense. There was to be a central roadway for slow-moving traffic (remember that horses were still in common use), flanked by roadways in each direction for fast vehicles. Outside these were to be interurban tracks, and the rest of a broad right-of-way was to be leased to commercial establishments so that the highway could be self-supporting.[9] The novelty of the scheme, plus quite unfounded mistrust of Mr du Pont's motives, prevented its adoption. He did build a road at his own expense the length of the state from north to south, but it was a standard two-lane concrete highway, still called the du Pont Road and even with its limitations much superior to anything Delaware had had before.

In the 1920s Hilaire Belloc, the English novelist, proposed that major cities be joined by what were in effect modern superhighways, except that they were to terminate at the city limits because Belloc assumed that their traffic would disperse easily into city streets.[10] The merits of such roads were recognised readily enough; the inhibiting factor was their cost. Some preliminary moves were made in the 1920s, as with New York's Metropolitan Parkways, but it took the Great Depression to create conditions where the attraction of stimulating employment overrode cost considerations.

SUPERHIGHWAYS

This was the motive behind the construction of the German *autobahnen* in the 1930s, the world's first national system of express highways specifically designed for motor-vehicle traffic – what our Federal Highway Administration officially terms 'Freeways' – roads with complete control of access, separate roadways, and no crossings at grade. In the United States the Merritt Parkway in Connecticut, the first part of the Pasadena Freeway in California, and the Harrisburg – Pittsburgh section of the Pennsylvania Turnpike were built at this same time.

The last-named was a novel enterprise. It was a make-work project, taking advantage of a completely graded railroad right-of-way that had been abandoned for fifty years, and it was planned to be supported through tolls. Expert opinion gave the scheme no chance of success. In particular, it was argued that trucks would take alternative routes rather than pay the tolls. Instead, almost from its opening truck traffic took the Turnpike by choice, not just because it offered easier grades over the Alleghanies but still more because the elimination of frequent stopping and starting produced savings that outweighed the cost of the tolls.

To reach ahead, in 1957, when it was possible to go by toll road without interruption from New York to Chicago, the Indiana Toll Road Commission conducted test runs for trucks between Chicago and Jersey City, comparing travel by the toll roads with travel by U.S. Routes 22 and 30. Here are the results:

Table 3.2   Indiana toll road truck tests, Chicago–Jersey City, April 1957

Factor	Turnpikes	Routes 30 and 22	Turnpike savings
Elapsed time	64 hrs 49 min.	94 hrs 43 min.	29 hrs 54 min.
Travel Time	41 hrs 5 min.	52 hrs 22 min.	11 hrs 17 min.
Gasoline consumption	363.9 gals.	394.4 gals.	30.5 gals.
Speed per hour	40.93 miles	32.73 miles	8.20 miles
Gear shifts	777	3,116	2,339
Brake applications	194	890	696
Full stops	58	243	185

Source: Courtesy of Indiana Toll Road Commission.

The success of these first superhighways produced an outburst of highway construction as soon as the restraints imposed by the Second World War were removed. Some 12 000 miles of toll road were projected and about 3500 miles actually completed before the Interstate Highway Act put an end to further toll road construction.[11] Meanwhile California was building its own freeway network, financed by gasoline taxes. In Northeastern states especially the toll road appeared as the solution to handling heavy volumes of through traffic without increasing local tax burdens. A superhighway was a special service for which a user charge in the form of tolls could properly be imposed, so that out-of-state vehicles would carry their share of the costs. In the absence of more substantial federal aid than was then available the alternative to tolls was higher gasoline taxes, which would have borne most heavily on local traffic, much of which would make little or no use of the express highway. For example, it is possible for a car to be driven across Pennsylvania, Ohio, or Indiana with only one filling of the fuel tank inside the state. Tolls, on the other hand, meant that every user of the highway, local or long-distance, in-state or out-of-state, was sharing the cost on equal terms. This condition does not prevail in California, which is a big state, and California cars account for most of the freeway users. The capstone of this structure, the Interstate Highway System, was provided for in 1956 and its 42.5 thousand miles are substantially completed[12] – some, indeed, already deteriorating and in need of maintenance. We may have to rethink the subject of highway tolls. We have to add to the express highways something on the order of 350 000 miles of primary routes (U.S.), plus 600 000 miles of city streets and over three million miles of other state and local roads, a grand total of approximately four million miles of roads and streets in the United States, of which over 80 per cent are hardsurfaced.[13]

PHENOMENAL GROWTH OF MOTOR VEHICLES AND ITS RESULTS

The number of motor vehicles registered in the United States has climbed from nine million in 1920, including a million commercial vehicles to 160.5 million in 1983, including 35.5 million commercial vehicles. I believe it to be highly significant that even in the depths of the Great Depression vehicle registrations declined less than 10 per cent, from 26.5 million in 1929 to 24 million in 1932, and truck registrations shrank even less, from $3\frac{1}{2}$ million to $3\frac{1}{4}$ million in the same period.[14]

What has been the result of all this? The total consequences – economic, social, cultural – of 'automobility' have never been completely measured and I suspect that they are basically immeasurable. So let us look just at traffic and transport. Between 1950 and 1984 intercity passenger travel in the United States changed as is shown in Table 3.3.

These figures will repay some consideration. In these last 34 years the total volume of intercity passenger travel in the United States has

Table 3.3 Intercity passenger miles by mode of travel

	Auto-mobiles[1]	Motor coaches[1]	Total motor vehicles[1]	Railways, revenue passengers	Airways domestic revenue services	Total
Passenger miles by mode (in billions)						
1983 (est.)	1 400.0	26.1	1 426.1	10.9	244.7	1 681.7
1982	1 344.9	26.9	1 371.8	10.9	226.7	1 609.4
1981	1 344.0	26.9	1 370.9	11.8	216.0	1 598.7
1980	1 300.4	27.4	1 327.8	11.4	219.1	1 558.3
1979	1 322.4	27.7	1 350.1	11.6	228.2	1 589.9
1978	1 362.3	25.6	1 387.9	10.5	203.2	1 600.8
1977	1 316.0	25.7	1 341.7	10.4	177.0	1 529.1
1976	1 259.6	25.1	1 284.7	10.5	164.4	1 459.6
1975	1 170.7	25.4	1 196.1	10.1	148.3	1 354.5
1974	1 121.9	27.7	1 149.6	10.5	146.6	1 306.7
1973	1 162.8	26.4	1 189.2	9.3	143.1	1 341.6
1972	1 129.0	25.6	1 154.6	8.7	133.0	1 296.3
1970	1 026.0	25.3	1 051.3	10.9	118.6	1 180.8
1965	817.7	23.8	841.4	17.6	58.1	917.2
1960	706.1	19.3	725.4	21.6	34.0	781.0
1955	637.4	25.5	662.9	28.7	22.7	716.3
1950	488.3	26.4	514.7	32.5	10.1	507.3
Passenger miles by mode (percent)						
						%
1983 (est.)	83.2	1.6	84.8	0.6	14.6	100
1982	83.5	1.7	85.2	0.7	14.6	100
1981	84.1	1.7	85.8	0.7	13.5	100
1980	83.5	1.8	85.3	0.7	14.0	100
1979	83.2	1.7	84.9	0.8	14.4	100
1978	85.1	1.6	86.7	0.7	12.6	100
1977	86.1	1.7	87.8	0.7	11.5	100
1976	86.3	1.7	88.0	0.7	11.3	100
1975	86.5	1.9	88.4	0.8	10.9	100
1974	85.8	2.1	87.9	0.8	11.3	100
1973	86.7	2.0	88.7	0.7	10.6	100
1972	87.1	1.9	89.0	0.7	10.3	100
1970	86.9	2.1	89.0	0.9	10.1	100
1965	89.2	2.6	91.8	1.9	6.4	100
1960	90.4	2.5	92.9	2.8	4.4	100
1955	89.02	3.56	92.58	4.01	3.17	99.76[†]
1950	86.19	5.20	91.39	6.39	1.99	99.77[†]

[†] Remainder represents inland waterway travel. Not included after 1960.
[1] Includes intracity portions of intercity trips. Omits rural to rural trips, and intracity trips with both origin and destination confined to same city for local bus or transit movements, non-revenue school and government bus operations.

Source: Interstate Commerce Commission and Transportation Policy Associates, *Transportation in America*.

more than doubled. Automobile travel has increased almost three-fold, although its share of the total has declined somewhat. Bus travel has grown slowly; its share of the total has gone down. Rail travel over the whole period has declined, but with the qualification that it has remained strikingly stable since the early 1970s. The spectacular change has been in air travel. This, and not the highways, is where the shrinkage in long-distance rail travel has gone – something fully understandable in a country of continental distances.

It is undeniable that some of the proliferation of highway travel has been at the expense of the railroads (but not to their regret – they would be happy not to have any passenger traffic). Much of this growth, perhaps the bulk of it, consists of trips that would not otherwise have been made; if it had not been possible to go by car, the travellers would not have gone at all. Let me give a personal example. Since 1959, when we moved from Massachusetts to California, my family has driven from coast to coast 24 times, varying our routes to see scenic and historic spots and visit friends, so that we have been, by car, in 47 states and several Canadian provinces. To have tried the same thing by any mode of public transportation would have been prohibitively expensive and time-consuming; we would simply have stayed at home.

It is very easy and reasonably popular to talk of getting people out of their cars and on to public transport, and in some situations, especially in large cities, it is probably desirable. Doing it is something else again. If only 1 per cent of all intercity automobile riders were to switch to air and rail travel, Amtrak and the airlines would be swamped. This situation could of course be remedied in time, but there would still be those other 99 per cent on the roads. In any case, I do not see any massive change in American travel habits without a degree of compulsion that would be unacceptable to the American people.

The freight picture is more complicated. Table 3.4 shows the figures for the 1950–83 period.

The total volume of intercity freight movement has almost doubled, and all carriers have gained. The proportion carried by road and by inland waterways has remained constant; the pipeline share had increased to just about the extent that the rail share has gone down. The greatest gain has been in air freight, although it is still a miniscule portion of the whole. Competition between railroads and trucks remains intense, but there have been significant changes in just the last few years. The deregulation of railroad rates in 1980 gave the

Table 3.4   Intercity freight movement by mode

	Motor trucks (1)	Rail- ways (2)	Inland waterways (3)	Pipe- lines	Domestic airways	Total
	Ton-miles, in billions					
1983 (est.)	551.0	838.0	356.0	582.0	5.500	2 330.0
1982	525.0	810.0	351.0	571.0	4.900	2 262.0
1981*	540.0	924.0	410.0	565.0	5.090	2 446.0
1980*	555.0	932.0	407.0	588.0	4.840	2 487.0
1979*	608.0	927.0	425.0	608.0	4.640	2 573.0
1978*	599.0	868.0	409.0	586.0	4.750	2 466.0
1977*	555.0	834.0	368.0	546.0	4.180	2 307.0
1976*	510.0	800.0	373.0	515.0	3.900	2 202.0
1975	454.0	759.0	342.0	507.0	3.730	2 066.0
1974	495.0	852.0	355.0	506.0	3.910	2 212.0
1973	505.0	858.0	358.0	507.0	3.950	2 232.0
1972	470.0	784.0	338.0	476.0	3.700	2 072.0
1971	445.0	746.0	315.0	444.0	3.500	1 954.0
1970	412.0	771.0	319.0	431.0	3.300	1 936.0
1969	404.0	774.0	302.0	411.0	3.200	1 894.0
1968	396.3	756.8	291.4	391.3	2.900	1 838.7
1965	359.2	708.7	262.4	306.4	1.910	1 638.6
1960 (4)	285.5	579.1	220.3	228.6	0.890	1 314.3
1955	223.3	631.4	216.5	203.2	0.490	1 274.9
1950	172.9	596.9	163.3	129.2	0.300	1 062.6
	Ton-miles, percent by type of transport					%
1983 (est.)	23.62	35.93	15.26	24.95	0.24	100
1982	23.21	35.81	15.52	25.24	0.22	100
1981*	23.10	37.90	16.50	22.40	0.21	100
1980*	22.30	37.50	16.30	23.70	0.19	100
1979*	23.70	36.10	16.50	23.70	0.17	100
1978*	24.40	35.20	16.30	23.80	0.27	100
1977*	24.10	36.10	15.90	23.70	0.18	100
1976*	23.20	36.30	16.90	23.40	0.20	100
1975	22.00	36.70	16.60	24.50	0.18	100
1974	22.40	38.50	16.00	22.90	0.18	100
1973	22.60	38.50	16.00	22.70	0.18	100
1972	22.70	37.70	16.40	23.00	0.18	100
1971	22.77	38.17	16.12	22.72	0.18	100
1970	21.30	39.72	16.49	22.29	0.17	100
1969	21.31	40.84	15.98	21.68	0.16	100
1968	21.55	41.16	15.85	21.28	0.16	100
1965	21.92	43.25	16.01	18.70	0.12	100
1960 (4)	21.72	44.06	16.76	17.40	0.07	100
1955	17.51	49.53	16.98	15.95	0.04	100
1950	16.27	56.17	15.37	12.16	0.03	100

* Revised.

(1) Ton-miles between cities and between rural and urban areas included whether private or for hire. Rural-to-rural and city deliveries are omitted. (2) Revenue ton-miles. (3) Does not include coastal and intercoastal ton-miles. (4) 1960 and later years include Alaska and Hawaii.

Sources: Transportation Policy Associates. *Transportation in America*. Tables 3.3 and 3.4 are taken from Motor Vehicle Manufacturers Association of the United States Inc., *Motor Vehicle Facts and Figures* (1984) pp. 56 and 57.

railroads what they needed most: freedom to compete without the burden of an obsolete regulatory system, created for transport as it was before the arrival of the motor vehicle and the building of hard-surfaced highways. Over these thirty years also, there has been a substantial integration of rail and road freight movement in the form of piggy-backing of truck trailers and containers, or, in formal terms, Trailer on Flat Car (TOFC) and Container on Flat Car (COFC).

One feature of the impact of highway transport on American life can be singled out because it involves passenger cars, trucks, and buses in a major transformation: namely, the transformation of our cities by decentralisation. Suburban expansion began with commuter rail lines, but the great explosion of Suburbia in our day is directly a product of the motor vehicle. The outward movement of industry is just as important, perhaps more so. The motor truck has made it possible, in fact desirable, for light- and high-technology industries to move to areas where there is ample land to build modern one-storey factories that can make maximum use of materials-handling equipment and to provide parking space for the workforce to come and go by automobile. So we have regions like Route 128 around Boston, Massachusetts, and Silicon Valley in the neighbourhood of San Jose, California. To an increasing extent people can not only live outside central cities but work outside them as well, and this in turn draws retail business out to the suburban shopping centers.

This process has had its drawbacks. One of the great unsolved problems of our time is the decay of city centres. I happen to believe that the solution does not lie in pushing people and business back into areas that they had good and sufficient reason for leaving; the crowded, highly centralised city of the past does not have to be enshrined as the model for the future.

We can, and we should, have a balanced system of transport in the United States, provided we really mean 'balanced'. (I have known the term to be used as if it was a synonym for 'anti-highway'.) Our modes of transport are competitive and ought to be, but they are also complementary. Passenger travel by air or train practically always has a road trip at each end. The same is true of much freight movement; as I have said, we are even integrating rail and road shipment through piggy-backing. However our national transportation system may develop, the segment represented by the motor vehicle and the road is going to be a vital part of it.

CONCLUDING SUMMARY

The positive consequences of the revolution in highway transportation can be summed up thus:

First, it provided an alternative, indeed the only viable alternative, to the dominating position that railroads had come to hold in overland transport. In short it gave travellers and shippers a choice.

Second, it provided a new mode of transport that for the first time in history made flexible, individual, low-cost mobility available to ordinary people; the ability to travel at will was no longer a luxury restricted to the wealthy.

If motorised road travel had not come into being; or had been limited in its function, we would have to consider the result not just in terms of the effect on other modes of transport, but in terms of the innumerable trips that would never have been taken and the innumerable shipments that would never have been made.

There are problems; I know of no new technology that has not brought problems with it. We have to balance the gains against the losses, and in this particular case I have to conclude that the gains come ahead.

Notes

1. Herodotus, *The History*, Book 8. See also *The History of Herodotus*, trans. by George Rawlinson, Everyman Ed. (New York, 1930) vol. 2, p. 251.
2. G. R. Taylor, *The Transportation Revolution* (New York, 1931) p. 132.
3. Late nineteenth-century Scottish investors in western American land were even more cautious. They wanted their land to be within ten miles of railroad or navigable waterway. See W. Turrentime Jackson, *The Enterprising Scot* (Edinburgh, 1968) p. 25.
4. See, for example, Eugen Diesel, Gustav Goldbeck, and Friedrich Schildenberger, *Vom Motor Zum Auto* (Stuttgart, 1957).
5. J. B. Rae, *The Road and the Car in American Life* (Cambridge, Mass., 1971) p. 32.
6. Malcolm L. Willey and Stuart A. Rice, 'The Agencies of Communication', *Recent Social Trends in the United States* (New York, 1931) pp. 172, 173, 177.
7. Ibid. pp. 169–70.
8. Mark S. Foster, *From Streetcar to Superhighway*, pp. 20–24.
9. J. B. Rae, 'Coleman du Pont and His Road', *Delaware History*, vol. 16, no. 3 (Spring – Summer, 1976) pp. 171-83.

10. Hilaire Belloc, *The Road* (London, 1924).
11. Rae, *The Road and the Car in American Life*, op. cit. p. 181.
12. For the evolution of the Interstate Highway System see Mark Rose, *Interstate: Express Highway Politics, 1941–1956* (Lawrence, Kan., 1979).
13. *Motor Vehicle Facts and Figures, 1984*, p. 84, Figures from U.S. Federal Highway Administration, *Highway Selected Statistics, 1982*.
14. Ibid. p. 19; Rae, *The American Automobile*, p. 109.

4 The Early Growth of Long-Distance Bus Transport in the United States

Margaret Walsh

Bus transport in the United States started in the second decade of this century when numerous entrepreneurs in all parts of the country operated local services between near-by communities using automobile sedans. Encouraged by their early success, ambitious pioneers built up longer networks by connecting their routes to those of like-minded venturers and by acquiring more reliable and comfortable vehicles. For those who could meet the requirements of state government regulations and withstand the competition from both railroads and other bus operators, prospects looked good. By the late 1920s the possibility of national lines suggested increased business; but the onset of the Great Depression forced reorganisation in the burgeoning and highly competitive industry. Many small carriers went out of business when passengers and revenue declined, while the larger companies had to restructure their operations to reduce costs at the same time as improving services. Federal regulation of motor carriers in the mid-1930s strengthened the position of surviving and larger companies who were able to retain their trade. By the end of the decade long-distance bus travel was established as a permanent and important part of the nation's economy.

ORIGINS · LOCAL SERVICES LINK UP

At the turn of the century Americans who wished to travel between cities either for work or for pleasure had limited options. The steam railroad offered the best, the most reliable and the fastest means of transport. Electric railways provided reasonable intraurban and short-distance intercity travel. They also offered some longer lines, but only in certain parts of the country. The coach or carriage was neither a competitive nor a comfortable alternative given the deplor-

81

able state of the nations's highways; and though bicycles were popular in both town and country, they, too, were hampered by poor road surfaces. It took the mass production and ownership of cars, together with increased attention to road construction, to bring the major breakthrough in travel in the 1920s. And alongside the rapid spread of the popular and individualistic auto came the slower, but no less significant, growth of bus transport. Not only did buses replace trams and trolleys in urban mass transit: they also opened up new avenues of intercity travel both to those Americans who could not afford cars and to those car owners who preferred to leave distance driving to others.[1]

No particular date marks the beginning of the American intercity bus industry because so many individuals were attracted to it at about the same time by the large profits available to those who could carry fare-paying passengers over public highways. These ubiquitous bus pioneers came from all walks of life. Few knew much about transport or about business, but they were willing to chance their hand in a new venture which had low entry costs. Frequently driving used vehicles, these 'carriers' concentrated on local services operated on a consumer-demand basis with the driver taking cash fares. There were no formal schedules or routes, though some 'bus owners' who carried men to work, began and finished their journeys either at hotels or stores. Others, however, loaded anywhere on the street where likely passengers stood waiting. Word-of-mouth or newspaper advertisements announced the existence of the new service, but a regular commitment was not guaranteed. If bus men did not get a payload, they frequently did not start until they did; and those who travelled on the early buses were satisfied with reaching their destination rather than enjoying a fast or comfortable journey.[2] The early 'jitney' or taxi-cab style days were exciting and promising, but were not particularly stable.[3]

The experiences of pioneers in the two states which led the way in long-distance bus travel, California and Minnesota, suggest the flavour, problems and possibilities of this new form of transport. In northern California Wesley E. Travis, who had previously worked in subcontracting the postal service in the Mountain and Far West and then in the taxi-cab business in Chicago, in 1907 set up another taxi service in San Francisco. Following his early success, he incorporated this enterprise as the California Taxicab Company in 1909 and gradually acquired forty taxicabs. But these vehicles were unsatisfactory both in terms of size and reliability, so he decided to improve their design and construction. Soon other carriers, such as the auto-stage operators of Stockton, wanted to buy his new elongated vehicles and this interest prompted him to investigate the possibilities of interurban as well as intraurban road transport.[4]

In 1912, ex-miner A. L. Hayes started an auto-stage line in southern California to take sightseers from San Diego to Oceanside. This proved to be so popular that he began another service to El Centro and then gradually built up longer routes and more regular schedules, later extending his Pickwick Stages to Los Angeles, a distance of 132 miles, where he joined up with another bus pioneer, Charles Wren. Wren had started off in suburban 'jitney' transport in 1913 with a service between Los Angeles and Venice, some 14 miles away. Persuading other city-taxi and longer-line operators to use his depot as a terminal, he moved into intercity transport with a route to San Fernando, twenty miles to the north. Shortly after this, he extended this line to Santa Barbara, ninety-nine miles away. By 1918 he had organised a number of rent-car drivers into a small auto-stage association operating between Los Angeles and San Francisco. Following consolidation with Hayes's San Diego line, Pickwick Stages became a state network of sorts using the coastal route.[5]

In the Iron Range district of northern Minnesota early bus operators started up business by carrying men between mining locations. In Hibbing, in 1914, Eric Wickman, a former diamond driller, entered the bus trade when he could not sell the first car shipped to his Hupmobile agency. Acquiring two partners, Andrew G. Anderson and Arvid Heed, they built up a regular service to the near-by community of Alice. The entrepreneurs then bought another auto and started a new route to the mines at Mahoning. As trade was brisk and their carrying capacity inadequate, they hired a local blacksmith to build elongated bodies welded onto truck frames. By 1915 they needed more capital and more drivers. So they acquired new partners and formed the Mesaba Transportation Company. Soon the firm was operating a fleet of eighteen buses with routes extending as far as Grand Rapids, forty miles away.[6]

These California and Minnesota experiences were replicated throughout the United States in the years between 1910 and 1920. Men from all walks of life who had touring cars carried passengers short distances for reasonable fares. If they worked hard and set up fairly regular services, either undercutting or amalgamating with near-by competitors, then they made enough money to expand and acquire more vehicles to develop these local routes. In most areas the possibility of operating over longer distances was prevented by the slow speed of travelling on dirt-top roads, in the rural areas in particular, and the frequent breakdowns. It was more difficult for drivers to make repairs on the road if they were far from their base. Any attempts to establish longer-distance services took the form of unco-ordinated links with the local services of other entrepreneurs. Passengers were left with the discomfort of long waits.

STATE LEGISLATION, TECHNICAL DEVELOPMENT AND BETTER ORGANISATION HASTEN CHANGE

But change was on the horizon. It came in two main ways: externally in the shape of state government regulations and internally through improvements in mechanical engineering and business organisation. Prompted by the need to finance better roads and to ensure public safety, and increasingly lobbied by railroad interests anxious to curb the new freight and passenger competition at an early stage, state governments moved to regulate motor carriers. By 1920 eleven states had imposed some control on bus operators. In the next five years twenty-six more states followed suit, making three-quarters of the total, including those most heavily populated. Regulations varied widely, but most states insisted that buses, as public utilities, should be subject to laws protecting passengers from unnecessary and inefficient lines, high charges, unsafe vehicles and inexperienced drivers. Many of the new requirements, such as those ensuring that drivers had licences and insurance coverage, that buses had brakes and lights, and that companies kept to advertised schedules, were difficult to enforce, particularly in rural areas. But a start had been made to ensure that bus lines observed basic standards.[7]

Within the bus industry advances in mechanical engineering and competition, both between small independent carriers and between buses and trains, stimulated improvements in motor-coach equipment and in operating practices. In terms of design and technology the new buses of the early 1920s were much superior to the improvised elongated auto sedans of earlier years which had been constructed on truck chassis and seat only up to seventeen persons. In 1921 the Fageol brothers of Oakland, California, produced the safety coach, a low-slung twenty-two-seater with sedan doors all along the side and equipped with a four-cylinder 60-horsepower airplane-type motor. The following year the White Company of Cleveland, Ohio, moved from a commercial chassis to one specially designed for bus use. Other auto and truck manufacturers like Pierce-Arrow of Buffalo, New York, General Motors of Pontiac, Michigan, and Mack Motor Truck Company of New York City also started to specialise in bus production, while several bus operators constructed motor coaches to meet their particular needs.

When larger and more comfortable vehicles with more powerful engines were called for, Fageol adopted a six-cylinder 100-horsepower motor, stretched the wheelbase and widened the body of his safety coach to accommodate twenty-nine rather than twenty-two passengers. Other leading bus manufacturers made similar design changes and worked to improve safety and comfort through introducing air brakes, sturdier frames, heating and ventilating systems, interior baggage racks, and even lavatory and toilet facilities. By the mid-1920s some ambitious firms were experimenting with two-level (or duplex) vehicles, with seating capacity for fifty-three passengers, and were designing night coaches which would provide sleeping and seating accomodation for twenty-six people. This general quest for improved vehicles in its turn brought systematic maintenance checks and more repair facilities in well-designed garages and terminals.[8]

Changes internal to the bus industry were equally important in contributing to the growing popularity of passenger travel in the 1920s. Modern business practices similar to those adopted by the railroads were essential to ensure smooth operation, profitability and customer satisfaction. Ambitious bus entrepreneurs in companies like California Transit and Pickwick Stages in California, or Mesaba Transportation and Jefferson Highway in Minnesota, thus began analysing traffic flows and planning routes and connections more carefully and developing rational fare structures. They installed new accounting procedures and created divisional lines of control. They then published schedules regularly and advertised services widely, taking care to draw attention to the comfort of the buses, the safety of the drivers, the amenities at the terminals and the ease of purchasing tickets.[9]

Bus entrepreneurs were anxious to systematise their procedures not only to increase profits but also to compete more effectively against steam and electric railways. In this they were so successful that many railway companies decided to use buses themselves or to work in collaboration with bus concerns rather than to try and limit their operations through restrictive legislation and heavy taxation. Accepting the economic facts that buses offered advantages of greater flexibility and cheaper running costs, especially for short and immediate length journeys and for routes in sparsely populated areas, the number of railroads operating buses rose remarkably in the mid-1920s. There were only three steam railroads running them in January 1925, but by the end of that year thirty-one companies were using 375. Encouraged by the successful example of major pioneers like the Great Northern, the Pennsylvania, and the New York, New Haven & Hartford Railroads, other rail companies quickly adopted buses on a system-wide basis. By 1929 sixty-two steam railroad companies ran 1256 buses over 16 793 miles of route.[10]

Railroads might attempt to retain their passengers by setting up bus auxiliaries in the late 1920s, but the larger bus companies, now with a decade of operating experience, looked further afield and threatened long-distance rail services. They announced their intention of providing nationwide routes using improved buses and interline connections. By 1927 the major Californian companies, Pickwick Stages and California Transit, had routes which stretched north to Portland, Oregon, south to San Diego, California, and east into the Plains and south-western states of Arizona and New Mexico. In the centre of the continent the Minnesota-based Motor Transit Company, having bought numerous small lines, started to acquire operating subsidiaries beyond the Middle West. The California Transit Company, reorganised as the American Motor Transportation Company, headed by Wesley Travis, had the distinction of putting together the first coast-to-coast bus service in 1928. For a one-way fare of $72.00 the adventurous traveller who could endure a journey of five days and fourteen hours incorporating 132 stops, could travel by road from Los Angeles to New York on the longest bus line in the world. Motor Transit shortly thereafter, in 1928, offered a transcontinental service through co-operative arrangements with Pickwick Stages.[11]

But Motor Transit needed to make more systematic arrangements and stringent controls were necessary to build up regular transcontinental passenger services. A national bus company had to be forged. In 1929, having put together a marketable portfolio of motor transport stocks, Midwestern investment bankers consolidated a sprawling bus empire consisting of four fully owned operating, three fully owned ancilliary and six partially owned operating companies with routes in forty-one out of the forty-eight states. This expansion carried the company from a total of five million coach miles in 1927 to 20 million in 1930. Early the following year Motor Transit officially adopted the name already used by most of the recently acquired

Table 4.1 The intercity bus industry 1925–1940

Calendar year	Number of companies	Number of vehicles	Route miles	Bus miles (millions)	Passenger miles (billions)
1925	3 610	21 430	218 601	n.a.	n.a.
1926	4 040	22 800	250 396	960	n.a.
1927	3 950	16 750	233 829	1 030	n.a.
1928	3 610	16 050	251 999	1 170	n.a.
1929	3 910	16 140	229 936	1 190	7.5
1930	3 520	14 090	318 715	1 230	7.1
1931	3 140	14 220	354 745	1 020	6.7
1932	2 760	13 640	364 676	920	6.3
1933	2 370	12 230	338 018	820	6.4
1934	2 210	10 660	321 449	800	7.1
1935	2 120	11 160	330 216	960	7.6
1936	2 130	11 260	312 825	1 110	9.2
1937	1 790	11 940	315 668	910	10.0
1938	1 880	11 700	320 640	880	9.0
1939	1 990	11 600	317 298	860	9.6
1940	1 830	12 200	313 136	820	10.1

n.a.–not available

Source: Burton A. Crandell, *The Growth of the Intercity Bus Industry* (Syracuse, N. Y.: Syracuse University, 1954), appx A, table A–2, 280–2.

companies and became the Greyhound Corporation. There were high hopes for the future. In the space of a few years entrepreneurs had successfully introduced a new component into the transport sector and were offering economic, speedy and safe travel on a nationwide basis.[12]

PROBLEMS OF THE 1930s

The adolescence of the intercity bus industry would, however, be more painful than its birth pangs and infancy. Coming of age in the major depression of the 1930s, all bus companies faced the prospects of declining traffic. Table 4.1 documents this. Many small carriers collapsed when passengers and revenue fell, while the newly merged large corporations, experiencing both a capital and cash flow shortage, had to reorganise both their financial and administrative structures and had to ensure system-wide economies in order to survive.

Having weathered their first national crisis in the early 1930s, motor coach entrepreneurs then had to comply with federal regulations. Still more economies and major improvements in equipment and operating practices, along with widespread advertising campaigns, were necessary to come through the recession of the late 1930s. But by the end of that decade bus travel in the United States was well established.

The impact of the Great Depression was felt throughout the intercity bus industry in the early 1930s as passenger traffic failed to continue its upward momentum. Though the decline in business was not as severe as that experienced in many branches of manufacturing, bus companies had to face shrinking demand. Certainly large operators had analysed traffic flows when establishing new routes or when buying existing lines, but they had generally assumed that they could persuade the public of either the necessity or the enjoyment of bus travel. Workers could get to their existing jobs more easily or could find better jobs in new locations. Women had more opportunities to go shopping and families were able to visit each other regularly. Furthermore, buses were admirably suited to serve the growing interest in tourism and to provide for holiday traffic. But with unemployment rising and incomes falling, fewer people rode distances to work. Fewer still rode the buses for social and recreational reasons. Bus operators would have to reassess market potential.

The number of intercity bus companies fell notably between 1929 and 1932 from 3910 to 2760 (See Table 4.1.) Many small and weak firms collapsed because their unsound accounting methods and inadequate maintenance facilities could not withstand the strain of competition for decreasing passenger traffic. Their larger and more ambitious counterparts, now in consolidations underwritten by investment financiers, were also in difficulties. Unable to repay the short-term loans used to float the mergers which created them, some companies declared bankruptcy, while others had to re-finance and then to apply themselves systematically both to reducing costs and increasing business.[13] The manœuvres undertaken by the newly formed Greyhound Corporation illustrate the problems of surviving the early years of the Depression.

THE GREYHOUND CORPORATION

Following the Stock Market Crash of October 1929 Greyhound was dangerously overcapitalised. Much of the Corporation's rapid expansion in 1928 and 1929 had been financed by short-term loans which could not be repaid as earnings fell. Two recapitalisation schemes were essential to meet current obligations in 1930 and in 1933. On the first occasion, before the Depression reached its nadir, Greyhound's major investment broker, Glenn Traer, managed to persuade bankers to take a $4 000 000 issue of three-year notes. But the corporation's financial position did not turn around in the next three years. Low profits in 1930 and 1931 and a net loss in 1932 meant that there was no money to repay the notes as they fell due in spring 1933 or to service the current debt of $1 640 000. More ingenuity and salesmanship were needed if Greyhound was to stay in business.

Turning first to the floating debt, Traer and top Greyhound officials, Eric Wickman and Orville Caesar, persuaded General Motors, who had been manufacturing buses for Greyhound since 1930, to take over a substantial portion. They were then able to get extensions on the remainder. Next they arranged a new issue of five-year notes in exchange for the maturing three-year notes. These measures, however, were insufficient to resolve the company's financial problems. The capital stock situation remained very unhealthy. Greyhound lagged badly in paying its cumulative preferred dividends and the common stock had never yielded any dividends. Recapitalisation created a new common stock in exchange for shares of the second preferred stock and the old common stock, while retaining the first preferred at par. Some stockholders took a loss under the rearrangements, but that alternative was deemed preferable to the company's demise.[14]

The nation's largest bus concern was still not out of its quagmire. The company had to make more money if it was to progress. Divisional restructuring promised savings when the two uneconomic companies in the system, Western Greyhound and Southland Greyhound, were reorganised into a sounder unit, Southwestern Greyhound.[15] Careful attention to maintenance facilities and to improving equipment also reduced operating costs. On the more positive side, company managers sought out and won new business by, for example, providing transport to and organizing tours at the Chicago World's Fair in 1933 and 1934. They also conducted a huge mass-media campaign which included scenes on a Greyhound bus in the 1934 film, *It Happened One Night,* starring Clark Gable and Claudette Colbert. Bus passengers travelling on Greyhound increased in 1934 and 1935 and prospects started to look better.[16] The experiences of Greyhound during the troubled years of the 1930s may not be typical; but they do suggest that considerable attention would have to be paid to better business methods, sounder financing and innovative marketing if the bus industry was to grow.[17]

LEGISLATION FAVOURS EXISTING OPERATORS, ESPECIALLY THE LARGER ONES

While bus companies were still struggling to combat the economic impact of the Depression, they were faced with further changes imposed by law. By the mid-1920s most states had regulated intra-state motor-carrier operations, not only on the grounds of public safety and financing road construction, but also with a view to controlling economic issues like the degree of competition and the level of fares. When, however, Supreme Court decisions rendered state regulation of interstate motor traffic ineffective in 1925, calls for federal legislation became frequent and loud. But the process of enacting this legislation was slow and painful. Congressional hearings in 1926, 1928, 1930, 1932 and 1934, and investigations by the Inter-state Commerce Commission starting in 1926 and 1930, convinced or persuaded enough politicians that the new motor-carrier industry, particularly the trucking section, should not yet be regulated and no fewer than forty bills failed to pass both houses of Congress between 1925 and 1935. Most bus operators, the railroads and tax-payers represented by state associations, favoured some regulations in the interests of reducing cut-throat competition, but little headway could be made while diverse interest groups fought over specific issues and methods of administration.[18]

Indeed initial federal guidelines emerged from the flurry of legisla-tion which was passed in the First Hundred Days of Franklin D. Roosevelt's Administration in 1933. Under the National Industrial Recovery Act, the motor-bus operators met government mediators and representatives of labour and consumers and drew up a code of self-regulation. Companies signing the code agreed to respect exist-ing state regulations, to eliminate destructive competitive practices, to observe maximum hours and minimum rates of pay for workers and to promote the fullest use of the industry. This Motor Bus Industry Code, however, had little impact on the shape of long-distance bus transport in 1933 and 1934. By this time depressed business conditions had eliminated sub-standard operators and most of the active firms had already adopted many of the suggested administrative practices and were following safety provisions.

Though the trade association, the National Association of Motor Bus Operators, favoured the principle of self-regulation rather than regulation by a government agency, the code had little chance to be effective, for in 1935 the National Industrial Recovery Act was declared unconstitutional.[19] Later that year, however, the long-awaited specific federal regulation came into effect with the Motor Carrier Act, alternatively known as Part II of the Interstate Commerce Act.

The objectives of the Motor Carrier Act were to prevent wasteful and destructive competition within the bus and truck industries in particular, and in the transport sector in general, and to promote and protect the public interest. The Interstate Commerce Commission, as regulatory agency, would achieve these aims by exercising three main controls. In the first place entry into the bus industry would require a certificate of public convenience and necessity which could be suspended, changed or revoked. Mergers and issues of securities also had to be approved. Second, bus operators had to conform to regulations governing safety, insurance, finance, accounting and records. For example, drivers had to be qualified, certain safety devices were made compulsory, insurance protection was required and companies had to comply with specified accounting procedures. Third, carriers had to publish and to adhere to rates and fares and they had to give 30 days' notice of any changes. The Interstate Commerce Commission could suspend these changes and could prescribe maximum, minimum and actual rates to be charged. Long-distance bus transport had now entered a new regulatory era.[20]

The terms of the Motor Carrier Act worked in favour of large companies and those with existing routes. Certification protected most companies already in business while the establishment of operating standards did not significantly affect those with adequate capital. Although the newly established Motor Carrier Bureau[21] did not follow a consistent policy in interpreting the 1935 Act, officials seemed to lean towards a course of regulated monopoly by limiting competition between long-distance bus operators. Existing companies contested applications for competitive routes, and on establishing that the service was adequate, new applications were often turned down. Mergers of existing carriers were allowed when it was clear that the volume of traffic for two carriers were inadequate. Bus subsidiaries of railroads could not acquire competing motor carriers if the sole purpose was to eliminate competition; but if traffic flow was light only one company would be certified. Small operators did continue to offer service, but the large firms dominated. National Trailways, an association of independent bus firms with each component retaining individual membership, offered transcontinental services; but Greyhound had no serious rival. And Greyhound proceeded to clarify its position under the Motor Carrier Act, grouping its twenty-nine companies into seven operating divisions. These mergers, known as 'The Greyhound Mergers of 1936', simplified the corporate structure at the same time as consolidating and strengthening Greyhound's position.[22]

Greyhound's continuing prominence in bus transport and its ability to grow during depressed economic conditions in the late 1930s, stemmed as much from the continued improvements in operating practices as from mergers and the changing legal environment. Bus officials hoped that increased passenger traffic would follow from improved internal management and better public services. They also realised, however, that they would have to stimulate bus travel through a systematic advertising campaign in major journals and newspapers, supported by travel brochures and window displays. Business needed to be encouraged.

Greyhound managers pursued various avenues of improvement simultaneously. Equipment was annually updated. The introduction of the aluminium and steel cruiser in 1936 and the diesel motor in 1938 constituted the two outstanding technical innovations, but many small changes were made to engine parts, particularly transmissions and brakes. In the maintenance division the regular servicing carried out at terminals was now supported by periodic overhauls undertaken at large garages specifically built at key points for that purpose. Many modern terminals were constructed not only to provide repair facilities, but also to offer those amenities such as waiting-rooms, cafeterias, baggage-checking, improved washrooms and travel bureaus, now demanded by passengers. Steps were also taken to improve one of the negative facets of bus travel, namely those intermediate rest stops run by concessionaires. Indeed in 1937 Greyhound started buying and building its own 'Post Houses' to be run by trained managers. As for the travelling itself, better interdivisional co-ordination brought superior timetabling arrangements, while more travel agencies in key cities provided full information on both the new schedules as well as the planned and expense-paid tours.[23]

Where Greyhound led, other long-distance bus operators tended to follow. As rates and fares were now subject to the authority of the Interstate Commerce Commission, competitors could no longer win passengers by cutting fares. Nor could they offer new routes without government permission. Instead they had to provide better services and for the most part this meant improving the safety, comfort and speed of travelling. Given this greater care and attention, bus operators increased their share of the nation's intercity travel from 3.6 per cent of the total in 1929 to 4.3 per cent a decade later.[24] Passenger bus mileage had gone up from 7.5 billion to 9.6 billion in the same period. (See Table 4.1) Though private automobiles accounted for the bulk of long-distance travel in the Depression years, buses were making inroads into the public carrier sector. Within the space of a quarter of a century a new public transport facility had grown to maturity and was able to offer passenger service comparable to, and in many cases superior to, those offered by the long-established railroad.

Notes

*The author wishes to extend many thanks to the American Philosophical Society, the Wolfson Foundation, the Nuffield Foundation and the Universities of Birmingham, Kansas and Wyoming, without whose generous financial support, the research for this paper could not have been undertaken.
 1. For general discussions of the condition of roads and the prospect of highway travel in the late nineteenth and early twentieth centuries see John B. Rae, *The Road and the Car in American Life* (Cambridge, Mass.: Massachusetts Institute of Technology Press, 1971) pp. 22–39; American Association of State Highway Officials, *A Story of the Beginning, Purposes, Growth, Activities and Achievements of the American Association of State Highway Officials* (Washington, D.C.: A.A. S.H., 1964) pp. 31–40; Charles L. Dearing, *American Highway Policy* (Washington, D.C.: The Brookings Institution, 1941), pp. 29–59, 224–40. For an overview of steam and electric railways see John F. Stover, *American Railroads* (Chicago: University of Chicago Press, 1961), and George W. Hilton and John F. Due, *The Electric Interurban Railways in America* (Stanford, Calif.: Stanford University Press, 1960).
 2. For general information on early bus activity see Albert E. Meier and John P. Hoschek, *Over the Road. A History of Intercity Bus Transportation in the United States* (Upper Montclair, N.J.: Motor Bus Society, 1975) pp. 1–9; Burton B. Crandall, *The Growth of the Intercity Bus Industry* (Syracuse, N.Y.: Syracuse University, 1954) pp. 6–13; Ford K. Edwards, *Principles of Motor Transportation* (New York and London: McGraw-Hill, 1933) pp. 5–9. Detailed information is available in local newspapers, as for example, *The Hibbing Daily Tribune*, 1915–27.

3. In California auto owners pulled up at trolley stops and offered rides for a 'jitney' or nickel. The name 'jitney' was frequently used before the manufacture and use of longer vehicles which might be called 'buses' or 'motor coaches'. In 1915 the newly incorporated Mesaba Transportation Company was described as a Jitney Company, *Hibbing Daily Tribune*, (18 Dec 1915).

4. Carl H. Gohres, 'History of Pacific Greyhound Lines', pp. 2–8, Box 3, Greyhound Corporation Records, University of Wyoming, Laramie (hereafter cited as G.C.R.); W. E. Travis, 'Statement', Folder 'Greyhound History: Pacific Greyhound Lines', Box 1, G. C. R.

5. Gohres, 'History of Pacific Greyhound Lines', pp. 18–20; John P. Hoschek, 'Greyhound's Western History', *Motor Coach Age*, vol. 12, no. 2 (1960) p. 5; Edwards, *Principles of Motor Transportation*, pp. 5–6.

6. F. H. Schultz, 'Greyhound The Greatest Name on the Highway' (Thesis, n.p. n.d.), 5–7, Box 4, G.C.R.; C. G. Schultz, 'Notes on Son's Thesis', Folder 3, Box 3, G.C.R.; 'Jitney into Giant', *Fortune*, vol. 10, no. 2 (Aug 1934) pp. 42–3, 110; *Hibbing Daily Tribune* (25 June 1946); Hilda V. Anderson, 'A History of the Beginnings of the Bus Industry with Grass-Roots in St. Louis Country' (unpublished MSS., *c*. 1954) pp. 5–13.

7. General discussions of the state regulation of motor carriers can be found in Shan Szto, *Federal and State Regulation of Motor Carrier Rates and Services* (Philadelphia: University of Pennsylvania, 1934) pp. 9–189; Crandall, *The Growth of the Intercity Bus Industry*, pp. 38–97; John J. George, *Motor Carrier Regulation in the United States* (Spartanburg, S. C.: Bond & White, 1929) pp. 1–213; Emory R. Johnson, *Government Regulation of Transportation* (New York: D. Appleton-Century Co., 1938) pp. 510–27. More detailed information is available either in trade journals like *Bus Transportation*, vols 1–4 (1922–5) and *Railway Age*, vols 70–9 (1921–5) or in state publications and local newspaper reports of state commission hearings, as, for example, Minnesota State Warehouse and Railroad Commission, Auto Transportation Company Division, *Biennial Reports* (1926), (1928), (1930) and *Minneapolis Journal* (1925).

8. Meier and Hoschek, *Over the Road*, pp. 17–20, 22–5, 43–6; Gohres, 'History of Pacific Greyhound Lines', 4–5, 7, 12–13, 25–7; Eli Bail, *From Railway to Freeway; Pacific Electric and the Motor Coach* (Glendale, Calif.: Interurban Press, 1984) pp. 51–2, 55; Anderson, 'Beginnings of the Bus Industry', pp. 12–15; 'The Fageol Safety Coach Story', *Bus Review*, vol. 1, no. 3 (1968), 4; *Minneapolis Journal* (3 Feb 1924, 1 Feb, 1925, 6 Feb 1927), 25 Sep 1928. *Bus Transportation*, vols 1–3 (1922–4) contains sections advertising buses manufactured by different companies.

9. For general information on business practices see *Making Bus Operations Pay* (New York: McGraw-Hill, 1932). For more specific insights see the trade journals, *Bus Transportation*, vols 1–8 (1922–9), *The Truck and Bus Owner*, vols 1–2 (1922–4) and *Travel by Bus*, vols 1–3 (1924–6).

10. The best discussion of railroad involvement with buses are found in the trade journal *Railway Age*, vols 76–87 (1924–9). Statistics on railroad-operated buses are available in *Bus Facts* for 1927–30 (National Association of Motor Bus Operators, Washington, D.C., 1927–30). For a

general discussion of the bus and the electric interurbans which turned to buses and are for greater scale than the steam railways and were using over 10 000 of them in 1929, see Hilton and Due, *The Electric Interurban Railways*, pp. 226–39.

11. Gohres, 'History of Pacific Greyhound Lines', pp. 8–17, 20–3; Bail, *From Railway To Freeway*, p. 59; Margaret Walsh, 'Tracing the Hound. The Minnesotan Roots of the Greyhound Bus Corporation' (unpublished paper); *Highway Traveler*, vol. 1, no. 2 (June 1929), 10–11, 28–9; vol. 2, no. 1 (Feb 1930) pp 16–17, 29; *Railway Age*, vol. 82, no. 4, 22 Jan 1927, pp. 329–31; vol. 85, no. 25, 22 Dec 1928, 1261–5.

12. 'The Greyhound Corporation', pp. 1–10, Greyhound History Folder 1, Box 3, G.C.R.; F. H. Schultz, 'Greyhound, The Greatest Name on the Highway', pp 11–15; Walsh, 'Tracing the Hound', pp. 14–17; 'Jitney into Giant', p. 42.

13. For a general discussion of the problems faced by the larger bus operators in the early 1930s, see *Bus Transportation*, vols. 9–13 (1930–4).

14. F. H. Schultz, 'Greyhound, The Greatest Name on the Highway', pp. 17–22; 'Jitney into Giant', pp. 34–43, 110, 113–14, 117; Greyhound Corporation, *Annual Reports*, to Stockholders, 1929–1932; 'A Brief History of the Greyhound Lines', *Mass Transportation* (1956 Sep) p. 56; Oscar Schisgall, *The Greyhound Story. From Hibbing To Everywhere* (Chicago: J. C. Ferguson, 1985) pp. 29–32.

15. Both Western Greyhound and Southland Greyhound had too many routes and low passenger traffic and both faced competition from the railroads who were cutting rates. F. H. Schultz, 'Greyhound, The Greatest Name on the Highway', p. 22.

16. Greyhound Corporation, *Annual Reports* to Stockholders (1934 and 1935); *Highway Traveler*, vol. 4, no. 4 (Aug–Sep 1932) p. 41, vol. 6, no. 2 (Apr–May 1934) p. 34; Carlton Jackson, *Hounds of the Road. A History of the Greyhound Bus Company* (Bowling Green, Ohio: Bowling Green Popular Press, 1984) pp. 45–48; Schisgall, *The Greyhound Story*, pp. 39, 41.

17. The structure of the bus industry in the early 1930s consisted of a majority of operators who owned a handful of buses and a very small number of larger firms, of whom Greyhound was the leading concern. Many operators would not have faced major problems of corporate financing, but they would have had recourse to loans from local banks.

18. Margaret Walsh, 'Federal Policy and the Intercity Bus Industry in the 1930s' (unpublished paper), pp. 4–12.

19. 'Code of Fair Competition for the Motor Bus Industry' in United States National Recovery Administration, *Codes of Fair Competition*, nos 58–110, vol. II (Washington, D.C.: United States Government Printing Office, 1934) pp. 110–117; *Bus Transportation*, vols 12–14 (1933–5), contains discussions of the bus industry's attitude to the Code; Crandall, *The Growth of the Intercity Bus Industry*, pp. 123–6; Szto, *Federal and State Regulation*, pp. 267–77.

20. Motor Carrier Act, 1935. Public Laws of the United States of America passed by 74th Congress (1935–6), *Statutes at Large*, vol. XLIX (Washington, D. C.: United States Government Printing Office, 1936)

pp. 543–569; Warren H. Wagner, *A Legislative History of the Motor Carrier Act, 1935* (Denton, Maryland: Roe Publishing Company, 1935); James C. Nelson, 'The Motor Carrier Act of 1935', *Journal of Political Economy*, vol. 44 (1936) pp. 471–97; I. L. Sharfman, *The Interstate Commerce Commission: a Study in Administrative Law and Procedure*, part 4 (New York: The Commonwealth Fund, 1937) pp. 102–22; Parker McCollester and Frank J. Clark, *Federal Motor Carrier Regulation* (New York: Traffic Publishing Company, 1935) pp. 88 ff.; William J. Hudson and James A. Constantin, *Motor Transportation. Principles and Practices* (New York: Ronald Press, 1958) pp. 476–82.

21. The Motor Carrier Act was to be administered by the Interstate Commerce Commission. The Federal Coordination of Transportation had recommended, in his third report, *Report of the Federal Coordinator of Transportation, 1934* (74 Cong. 1 Session, 1935) Document 89, that the Interstate Commerce Commission should be reorganised to deal with regulation of motor and water carriers, but Congress failed to act on this suggestion. Nevertheless it was clearly understood that special divisions for handling each kind of transportation were essential if regulation was to be effective. See Wagner, *A Legislative History*, pp. 13–14 quoting Senator Wheeler, 79 Cong. Rec. 5650, 5656, 5657. On 1 October 1935 the Interstate Commerce Commission did reorganise, reducing the number of its divisions from seven to five and establishing Division 5 as the Bureau of Motor Carriers.

22. F. H. Schultz, 'Greyhound, The Greatest Name on the Highway', pp. 31–5; *Railway Age*, vol. 100, no. 17 (25 Apr 1936) 692–6, vol. 100, no. 21, (23 May 1936) pp. 837–8, 844–5; *Mass Transportation* (Sep 1956) p. 57. Crandall, *The Growth of the Intercity Bus Industry*, pp. 166–220, provides the best summary of those early Motor Carrier Cases which affected buses. Interstate Commerce Commission, *Annual Reports*, from 1937, lists the most important decisions made by the Bureau of Motor Carriers, but frequently these decisions are concerned with trucks.

23. For details of improvements made by Greyhound see *Highway Traveler*, vols 7–12 (1935–40). For a summary of some of these improvements see 'Greyhound Still Growing', *Fortune*, vol. 30, no. 3 (Sep 1944) pp. 121–5, 236–9.

24. Private automobiles were responsible for 82.7 per cent of the 186.6 billion intercity miles travelled in 1929. In 1939 they accounted for 87.5 per cent of the 268.1 billion miles travelled. Of the 17.3 billion intercity miles undertaken by public carriers only (railroads, intercity buses, airlines and waterways) in 1929, buses were responsible for 21.2 per cent. In 1939 the respective figures were 12.5 billion miles and 34.7 per cent. Intercity buses increased their mileage in the 1930s, while their major public carrier rival, the railroads, were in serious difficulties and experienced falling passenger traffic. (*Bus Facts*, vol. 21 (1952) pp. 7, 10.)

5 Diesel Trucks and Buses: Their Gradual Spread in the United States

James M. Laux

In the United States trucks and buses adopted the diesel engine much later than occurred in Europe. The fact of this delay is well known in automotive circles, but behind it are some interesting developments that deserve exploration.

Actually the Americans have been slow to adopt many important automotive innovations. The automobile itself came from Europe, as did pneumatic and later radial tyres, the direct drive, four-wheel drive, fuel injection, etc. Whether American auto makers adopt a technical improvement depends on many factors, but their under-standing of what the American market might want almost always receives a higher priority than technical elegance for its own sake.

ORIGINS AND EARLIER POPULARITY IN EUROPE

The diesel engine was conceived and brought to fruition by the Paris-born German Rudolf Diesel in the 1890s. Although Diesel projected an automotive market for his engine, its main use at first was to power workshops and to turn turbines in electric power stations. Early diesels' great weight and size precluded their use for automotive purposes. The typical diesel of 1913 weighed 250 kg per horsepower. This figure would have to be cut twenty-fold before the diesel engine could be considered seriously for road transport. The first early push toward such a reduction in weight came from the use of diesel engines in submarines. These engines weighed about 25 kg per horsepower. One complexity that added weight and took power was the compressor that provided compressed air to blast fuel into the cylinders. Shortly before the First World War, however, the German Prosper L'Orange and the Briton James McKechnie sep-arately worked out laboratory experiments that dispensed with such air compressors. After the First World War diesel engines found wider markets in small ships and boats and in small railway

97

locomotives. Serious efforts also began to apply diesels without air compressors to automotive use, first of all in Germany. There the MAN and Benz companies began selling small, high-speed diesel engines for agricultural tractors and trucks in 1924. Two years later the Bosch company of Stuttgart perfected an improved fuel-injection system that raised the reliability and efficiency of high-speed diesels and was the major breakthrough to automotive use. By 1929 Germany had produced some 300 diesel trucks with engines that weighed abound 10 kg per horsepower. Swiss, British and French firms followed the German lead in diesel trucks and buses, with the Englishman Harry Ricardo making significant improvements in engine design.

Diesel engines for trucks and buses found a ready market in Europe because of this engine's fuel efficiency. Theoretically, the diesel at full load burned about three-quarters as much fuel as a gasoline engine, and with a light load consumed half or less. In addition, in the early 1930s diesel fuel cost from about one-quarter to one-half as much as gasoline in Western Europe. There, gasoline prices varied widely, ranging from 23 cents per American gallon in France and Germany to 27 cents in Britain and 65 cents in Italy. In the United States gasoline brought about 13 cents plus an average 3 cent state tax. So the relatively high cost of gasoline in Europe made an alternative fuel quite attractive.

Buses and heavy trucks dieselised first in Europe, for their fuel expense accounted for a higher proportion of operating costs than lighter vehicles. For these types of machines many operators determined that the fuel savings more than overcame the higher cost of the engine, and at first, the higher maintenance charges. Diesel buses in Britain rose from 5.4 per cent of the total in 1934 to 31 per cent in 1938. These 17 000 diesel buses outnumbered the 9000 diesel trucks in Britain at this time. Germany went further. There, the vast bulk of new trucks rated at loads over 3.5 tons produced during the 1930s used diesel power. A total of 51 000 diesel trucks and 7300 diesel buses (37 per cent) were on the road in 1938.[1] Ultimately the same factor of relative fuel cost operated in America but much more slowly.

RELUCTANCE TO ADOPT DIESELS IN THE UNITED STATES AND THE PIONEERS WHO OVERCOME THIS

In the United States between the wars extremely low gasoline prices made fuel costs a minor factor for most truck operators, few of whom engaged in long-distance work before the early 1930s. Consequently, Americans tended to ignore European developments in automotive diesels. Some experimentation did occur, but little was by the makers of heavy diesel engines for stationary or maritime use. Charles F. Kettering, head of research for General Motors, became interested in the diesel engines on his private yacht in 1928 and persuaded his company to begin a diesel project. Kettering favoured the two-stroke system and insisted that GM development be confined to this type, even on small high-speed automotive sizes where most European diesels used a four-stroke cycle. GM by 1934 began to sell medium diesels for railway locomotives, soon with tremendous success, but offered smaller ones suitable for its own buses and trucks only in 1938, ten years after Kettering had begun his experiments. Ford and Chrysler did almost nothing with diesel engines in the 1930s. Nor did most of the major independent truck-makers. These firms were dubious about investing in a new engine in the midst of a Great Depression when the market for it was problematical. They would let others take the risk. This caution, compared with automotive firms in Europe, is striking.

Still some American manufacturers, including several firms making agricultural and other off-the-road tractors, did look into diesels. International Harvester, an important producer of both farm equipment and trucks, began experimenting with diesel engines as early as 1916, but did not offer anything in this line until 1933. The Caterpillar company began its diesel development in 1925. Sparked by competition from European diesels when one of its salesmen encountered a Daimler-Benz in the Sudan in 1927, it introduced its first diesel-powered crawler tractor in 1931. This machine soon sold very well for large industrial type farming operations, for use on construction projects, and for logging duties. In these uses it ran many hours per year and its saving on fuel – it consumed less than half that of a gasoline tractor – could outweigh its 20 per cent higher first cost. Sales totalled 2000 by the end of 1933 and 20 000 by November 1936.[2] In 1937 of all diesel horsepower sold in the United States for any purpose, Caterpillar accounted for about 40 per cent.[3] This was the first major success for high-speed diesels in the United States. The company sold them through its network of high-quality dealers who implemented Caterpillar's policy of complete and speedy service after the sale.

International Harvester followed Caterpillar into the diesel market in 1933 with wheeled as well as crawler tractors but sold less than one-quarter as many by 1940. Why IH, the largest independent truck manufacturer, hesitated to enter the diesel truck market is unclear. Other tractor-makers, usually employing diesel engines bought from specialist makers, also entered this market in a small way, but farmers with small- and medium-sized holdings preferred gasoline tractors at this point. They were familiar with this type of engine and they did not use the tractor enough hours per year to overcome the higher diesel price.

Clessie L. Cummins pioneered highway diesels in the United States. By 1930 this self-educated mechanic had been running the Cummins Engine Co. for twelve years in Columbus, Indiana, making small and medium high-speed diesels. He sold some for fishing boats and yachts but had yet to make a profit. Just a few months after he developed his own fuel-injection system the Stock Market Crash of 1929 threatened to bring an end to his market and his business. As a desperate move to gain publicity and possible new clients, early in 1930 Cummins installed a marine diesel in a large automobile and drove it to New York City for a fuel cost of less than one-third that of gasoline. The general notice he won with this stunt brought him more advances of capital from his financial backer. Thereafter he engaged in several other exhibitions of diesel engines in cars, trucks and buses, to demonstrate the reliability and modest fuel consumption of his engines. He designed a new one specifically for automotive use and tried to sell it for trucks. To do so he had to overcome some disadvantages, real and perceived, of the diesel:

(1) The high price (diesel engines cost more than gasoline types of comparable power because of the very precise and expensive fuel-injection system, the more rugged construction, the need for heavier batteries and starter motors, and the short production runs);
(2) Starting difficulties, especially in cold weather;
(3) The diesel's heavier weight;
(4) It was noisy and produced ill-smelling smoke;
(5) Diesel fuel was not widely available;
(6) The engine required specially trained mechanics and some clients feared higher maintenance costs.

The diesel's primary advantage, low fuel cost, carried little weight if a gasoline truck's fuel expense amounted to under 15 per cent of the total operating costs. A study dating from 1931 found that gasoline costs for 149 trucks operated by 39 firms in the United States averaged 18.2 per cent of total operating costs, including depreciation. Another investigation a few years later estimated that gasoline costs for a typical American truck were 13 per cent, but reached 33.7 per cent for a European truck.[4] Cutting a gasoline cost of just 15 per cent by half bulked small compared with the initial high price of the diesel. This engine would appeal at first only to those whose trucks had high fuel costs because they operated many thousands of miles per year, and where the differential between the prices of the fuels was large.

SLOW PROGRESS WITH WEST COAST RAILWAYS IN THE LEAD

California matched these requirements. The railway network in the western states was less dense than in the east, distances among cities were greater, and roads frequently climbed long grades. Also, diesel fuel in this state cost about 5 cents per gallon (about one-third the price of gasoline) because it paid no state tax (until at least 1938), while all forty-eight states levied a tax on gasoline. The federal government charged a gasoline tax from June 1932 at a rate of 1 cent per gallon until 1940. Cummins sold his first truck diesel in May 1932 to a large chain of grocery stores in California. Three years later this firm's trucks used sixteen Cummins engines and found that for comparable mileage its diesel fuel cost was only one-sixth that of its gasoline trucks.[5] Also in 1932 a common-carrier trucking firm in San Francisco opened a route to Salt Lake City, 800 miles, when a surfaced highway was completed. Although rails linked these cities, this company's trucks carried perishable goods under refrigeration in less-than-railcar-load lots, and at a lower price. To cut its expenses the firm tested a Cummins diesel on this route and reduced fuel costs by 80 per cent. Thereupon it ordered four new trucks with Cummins engines.[6]

In 1933 Consolidated Freight Lines of Portland, Oregon, a firm that would grow to giant proportions after 1945, began replacing its gasoline engines with Cummins diesels to serve its routes to towns up to 500 miles from Portland.[7] Somewhat later, in 1938, the Colonial Sand & Stone Co. of New York City converted its large fleet of construction trucks to Cummins diesels, a move that eventually required 171 engines.[8] These trucks did not cover many miles per year, but operated long hours, many at low or idle speeds, where diesel economy was outstanding. These examples show the kinds of truckers to whom the diesel appealed in the 1930s.

Diesel fuel in the United States typically cost 8 cents a gallon at the pump in the mid-1930s and gasoline 17 cents. If a truck operated about 80 000 miles per year, and very few did, it could regain on fuel savings the $1200 premium for a replacement diesel engine in less than one year.[9] Most American trucks at this time, however, were operated as single units for short distances by farmers or were engaged in pickup and delivery services in towns. They fed rather than competed with railways. As such they accumulated under 20 000 miles per year and their owners believed diesels were irrelevant.

Although some of the smaller truck-makers offered new vehicles with Cummins engines in 1932, and eleven stood ready to supply them with new trucks by 1934, most of Cummins' sales in the 1930s went to replace worn-out gasoline engines. Through 1940 the Indiana producer dominated the diesel engine market for heavy, intercity trucks. Its sales rose from 133 truck engines in 1933 to about 650 in 1937. In the latter year this company made its first profit.[10] Clessie Cummins' early entrance into this market, his reliable engines and his high-quality service through local dealers gave his products a distinguished reputation comparable to that of Caterpillar in diesel tractors. At the end of the decade over 80 per cent of diesel trucks used Cummins engines, and about one-third of the estimated total of 7500 diesel trucks were registered in California.[11]

Cummins' opening up of the diesel automotive market encouraged other engine builders – Hercules, Buda and Waukesha – to enter it, and by 1936 some of the major truck manufacturers began offering diesel engines supplied by these independent firms or by Cummins as an option, primarily for export markets to compete with the European diesels they met there. Nevertheless, in 1937 about $4\frac{1}{3}$ times more diesel horsepower was produced for tractors as for trucks and buses.[12]

GENERAL MOTORS SELLS DIESEL BUSES AS WELL AS TRUCKS FROM 1938; BUT DIESEL PENETRATION IS STILL SMALL AND FAR BEHIND THAT OF LEADING EUROPEAN COUNTRIES IN 1941

The gradual penetration of diesel engines, the improving highway system that encouraged longer-distance trucking, and the precedent of European diesel trucks ultimately forced the major American truck manufacturers to take the diesel seriously. In January 1938 General Motors introduced its truck and bus diesel engine with a unique two-stroke design. It found a good market in buses, delivering 4193 diesel-powered ones through 1942. Dieselisation of buses in the United States had not begun until 1936 when the New Jersey Public Service Company ordered 27, using Hercules engines. At this time transit buses in the United States travelled about 36 000 miles per year, on average, compared with 10 600 for the average truck.[13] This bus mileage was enough to make a diesel competitive, for by 1939 most diesel buses cost less than 10 per cent more than gasoline types.[14]

Soon GM's diesel bus led this market. GM helped itself by obtaining indirect ownership of transit systems in a number of cities and selling them gasoline or diesel buses*. The Greyhound system, the leader in intercity bus transportation and also influenced or controlled by GM, was the largest client for GM diesel buses before the war. But GM faced competition in the diesel bus business, for the Mack Company especially furnished many in this period. Even so, of 14 535 transit and intercity buses produced in 1940 and 1941, just 2362 or 16 per cent used diesel engines. When the United States entered the war late in 1941 about 2700 of 57 580 common carrier

---

*The charge has been made that GM forced uncompetitive motor buses on the U.S. transit industry, eventually to weaken or destroy it and consequently expand the sale of private automobiles. See Bradford Snell, 'American Ground Transport', U.S. Congress, Senate, Comm. on the Judiciary, Industrial Reorganization Act, Hearings before the Subcommittee on Antitrust and Monopoly, 93rd Congress, 2nd Session (Washington, D.C., 1974); and Jonathan Kwitny, 'The Great Transportation Conspiracy', *Harpers*, 262 (Feb 1981) pp. 14–15, 18, 20–21, which is useless because it is undocumented. The argument is a *post hoc ergo propter hoc* one, without direct evidence of GM's motives but imputing to GM's management in the 1930s perfect vision of the 1960s. It ignores GM's profits from buses and especially the large market for its high-speed diesels in trucks and other equipment. It plays down the transit systems' need for capital to modernise, capital available for motor buses but not for alternate systems. It assumes that contemporaries were as convinced of the merits of electric transit as are those making the charges. The argument remains very far from proved. Recent investigations of it include David St Clair, 'Entrepreneurship and the American Automobile Industry', (unpub. diss. University of Utah, 1979); 'The Motorization and Decline of Urban Public Transit', *Journal of Economic History*, 41 (1981) pp. 579–600; and an unpublished MS. (1984) by the same author.

buses employed diesel power.[15] The prophets of diesel power still had some convincing to do.

In addition to General Motors, other companies that introduced diesel trucks in 1938 included Mack, with its own engine, International Harvester, using Cummins engines, and the Dodge division of Chrysler. The estimated 7500 diesel trucks in service by 1940 were only a tiny fraction of the 4.9 million trucks registered, but most of these were unsuitable for diesel power. More appropriate comparisons would be that the 7500 diesels were 5.4 per cent of trucks with five or more tons capacity registered, or 20 per cent of the 36 600 trucks of five tons or more capacity built from 1936 through 1940.[16] Of course the U.S. figures show a lag behind the estimated 63 000 civilian diesel trucks operating in Germany in 1940, the 1938 figure of 9000 in Britain and approximately 35 000 in France.[17]

The delay in U.S. dieselisation can be summed up as due to the low gasoline price, to the reluctance of major automobile producers to enter this field, and to improvements in high-powered gasoline engines. At the end of the 1930s gasoline and diesel fuel prices looked like this in various countries, measured in U.S. cents per U.S. gallon, including taxes:

	Gasoline	Diesel fuel
United States	17.5	9
Great Britain	30	26
Germany	50	25
France	30	24
Italy	185	121

Sources: U.S.A., *Survey of Current Business, Supplement 1951*, p. 169, G. B; W. Plowden, *The Motor Car and Politics* (London: Bodley Head, 1971) p. 264 n.; France, *Annuaire Statistique 1939*, p. 157, and author's estimate; Germany, *Statistisches Jahrb. 1939/40*, p. 333; Italy, A.N.F.I.A., *Automobile en Cifre* (Torino, 1982), pp. 134–5.

LARGE EXPANSION OF PRODUCTION DURING THE WAR, BUT NOT FOR TRANSPORT PURPOSES

During the Second World War American production of diesel engines rose to a level ten times higher than in 1940. Most of the high-speed diesels used for military purposes did not go to tanks or trucks, but to naval vessels, especially landing craft and other small types. One-third of the diesel horsepower produced in the highest wartime year, 1944, consisted of GM two-stroke engines,[18] some of which went to tanks supplied to Britain and Russia, as well as to naval uses and for crawler-type construction tractors. Thousands of Caterpillar diesel tractors maintained their high reputation in military service. But early in 1942 the U.S. Army decided to use gasoline engines for almost all of its road vehicles so as to simplify logistics. Diesel trucks, then, did not appear with the U.S. land forces, except for a few large tank transporters. Probably the main impact the Second World War had on the spread of diesel engines was first, to acquaint many young sailors with the operation and maintenance of this type of engine, and second, to expand American industry's capacity to manufacture high-speed types.

POST-WAR YEARS. MORE DIESELS FOR COMMERCIAL, BUT NOT FOR THE NUMEROUS SCHOOL, BUSES. LARGE DIESELS CAPTURE A MUCH LARGER SHARE OF INTERCITY FREIGHT

After the war diesel engines continued to penetrate the commercial bus market, especially the two-stroke GM type. In round figures a diesel bus in this period cost about $1000 more than a gasoline model. As it saved about $500 per year in fuel costs, the premium for the engine could be paid off in two years. The engine's smoke and noise were no longer considered a problem.[19] By the early 1950s about half the new commercial buses sold had diesel engines. This proportion rose gradually and by the end of the decade almost all new transit and intercity buses used diesel power. The total commercial bus fleet had become 54 per cent diesel by 1956.[20]

The number of school buses exceeded commercial buses in operation after the war. They expanded in numbers as rural school districts consolidated and urban districts began to provide bus service to pupils. Yet these buses remained wedded to gasoline power because their annual average mileage of under 10 000 precluded diesel engines through the 1950s. Because the fuel they burned was exempt from state and federal taxes, its cost was a smaller share of operating costs than for commercial buses.

As the trucking industry grew after the Second World War, diesel truck manufacturers saw their sales increase. Cummins, Mack, and

GM's Detroit Diesel division made the most of the automotive diesel engines in these years. Mack, White and International Harvester led in sales of diesel trucks, with Cummins supplying most of the engines for the second and third of these. The market for heavy trucks expanded much faster than the total. The proportion of intercity freight carried by trucks had been 10 per cent in 1941. In 1948 it surpassed this with a figure of 10.6 per cent, but then it rapidly doubled to 21.4 per cent in 1959, at which rate it remained nearly constant thereafter. In actual ton-miles this freight rose from 81 billion in 1941 to 115 billion in 1948, and to 279 billion in 1959. In a dozen years then, from 1948 to 1959, intercity truck freight rose 143 per cent. And these trucks rolled faster, from an estimated average of 42.8 m.p.h. in 1948 to 48 in 1959 and 51.8 in 1965.[21]

All these figures point toward more and larger trucks in intercity commerce, carrying heavier loads at higher speeds. In turn this meant higher fuel consumption per mile, and diesel engines provided one important way to cut these costs. Although the price of diesel fuel including taxes rose after the war to about three-fourths that of gasoline by the early 1950s, the inherent efficiency of the diesel brought more clients to this engine. A trade magazine in 1956 surveyed truck and bus line executives who used diesel power. It found that this engine's economy of operation, reliability and long life, in that order, persuaded them to choose it.[22] Expanding production runs of diesel engines did reduce the price differential with gasoline engines; so by the late 1950s some experts argued that the diesel was economic for intercity trucks (line-haulers) that travelled over 35 000 miles per year.[23]

The earliest figures available showing the sales of diesel trucks by detailed weight groups, 1951, indicate that diesel powered 23.2 per cent of the heaviest trucks sold in the United States in that year (class 7, with a gross vehicle weight – GVW – of 26 001 lb. or more). New class 6 trucks (GVW 19 501–26 000) lb.) were 5.1 per cent diesel in 1951, but thereafter fell below this share for class 6 for over a decade.

The first major jump in post-war diesel truck sales came in the mid-1950s. They rose from 10 500 in 1954 to 25 800 in 1956, or from 1 to 2.3 per cent of total truck sales. Almost all of this increase came in the heaviest trucks. It was stimulated by an economic boom in this period, but also by the opening of a series of intercity express toll highways, financed by state governments. By 1956 the most important of them linked New York City with Philadelphia, Pittsburgh, Cleveland and Chicago, 840 miles.[24] In that year came the passage of the Interstate and Defence Highway Law. This provided for federal government finance of 90 per cent of the cost of an expressway system extending some 40 000 miles to all of the major cities of the country. Already the toll expressways allowed huge trucks to travel swiftly among major cities. The Interstate system promised much more, and no new tolls. It would permit trucks to move freight faster than railways over long distances among all the major markets. Firms in line-haul trucking saw this as a great opportunity and began preparing for it by expanding their fleets with diesel machines. By 1960 10 440 miles of the Interstate network already were in use, including 2264 miles of the earlier toll roads, and at the end of 1965 over 21 000 miles, half the system, was finished.

These highways, along with a business boom in the mid-1960s, encouraged another upward bound in diesel truck sales, which rose from 28 400 in 1961 to 96 600 in 1966. Three-fourths of the increase went to the heaviest trucks, a new class 8 (over 33 000 lb. GVW), but diesels also tried to break through into the medium range, classes 5 and 6 (16 001–26 000 lb. GVW), where truck-makers sold 8921 diesels in 1965, their best showing of the decade, but this amounted to only 4 per cent in these classes. In the U.S. market most of the class 5 and 6 trucks engaged in some kind of low mileage P & D service (pickup and delivery, or 'peddle') in and around cities. Their owners were encouraged to switch to diesel by the success of these engines in heavy trucks and in buses; by the advantage of standardising on one type engine throughout a fleet, for many had diesel engines in their line-haul vehicles; and by arguments that the diesel fuel economy at low loads or idle would make it economic for them, and by the European precedent.

FOREIGN COMPETITORS INVADE PREMATURELY A GROWING MARKET

The Perkins Company of England played the European card. It had won great success with its small diesel engines over much of the world, and in 1959 decided to sell its diesels to American P & D trucks. It supplied engines in sizes up to 120 h.p. to some leading truck-makers as well as for replacement in older vehicles. By 1964 Perkins diesels took 9 per cent of the total U.S. automotive diesel market. American firms also entered this mid-size range: Ford with engines from Dagenham, England, Cummins with some made in its Scottish factory, Caterpillar, after more than 30 years' experience with its tractor diesels, now trying for trucks, and finally GM added a group of four-stroke medium diesels to its two-stroke line. (Charles Kettering, the stubborn apostle of the two-stroke design, had died in 1958.) Among the clientele were several large breweries, transit mix cement trucks and communities in the Boston area and in Chicago which used them in their refuse collection trucks.

Late in the decade disillusion set in when many truckers had trouble with these diesel engines in mid-weight trucks. The word 'disaster' was used. Most of the problems appeared to be a matter of misapplication and misuse, along with inadequate service facilities for some types, but it gave mid-size diesel engines a poor reputation in some quarters.[25] Consequently diesel truck sales in class 5 disappeared by 1970, and dropped by half in class 6. Both Perkins and Ford lost almost all their sales of automotive diesel engines in the United States; but Caterpillar rose to a prominent place in diesels for large as well as medium trucks, regularly from 1969 winning the fourth largest share of the market.

In 1972 domestic diesel truck sales reached 129 628, by far the highest total on record, but almost all (88 per cent) were for the heaviest trucks, class 8 (see Table 5.1). Diesels clearly had conquered the market for line-haul trucks, but years of marketing strategy and salesmanship had failed to dent seriously the medium P & D segment. Diesel buses were 68 984 or 78 per cent of a total 88 722 transit and intercity buses in service.[26]

THE SCHOOL BUS MARKET IS OPENED AT LAST

During these years when the commercial bus market remained flat, the school bus market grew ever larger to over 250 000 in operation by 1972. Diesel engines finally began to penetrate it. Again, in California first. In 1958 a school district in suburban Los Angeles, Palos Verdes, began to buy diesel buses, persuaded by an equipment superintendent with diesel experience and by a school bus assembler, Crown Coach, eager to supply diesel power.[27] In the neighbouring Los Angeles school district buses averaged almost 15 000 miles per year, well above the national average. This district began to shift its large fleet to diesels in 1963. Its managers believed they could economise on their costs with diesels, and the recent introduction of some new diesel automotive engines in the 130–220-h.p. range encouraged them to try. The nature of school bus service – operating at less than full load much of the time because of heavy traffic on their

Table 5.1 Factory sales of diesel trucks in the U.S. market and in proportion to total domestic sales in each GVW size

	Class 1 6000 & less	%	Class 2 6001– 10 000	%	Classes 3–5 10 001– 19 500	%	Class 6 19 501– 26 000	%	Class 7 26 001– 33 000	%	Class 8 33 001 & over	%	Total	%
1984	15 234	1	122 995	12	–		13 789	26	43 412	55	137 692	99	333 122	11.6
1983	25 565	2	89 161	12	–		9 703	23	29 468	55	78 008	98	231 905	10.2
1982	44 586	5	72 868	10	–		8 559	21	26 282	49	67 453	98	219 748	12.4
1981	47 789	8	20 393	3	–		19 731	29	27 712	60	93 769	97	209 394	13.8
1980	68 868	13	4 397	1	–		12 337	16	30 958	60	98 002	96	214 562	14.7
1979	31 894	3	653	0	–		14 404	11	23 121	54	157 762	95	227 834	8.3
1978	34 751	3	840	0	–		12 229	8	22 597	60	149 498	94	219 915	6.4
1977	2 386	0	975	0	–		11 142	7	15 695	58	135 361	95	165 559	5.2
1976	–		1 498	0	–		5 045	4	9 053	45	91 620	92	107 216	3.9
1975	–		1	0	159	1	3 517	3	8 992	42	59 936	85	72 605	3.6
1974	–		–		41	0	2 704	3	10 333	36	131 624	87	144 702	5.9
1973	–		–		302	1	3 197	2	15 164	40	133 496	89	152 159	5.5
1972	–		–		220	0	3 196	2	11 922	32	114 290	90	129 628	5.6

Source: Calculated from Motor Vehicle Manufacturers' Association of the United States, *Motor Vehicle Facts & Figures '84* (Detroit, 1984) pp. 10–11.

routes and waiting for passengers to board – suggested that fuel savings would be large. They were. In 1964–5 the district's gasoline buses cost $0.149 per mile for fuel, oil, maintenance labour and parts. For its diesel buses the figure was $0.077, or half,[28] a saving of $1000 per year for a diesel bus. Thereafter, other west-coast school bus operators turned to diesel power and for a time 85 per cent of new school buses delivered in that region mounted diesel engines.[29] Nevertheless, only in the 1970s did the trend toward diesel buses spread eastward.

THE EFFECTS OF THE OIL SHOCKS AND MORE COMPETITION, FOREIGN AND DOMESTIC

Late in 1973 the characteristically low petroleum prices in the United States suddenly ended with the OPEC-sponsored price increases. In 1972 gasoline and diesel fuel prices were still at remarkably low levels, gasoline selling at a national average wholesale price with taxes of 24 cents per gallon, only 150 per cent above the 1940 price. The diesel price was just 2 cents less. (Large trucking firms did not pay much more than the wholesale price for their fuel.) The price revolution in petroleum brought an increase of just over 70 per cent for their fuel.

The sharply higher gasoline prices after 1973 did not immediately bring larger sales of diesel vehicles in the United States. In fact, after a peak in 1973 and 1974, resulting from orders placed at the top of the business cycle in 1973, sales of U.S.-made diesel trucks to the domestic market dropped disastrously, by half, in 1975. The recession of 1974–5 brought this fall, and it affected diesel truck sales more than trucks generally, for most were expensive capital goods whose purchase could be delayed. Then, on the way up from the trough of the recession, diesel truck sales did attain new records annually from 1977 through 1979. Not only did they rise from 87–90 per cent to 95 per cent of new class 8 trucks, but to higher levels in smaller sizes – to 60 per cent in class 7 in 1978, and to 11 per cent in class 6 in 1979. Sales of imported diesels in these two classes raised the proportions slightly higher. By the late 1970s it seemed clear that oil prices would remain high for the foreseeable future; the diesel offered a way to limit the higher operating costs.

U.S. manufacturers began to meet foreign competition in the class 6 and 7 markets in the 1970s. In 1969 Mercedes-Benz had cautiously ventured into the eastern United States with class 6 P & D models. It

sold a few hundred of them annually as it established a solid marketing organisation. When the U.S. devaluations of 1971–3 made German goods more expensive, Mercedes shifted the source to its
Brazilian operation, which sold a record 901 trucks in the eastern
United States in 1974, one-fourth of the market for class 6 diesels that
year. In 1979 Mercedes delivered its 10 000th truck and considered
itself well enough established to erect an assembly operation in
Hampton, Virginia, in 1980. In the mid-1970s the Swedish Volvo
company came into the market, selling a few hundred a year of
29 000 and 35 000 GVW trucks. To expand its marketing, it arranged
to have the Freightliner company sell Volvo trucks. Then in 1978
came Iveco, the truck-building subsidiary of Fiat, with German,
French and Italian factories. A fourth major European diesel truck-
maker, Renault, turned up in 1979, hoping to sell its Berliet and
Saviem vehicles in the U.S. market by an arrangement with Mack.
Renault bought 20 per cent of the equity of Mack, which agreed to
sell the French-made class 6 and 7 trucks. All these foreign producers, with the notable exception of the British, apparently believed
that the U.S. market had left a niche open for them.

   The opportunity seemed even greater when the second oil shock of
1979 brought even more pressure for fuel economy.

Table 5.2

	Wholesale prices in cents per gallon		Taxes levied on both fuels
	Gasoline	Diesel fuel	
1972	12.70	10.61	11.3
1975	30.27	26.09	11.65
1982	92.90	91.95	13.1

Sources: Gilbert Jenkins, *Oil Economists' Handbook 1984* (London: Ap-
plied Science Publications, 1984) p. 30; U.S. Department of Transportation,
Highway Statistics Summary to 1975 (Washington, D.C., 1977) pp. 39–41;
Statistical Abstract of the United States 1984, p. 612.

Especially noteworthy in Table 5.2 was the failure to raise taxes on
motor fuel more than a trifle, although in 1983 the federal tax on both
types rose by 5 cents. Then in August 1984 the federal tax on diesel
fuel was raised another 6 cents. This brought the total tax on diesel to
a little over 24 cents and on gasoline to just over 18 cents. But again,

a recession, extending from 1979 through 1982, sharply reduced sales of all trucks except class 7. In 1980 domestic sales fell to just 42 per cent of the 1978 total. Diesel trucks followed a similar but lagging curve. Sales of classes 6, 7 and 8 dropped to 52.4 per cent of the 1979 peak. With economic revival in 1983 and 1984 diesel models gained market share, but both diesel and gasoline trucks in class 7 took sales away from class 8, which remained far below its earlier levels until 1984.

Two special difficulties weighed on the heavy line-haul trucks until the latter year. The deregulation of interstate trucking in these years brought fierce competition. Some fifty truck fleets went bankrupt in 1979–82, which dumped large numbers of class 8 vehicles on the market. Second, there was a continuing controversy among the state and federal governments about maximum length and weight of trucks, which forced postponement of purchase decisions.

The most novel effect of the second oil shock was the appearance of diesel engines in light trucks and vans up to 10 000 GVW. Most trucks of this size were operated by individuals as personal vehicles, so these machines really should be considered part of the passenger car sector rather than in the truck market. When U.S. manufacturers began to supply small diesel engines for cars, 1977, they also fitted them to light trucks, where domestic diesel sales rose to 117 454 by 1982. The Japanese soon began to exploit this market also.

After decades of hesitation, diesel engines also penetrated the United States' Army. This type had been under consideration since the late 1930s, but always the decision had gone to gasoline. Then, in 1956, a major tank program prescribed a diesel engine made by Continental Motors. The diesel's fuel economy won over the doubters, for it gave the tank twice the range. The Army employed a multi-fuel engine in heavy trucks in the 1960s, and finally, in the 1970s and 1980s, went to diesel entirely for personnel carriers and most trucks.

TRENDS IN THE 1980s

By the early 1980s commercial buses were largely dieselised, 88.4 per cent in 1981,[30] and diesel school buses in 1985 amounted to an estimated 16 per cent of the total.[31] There are, however, widely varying reckonings of the school bus fleet.

As for trucks, the estimated 1 406 282 diesels registered in 1981 came to 4 per cent of the total of 34 451 110, but a breakdown by weight classes is not available. Surely there were few gasoline models in class 8 still working in the mid-1980s, for the fuel cost for line-haul trucks now reached about 35 per cent of operating expenses. One can estimate that about half of class 7 is diesel, but for class 6 only some 15 per cent employ diesel engines. The European producers continue to contest in classes 6 and 7, with Volvo even competing against the Americans' greatest strength, class 8. Mercedes-Benz in 1981 acquired Freightliner to improve its marketing. Renault bought majority control of Mack and its diesel truck sales through Mack rose to surpass Mercedes and took about 7 per cent of the entire class 6 market in 1984. The Japanese Hino and Isuzu firms also are mounting assaults.

It appears that there is plenty of room for a rising share of diesel trucks in class 7 and especially class 6 trucks, and in school buses, but the slowly falling price of petroleum from its high of 1981 and the failure of Americans to rush to buy diesels for these uses suggests that we will see a gradual rather than a strong trend to diesel in these segments. In 1981 gasoline consumption in the United States was only 6 per cent higher than in 1972, but diesel motor fuel had risen 109 per cent over this decade to 17.5 per cent of the gasoline level.[32] Clearly a trend was present.

Was there also a trend in intercity freight traffic? From 1965 to 1982 the total ton-miles of traffic rose 37.5 per cent as railways lost 7 per cent to settle at 36 per cent. The offsetting gain did not come from trucks but from oil pipelines which rose 6 per cent to nearly 25 per cent, surpassing trucks at 22.29 per cent and waterways at 16.65, each of which gained only a trifle more of the market.[33] All these forms of freight traffic used petroleum as a fuel. Presumably, the line-haulers' switch to diesel saved them from losing market share.

In conclusion the diffusion of the diesel engine in the U.S. automotive market has been a slow process, conditioned by two major factors since 1945, the price of gasoline and the growth of line-haul trucking. The alleged U.S. lust for change and novelty is not exhibited here. One thing that surely would have speeded the transition of diesel engines in the 1970s and 1980s would have been to tax motor fuel at a rate comparable to that in European states.

Notes

1. *Oil Engine* (London), 6 (Jan 1939), 270–1; *Statistisches Jarhbuch* (1938) p. 232.

2. Caterpillar Tractor Co., *Fifty Years on Tracks* (Peoria, 1954) p. 40; *Business Week* (12 Dec 1936) p. 16.
3. 'The Cat', *Fortune* (May 1938) p. 99.
4. U.S. Department of Commerce, Domestic Commerce N.66, *Motor Truck Freight Transportation* (Washington, 1932) p. 52; Orville Adams, *Elements of Diesel Engineering* (New York: Henley, 1936) p. 348.
5. *Diesel Power*, 13 (Oct 1935) p. 168.
6. *Ibid.* 11 (Jan 1933) 14, pp. 17–18.
7. *Ibid.* 12 (May 1934) p. 244, and 14 (Sep 1936) p. 595.
8. *Diesel Power & Transportation*, 17 (Jan 1939) p. 31.
9. 'Diesels on Wheels', *Fortune*, 10 (Dec 1934) p. 116.
10. C. L. Cummins, *My Days with the Diesel* (Philadelphia: Chilton, 1967) p. 159; *Automotive Industries*, 77 (3 July 1937) p. 20.
11. *Business Week* (7 Sep 1940) p. 36; *Diesel Power & Transportation*, 18 (July 1940) pp. 581–2.
12. *Business Week* (16 Apr 1938) pp. 45–6.
13. Motor Vehicle Manufacturers' Association, *Facts & Figures '82* (Detroit, 1982) p. 54.
14. *Transit Journal*, 83 (Sep 1939) pp. 376–77.
15. *Bus Transportation*, 20 (Jan 1941) pp. 45, 47; 21 (Jan 1942) pp. 15, 18, 21.
16. *Roads & Streets*, 83 (July 1940) p. 61; Automobile Manufacturers' Association, *Auto Facts & Figures 1941* (Detroit, 1941) p. 7.
17. For Germany, author's estimate based on *Statistisches Jahrbuch 1938*, pp. 232, *1939/40*, p. 237, and *1941/42*, p. 261. For Britain, *Oil Engine*, 6 (Jan 1939) p. 271. For France, author's estimate.
18. 'GM Diesel', *Fortune* (July 1948) pp. 78–9.
19. E. N. Hatch, 'Diesels for Buses', *SAE Journal*, 57 (Apr 1949) pp. 25–6.
20. U.S. Bureau of Public Roads, *Highway Statistics 1956* (Washington, D.C., 1958) p. 16.
21. U.S. Department of Commerce, *Historical Statistics of the United States* (Washington, D.C., 1975), vol. 2, pp. 707, 718.
22. *Diesel Power*, 24 (Aug 1956) pp. 34–5.
23. *Business Week*, 12 Dec 1959, p. 142.
24. John B. Rae, *The Road and the Car in American Life* (Cambridge, Mass: M.I.T. Press, 1971) pp. 175–7.
25. *Diesel Equipment Supervisor*, 58 (Feb 1980) p. 19.
26. U.S. Department of Transportation, *Highway Statistics 1972* (Washington, D.C., 1974) p. 36.
27. *Diesel Equipment Supervisor*, 41 (Apr 1963) pp. 30–2.
28. R. Davis, 'School Fleet Dieselization Program', *SAE paper 650711* (1965) pp. 1–5.
29. *Diesel Equipment Supervisor*, 43 (Feb 1965) pp. 41–5.
30. Motor Vehicle Manufacturers' Association, *Motor Vehicle Facts & Figures '83*, pp. 24–5.
31. *School Bus Fleet*, 28 (Feb 1983) pp. 12–13, 35–6.
32. OECD, *Energy Statistics 1971–1981* (Paris, 1983) pp. 137, 155.
33. *Statistical Abstract of the United States 1984*, p. 607.

6 The Automobile and the City in the American South

David R. Goldfield and Blaine A. Brownell

In the American South the automobile was never a luxury: it was a way of life. It was an inextricable part of the family patrimony and frequently, through various permutations, did indeed pass down through generations. The auto's greatest impact on the South probably occurred in the region's rural reaches where the new technology broke the chronic isolation and loneliness of life, made urban life and its attractions and distractions more readily available to country residents, and provided a means to get to that mill job while still retaining the family farm. The auto encouraged or, rather demanded, decent roads. In fact, the Good Roads Movement that energised Southern states during the 1920s attained its most loyal constituency in the rural districts. Good roads, in turn, encouraged the establishment of better services and industry. By bringing farms closer to market towns and cities, improved roads and the motor vehicles upon them transformed Southern agriculture in certain areas from the historically soil-leeching staple crop cultivation to dairying and truck farming.

Above all, the automobile facilitated the choice of an alternative to the sometimes-grinding poverty of rural life. Commuting, an activity many associate with the affluent suburban lifestyle, became a rural standard by the 1930s as men and women endured sometimes lengthy journeys to jobs that would supplement meagre family farm incomes. A 1948 South Carolina labour market survey revealed that nearly one in five mill workers commuted more than fifteen miles each way to their jobs. It was little wonder, given the general poverty of the rural South, that the relatively expensive, but necessary, automobile received great personal love and attention, akin to the proverbial cowboy and his horse, for in both relationships repair shops and spare parts were in short supply on the open range, and self-sufficiency required mechanical knowledge and imagination.[1]

This intimate connection between man and car eventually became an integral part of Southern folk culture when a group of country boys in North Carolina began to modify the engines of their precious

115

vehicles and race them on dirt tracks and at county fairs during the late 1940s. Thus, stock-car racing was born as were such regional folk heroes as Junior Johnson and Richard Petty. Today, stock-car racing is a national sport and while many still express puzzlement over the entertainment of watching cars going around in circles to the accompaniment of ear-splitting drones and gasoline smells, they do not understand that this is all part of a folk ritual centred on the sacred automobile.

Of course, Americans generally, not merely in the rural South, have evolved special relationships with the automobile that transcend their mere utility. The *Middletown* studies of Robert S. and Helen M. Lynd suggest the importance of the automobile to Muncie, Indiana, even during the Great Depression of the 1930s when motor-vehicle registrations actually increased proportionate to the town's population. Muncie's working class, the Lynds reported, 'want what Middletown wants, so long as it gives them their great symbol of advancement – an automobile. Car ownership stands to them for a large share of the "American dream"; they cling to it as they cling to self-respect. . . .' Accordingly, scholars have attached almost mystical qualities to the automobile. Urban historian, Lewis Mumford has speculated that the automobile 'appeared as a compensatory device for enlarging an ego which had been shrunken by our very success in mechanization'. A symbol of freedom, a means of escaping the growing limitations of an industrial society, the motor car in Mumford's view became virtually a religion, 'and the sacrifices that people are prepared to make for this religion', he wrote, 'stand outside the realm of rational criticism'.[2]

While the rest of the country shared the South's attachment to the automobile, Southerners, as has been their wont in numerous aspects of American life, exaggerated this love affair and not only planned their work and play around the auto, but their farms, as noted earlier, and more pertinent for this chapter, their cities as well. Although the auto's impact may have been most significant in the countryside of the American South, there is little gainsaying its urban dimension in the region. For one thing, city and country were inextricably entwined in the South of the early auto age, from the roads bringing produce and economic sustenance to the customs carried by rural migrants who comprised the majority of Southern urban populations. For another, Southern cities were typically small: the era of the great cities that accompanied industrialisation in the late nineteenth century generally bypassed the South as the region's towns and cities persisted on a commercial economic base. So the Southern city as it grew in the twentieth century would be more malleable and sensitive to technological changes than Northern cities, many of which had assumed their major spatial proportions by the early twentieth century before the halcyon era of automobile transport.[3] Chicago urbanist, Homer L. Hoyt, in his classic study on urban land use a half-century ago, observed that the 'location, shape, and size of our cities has . . . been a function of the form of transportation prevailing during a city's main period of growth'.[4] Thus, the Southern city, achieving maturity by the middle and latter quarters of the twentieth century, has grown up with the automobile and its impact was bound to be far-reaching.

ATLANTA, THE EXEMPLAR

The economic impact of the automobile on the Southern city was apparent as far back as the early decades of this century. Not surprisingly, Atlanta took the early lead in securing for itself the economic benefits of the new technology. The city was the spiritual headquarters of a regional booster ethic called the 'New South Creed' initially framed and promulgated following the Civil War by a local newspaperman, Henry W. Grady. The 'Creed' doted upon economic development as the South's deliverance from the chronic poverty and ignorance that had, in large part, led to the disastrous defeat during the Civil War. Atlantans, in fact, became quintessential boosters, impressively rebuilding their ravaged city quickly and establishing it as a major rail transport nexus. It was Atlanta's strategic location as the vital transport link between the coastal South and the trans-Appalachian South that first drew the attention of Northern automobile manufacturers. That and the vigorous and imaginative marketing of Henry Grady's heirs were instrumental in Atlanta's becoming the Southern distribution centre for the automakers.

The combination of location and lobbying lured the National Association of Automobile Manufacturers to unveil their 1910 model cars in Atlanta in November 1909. Atlantans stressed the compatibility of new technology and the old South by distributing lapel buttons depicting a bale of cotton framed in an automobile wheel. The visiting manufacturers evidently got the picture and decided upon Atlanta as the Southern distribution centre for their product. Automobile dealerships appeared almost immediately along the city's fashionable Peachtree Street in order to take advantage of proximity to the city's wealthier residents. Within a decade the nascent automobile industry had precipitated a startling multiplier effect on the city's economy, soon to be duplicated on a smaller scale in other cities throughout the South. Automobile dealerships required expansive physical plants that spurred the local building construction industry, as well as a host of subsidiary economic activities to support the automobile, such as tyre and battery dealerships and service stations, all of which added to the employment base of the city.[5]

The automobile not only helped to touch off a building boom in the city, but outside it as well. Prior to 1910 suburban development in the South, as elsewhere, was closely tied to the extension of electric streetcar lines. In fact, suburban developers and traction company entrepreneurs were frequently one and the same. Among Atlanta financier Joel Hurt's numerous economic activities included a trolley franchise and the middle-class streetcar suburb, Ansley Park. Charlotte's Edward Dilworth Latta opened his trolley line extension and his new suburb, Dilworth, on the same day. Suburban home and land advertisements of the period always touted proximity to streetcar lines if they hoped to secure purchasers.[6]

By the 1920s automobile usage became more common in Southern cities. In 1920 Atlanta, New Orleans and Memphis each contained approximately 20 000 motor vehicles, respectively; by the end of the decade motor-vehicle registrations had risen to 64 000, 58 500 and 40 300 in each of the three cities, or an average of about one vehicle for every 1.5 families. Developers were no longer constrained by trolley tracks and a divergence of transport entrepreneurship and land development began to occur.[7] The rapid expansion of possible development sites touched off a suburban land boom that fed the prosperous Southern urban economies of the 1920s.

Little wonder that the urban press, the traditional oracle of the booster philosophy, extolled the virtues of automobiles in terms that their antebellum predecessors had reserved for railroads. 'The operation of automobiles', the Nashville *Tennessean* intoned, 'is one of the great blessings that has come to us who live in this age, and its influence on our national life is good and so powerful and far-reaching as to challenge our imagination as to what its final results will be.' Four years later, in 1924, the *Tennessean* remained impressed: 'The far-reaching effects of modern inventions are nowhere better demonstrated than in the automobile and the way in which it has directed our development into new channels and molded our lives in a way unimagined twenty years ago.'⁸ The automobile comported well with the Southern concept of progress, the utilisation of new-fangled technologies, without upsetting the traditional free enterprise, biracial context of regional life.

THE COSTS AND DISADVANTAGES, REAL OR IMAGINED

But the motor car was not an unmitigated success in the urban South. The automobile and its use altered behavioural patterns, expanded the urban economy and drastically changed land use. Despite the boosters' inherent belief that their society could easily absorb the changes wrought by automobility, others were not so sure, others who saw more sinister implications in the widespread popularity of the new machine. For one thing automobiles added greatly to the already-considerable danger of chaotic Southern urban streets. The director of the local safety council in Birmingham dramatically declared in 1925 that automobile accidents were 'probably the greatest menace in the country and certainly the greatest in Birmingham . . . '. The Executive Committee of the New Orleans Safety Council recommended to the municipal government that the death toll from motor-vehicle accidents constituted 'so appalling a loss of life as to justify an immediate study and the application of stern legal measures to improve so grave a condition'. In most instances city officials were only too willing to agree that the problem was serious and the need for action obvious. New Orleans Commissioner of Public Safety P. B. Habans reported in 1926 that 'The death rate in the United States, resulting from traffic, is appalling and our city has its share in the making of that rate.' The Birmingham Chamber of Commerce even found it necessary in 1926 to issue to the city's motorists an appeal to avoid running down any of the old soldiers in town for the annual reunion of Confederate veterans.⁹

Some critics of the automobile noted its visual and environmental pollution, an alarming development in a region where sense of place and commune with nature were essential elements of the folk culture. Nashville poet and essayist, Donald Davidson, noted with sadness in his epic against modern technology, *The Leviathan* (1938), that 'Cities that preserved the finest flavor of the old regime had to be approached over brand-new roads where billboards, tourist camps,

filling stations, and factories broke out in a modernistic rash among
the water oaks and Spanish moss.'[10] An anonymous poet in Atlanta
penned a troubling sonnet of a gasoline interloper in the once-
pristine Southern landscape:

> Give me a house by the side of the road
> Far from the town's turmoil;
> Where one inhales the reeking scent
> Of burning gas and oil;
> Where the lilting songs of mating birds
> Are hushed in every bower,
> By the rush and roar of motor cars
> Of ninety-odd horsepower.
>
> Give me a house by the side of the road
> Where midnight's quiet skies
> Are gashed with glarish motor lights,
> And pierced with frenzied cries;—
> Where I can rest at peace at night
> And never leave my bed,
> Except to rise at intervals
> And gather up the dead.[11]

It was inevitable, given Prohibition and the rise of organised crime
in the 1920s, that concerned citizens focused on the new technology
as a major accomplice, if not co-conspirator. The image of machine
guns blazing from sleek, black sedans is synonymous with many with
urban crime in the 1920s and 30s. The motor car made the 'getaway'
and the chase Hollywood staples by the 1920s and to some extent this
was only slightly larger than life itself. The Nashville *Christian
Advocate*, for example, directly linked the motor car to the rising
crime rate: 'With the use of an automobile for escape the highway-
man may flash his pistol into the face of his victim even in the
crowded streets at noonday. . . .' Another Nashville editor complained that 'There is something wrong with our Government and
social system when bloody thieves can rove our streets and highways
in swift automobiles, while the law-abiding citizen is compelled to go
unarmed and to be a defenseless spectator of their murderous
projects.'[12]

Given the strait-laced religiosity of the South, it was not surprising
that those who blamed the automobile for a new type of crime wave

frequently accused it of precipitating a moral decline as well. The Birmingham *Labor Advocate* attributed the decline of youth to 'night automobile rides, and the consequent and inevitable bottle of white lightning', while a black writer in Atlanta complained of 'pleasure seeking' among the city's black youth: 'They go automobile riding at nights and all day Sunday, throwing away their hard earnings for a few hours' pleasure.' The automobile, some felt, contributed to the decline of family life or exacerbated the consequences of lax parental control. An Atlanta judge concluded 'that a large percentage of cases are the direct result of too much automobile and too little parental control.' He also lauded an Atlanta grand jury's investigation in 1921 of the link between automobiles and moral degeneration in the city. A Nashville minister expounded on the same theme: 'The picture shows and the automobile call the people, young and old, from their houses and every year adds to the problem of conserving the family life.'[13]

Older courtship patterns, especially to the extent that they involved chaperonage, were difficult to maintain in an age of motorcar mobility, and the automobile was inevitably drawn into discussions of allegedly increasing sexual promiscuity. The Salvation Army in Nashville estimated that the majority of unwed mothers in the Army's 'maternity homes' could blame their condition on 'the predatory drivers of automobiles . . .'. The religious literature of the time contained more than its share of frantic references to the unseen horrors of 'parking with drawn blinds and lights out'; but such comments were not limited to the clergy. A statute in New Orleans went so far as to prevent the use of motor cars for 'immoral purposes'.[14]

These demurrers from the automobile's positive and popular image reflected concerns ranging from the real to the fanciful. For the most part, however, the critics imputed too much to the automobile and not enough to much broader currents of social and economic changes affecting the South. Writers as diverse as William Faulkner, Thomas Wolfe and Allen Tate questioned and worried over the advance of modern industry and technology into an essentially rural, insular culture. The automobile was undoubtedly a part of this advance, but it was, in the perspective of the Southern Renaissance novelists and poets, a superficial manifestation of a much larger problem confronting traditional Southern society.

CHANGING LAND USE AND THE GROWING DEMAND FOR PARKING SPACE

The more serious and far-reaching impact of the automobile lay in other directions, particularly with respect to the manner and extent it helped to reorder urban land use and economic base. The most immediate change occurred downtown. Automobile dealerships and supporting services tended to congregate along major thoroughfares such as Peachtree Street in Atlanta and Trade Street in Charlotte. Southern cities were not particularly well-endowed with paved streets outside the central business district and, typically, the wealthier citizens lived along or near these prominent thoroughfares. One result was the evacuation of these streets by the leading citizens. As early as 1921 the *Atlanta Constitution* reported the transformation of Peachtree Street due, in part, to the automobile:

> [F]or many years Peachtree . . . was known as the principal residence street of Atlanta. On it were located many handsome houses, but time and commerce have changed this street perhaps more than any thoroughfare in the city. Steadily encroaching upon residences, business houses such as automobile sales rooms and other high-class concerns, have forced the residences outward until at the present time all but a few of the handsome homes . . . have been dismantled and removed.[15]

It is likely that in any prosperous, expanding city such as Atlanta the residential exodus would have occurred, automobile or no; but the motor car both gravitated to these high-income central-city residential areas and facilitated the choice of moving to the suburbs.

A more direct impact of the automobile related to commercial land use in the central business district. Southern downtowns were typically congested prior to the widespread use of the motor car. No fewer than five streetcar lines converged on Rich's Department Store in Atlanta, for example. The automobile exacerbated a bad situation, contributing significantly to the alarming rise in death and injury mentioned earlier. Though downtown merchants initially hailed the automobile as a business booster, by the 1920s they were already having second thoughts as to its utility in swiftly and conveniently bringing customers to and from the downtown area. 'From the moment your car hits the edge of Peachtree', an Atlanta writer observed in 1929, 'it is touch and go, jostle and jump, "horn in" fast and then throw your wife abruptly through the windshield by suddenly braking. . . .' Many shared the opinion of the Birmingham *Age-Herald* 'that the automobile which was formerly such a source of convenience and means of swift transportation is rapidly becoming a serious handicap in downtown business centers'.[16]

Both a cause and a consequence of traffic congestion in the downtown was the lack of adequate parking space. 'About one-fifth of the automobile drivers are lucky enough to find parking space downtown,' a Memphis writer complained in 1929, 'and the other four-fifths burn up gasoline and lose time looking for it. Finding an unoccupied space large enough to park a car in the business district is like sighting an oasis in the desert.' Four years earlier a Nashville publication had half-jokingly recommended that, 'After it gets the good five-cent cigar . . ., what the country needs is more and better parking space.'[17]

Some downtown merchants began to complain loudly about the lack of parking and about the traffic congestion that threatened to render the central city even less accessible than before to potential customers. 'Lack of parking space', a Nashville business publication noted in 1926, 'causes purchasers to patronize suburban rather than downtown stores'.[18] Thus, the economic decline of the central business district which many attribute to the proliferation of suburban shopping malls after the 1950s actually originated at least a generation earlier. During the subsequent three decades cities made little progress in solving the downtown merchants' twin woes of inadequate parking and congestion; indeed, these remain problems to this day.

Some businesses were unwilling or unable to overcome the obstacles exacerbated by growing automobility. Some succumbed outright, while others closed; only to reopen in the suburbs, following the lead of their customers. The pattern is evident in Atlanta where the area of middle- and upper-class settlement expanded northward from downtown and beyond the city boundaries, and retail establishments followed suit. The reflections of Thomas H. Pitts, who closed his well-known cigar store and soft-drink bar at Five Points in the heart of downtown Atlanta and reopened his establishment in a suburban location, are revealing:

I think the real thing that did it was automobiles and more automobiles. Traffic got so congested that the only hope was to keep going. Hundreds used to stop; now thousands pass. Five Points has become a thoroughfare, instead of a center. . . . Now instead of having one community center, Atlanta has many.[19]

THE TOWN PLANNERS' RESPONSE

The fragmentation of the city and the disinvestment that implied was not only alarming to merchants but to city officials as well. Suburban expansion meant an eroding tax base, increased service costs to handle the spiralling traffic problem, and visual and environmental problems in urban neighbourhoods. The most common solution offered was simply to make more room for cars through the construction of additional roads, the paving and widening of existing streets and the provision of more parking space. The tax and aesthetic costs of devoting increasing amounts of urban space to the automobile frequently short-circuited comprehensive transport schemes, at least until the 1950s when the Interstate Highway Act enhanced the feasibility of such large-scale road projects within urban boundaries.

City planners in particular advanced transport plans as panaceas for the economic vitality of downtown, taking their cue from planner Daniel H. Burnham's elaborate circulatory scheme proposed for Chicago as early as 1909. Noted city planner, Harland Bartholomew visited several Southern cities during the 1920s and 30s, recommending a variety of transport modifications as centrepieces for his comprehensive city plans. By the 1940s, however, he had recanted, realising that improved road systems encouraged further fragmentation of the metropolis and attracted more traffic thereby increasing rather than alleviating the congestion. He recommended instead that Southern cities attempt to revitalise and strengthen inner-city neighbourhoods as the best plan for preserving both residential and retail land uses in the core area. His suggestions generally went unheeded until the 1970s when urban neighbourhood redevelopment became part of a national urban strategy.[20]

Other suggestions at solving the dilemma posed by the automobile and its uses failed either in implementation or in conception. One obvious proposal was to enhance the attractiveness of mass transit; first the streetcar, then the bus. Generally, however, cities or the private franchises that operated public transit service underestimated the great attraction of the automobile as a commuting competitor. Further, city officials tended to shield these companies from effective competition by limiting or prohibiting outright competing services. The short, controversial history of jitneys during the 1910s and 20s in

Southern cities provides a glimpse of how officials protected obsolescent and relatively expensive public transit.

JITNEYS

Jitneys were typically large touring cars that had been converted to provide space for five to seven paying passengers. In some cases even larger bodies were fitted on to truck chassis to produce a crude sort of motor bus capable of carrying as many as a dozen to twenty persons. They combined many of the virtues of the taxicab with a cost comparable to the streetcar fare. Jitney service was also especially concentrated along the principal avenues of traffic, and it siphoned off increasing numbers of commuters from the established streetcar lines. These vehicles, however, tended to be a less safe mode of transport than the streetcar. The reaction in most Southern cities was to throw a legislative *cordon sanitaire* around the streetcar companies by outlawing the automobile hybrid.

The jitney was particularly popular in working-class districts poorly served by public transit. When Birmingham officials moved to ban jitneys from the streets of that city, the *Southern Labor Review* charged: 'You don't hear of any effort on the part of our Civic bodies and city officials to forbid . . . automobiles from using the streets. Oh, No. . . . [T]hey only want to put the poor man's automobile, The Jitney, off the street and thus deprive the poor of the little pleasures of an automobile ride and a quick trip to and from work and incidentally force him to render a higher tribute unto Caesar, while the more fortunate ride in privately owned cars.' The Birmingham *Labor Advocate* also bitterly assailed the streetcar company, and endorsed the automobile as perhaps the most desirable alternative for the working man: 'The gas method is cleaner, as safe, is more rapid, and a better method than the old way of packing them in like sardines, and shaking them up to make the breakfast digest better, and the people who have to be taken to work and back again are eagerly accepting the new method of transport and refusing to be strap hangers, and dividend producers for watered stock.'[21]

The comments of the workers' press indicate the disdain which workers felt toward so-called 'mass transit' as well as the promise and desirability of an automobile alternative. Despite the short-term protection afforded to public transit by monopolistic practices, the long-term future boded ill for the transit companies. Further, the contempt for streetcar companies extended to the black community with even greater fervour. Blacks in Southern cities, because of their low economic status, had even fewer transit options than working-class whites. Every ride on a streetcar, and later, a bus presented the prospect of humiliation ranging from segregated seating to abusive operators and to infrequent stops in black neighbourhoods.

THE AUTOMOBILE, THE LIBERATOR

The automobile, expensive as it was, represented a liberation in numerous respects for blacks. As early as 1921 in Atlanta, a resolution adopted at an interracial meeting protesting the racial policies of streetcar companies urged blacks to 'Buy a car of your own and escape jim-crowism from street car service.' Some years later, a black correspondent suggested in a letter to the Atlanta *Independent*, a black newspaper, the formation of an 'automobile club that would take up women and children on the streets and who [sic] would give repair work to our own men.' Though to some degree the automobile and the suburban expansion it facilitated helped to harden lines of residential segregation in Southern cities, blacks shared with whites the association of automobility and freedom in many senses of that word. As Chapel Hill sociologist, Arthur Raper concluded in 1936: 'the opportunities afforded by the automobile provide a basis for a new mobility for whites as well as Negroes, based upon personal standards rather than upon community mores – upon what the individual wants to do rather than what the community does not want him to do.' The automobile was significant, in other words, because it provided 'the mechanical means for a greater degree of self-direction and self-expression'.[22]

So the public transit alternative to automobility use was botched or perhaps not even viable to begin with, considering the mystique enveloping the automobile as early as the First World War. Other efforts to relieve congestion and urban disintegration rank as fanciful or, at best, unlikely. Some suggested that the organisation of auto-mobile clubs would more fully insure motorists observance of the traffic laws and hence result in the safe operation of motor vehicles. One editor appealed to Birmingham citizens to prevent traffic congestion by not putting 'automobiles to unnecessary uses'. Atlanta businessmen organised a 'Be Careful Drive' in 1923 to alert citizens to the dangers of motor traffic and thereby reduce the accident rate. And the Nashville *Tennessean* proposed that the radio was 'destined to entirely counteract the influence of the automobile in calling people from the home'.[23]

It was not to be, of course. By the 1920s the automobile had become an integral part of Southern lifestyle both in the countryside and in the cities. Much more than a means of transport, it was a vehicle of freedom, access to employment, invitation to recreation (and in some cases even procreation) and in the less densely populated South especially, occasionally the difference between life and death – the time it took for a doctor or midwife to reach a stricken patient. Under these circumstances it was not surprising that the nostrums of city planners, as wrongheaded as they were in any case, the pleas of preachers and the lectures of editors had relatively little impact on the urban South's automobility and may even have contributed to the auto's increased utilisation.

By 1980 the South had more motor vehicles registered per capita than any other section of the United States. The region's cities, invariably low-density, sprawling affairs reflected the automobile's impact on land use. If public transit was ever a viable challenge to automobile hegemony, it is no longer so; most Southern cities simply lack suitable density thresholds to adequately support such service. The segregation of land use, strip development and the ubiquitous shopping mall, all within the framework of a virtually unfettered real estate market in the South, ensure the continued dominance of and dependence on the automobile. Houston is the illogical conclusion of such events, a city choking on its own traffic in both environmental and physical senses with no promising policies on the horizon. It has become apparent in the urban South, as elsewhere, that drivers are willing to endure much adversity in order to use their automobiles. In the South the most individualistic and privatistic region in the land, the automobile is part of the citizen's inalienable Southern rights. The immunity of the automobile to policy directives in the South may have less to do with the policies themselves than with the fact that it is difficult to alter an integral element of the regional folk culture, a folk culture that has secured numerous economic and psychic benefits for the region.

Notes

1. Anthony M. Tang, *Economic Development in the Southern Piedmont. 1860–1950: Its Impact on Agriculture* (Chapel Hill, 1958).
2. Robert S. and Helen M. Lynd, *Middletown in Transition: A Study in Cultural Conflicts* (New York, 1937), 26, pp. 265–7; Lewis Mumford, *The Highway and the City* (New York, 1964), pp. 244–5.

3. For a discussion of how Southern cities differed spatially from Northern cities, see David R. Goldfield, *Cotton Fields and Skyscrapers: Southern City and Region, 1607–1980* (Baton Rouge, 1982), pp. 29, 96.

4. Homer Hoyt, *One Hundred Years of Land Values in Chicago: The Relationship of the Growth of Chicago to the Rise in Its Land Values, 1830–1933* (New York, 1970), p. 32.

5. Howard L. Preston, *Automobile Age Atlanta: The Making of a Southern Metropolis, 1900–1935* (Athens, Ga, 1979), pp. 25–35, 50–70.

6. David R. Goldfield, 'North Carolina's Early Twentieth-Century Neighborhoods and the Urbanizing South', in Catherine W. Bishir (ed.), *Early Twentieth-Century Suburbs in North Carolina* (Raleigh, 1985).

7. Blaine Brownell, 'A Symbol of Modernity: Attitudes Toward the Automobile in Southern Cities in the 1920s', *American Quarterly*, p. 20.

8. Nashville *Tennessean*, 9 Aug 1920; 20 June 1924.

9. Perkins J. Prewitt, 'Making Birmingham Safe for Life and Property', *Birmingham*, 1 (May 1925) p. 13; Executive Committee of the New Orleans Safety Council to the Mayor and Commission Council of New Orleans, 29 Oct 1929 (City Archives, New Orleans Public Library); New Orleans Commission Council, *Official Proceedings of the Commission*, Council of the City of New Orleans, 16 Nov 1926, 2 (City Archives, New Orleans Public Library); *Birmingham*, 2 (Mar 1926) p. 6.

10. Quoted in George B. Tindall, *The Emergence of the New South, 1913–1945* (Baton Rouge, 1967) p. 257.

11. *Atlanta Life*, 18 June 1927.

12. Nashville *Christian Advocate*, 85 (25 July 1924) p. 933, Nashville *Tennessean*, 12 Aug 1921.

13. Birmingham *Labor Advocate*, 17 Dec 1921; Atlanta *Independent*, 14 July 1921; Nashville *Christian Advocate*, 82 (7 Oct 1921) p. 1254; the Rev. Richard L. Ownbey, 'The Church and Leisure', Nashville *Christian Advocate*, 88(14 Jan 1927) p. 42.

14. Nashville *Tennessean*, 16 June 1926; A. T. Robertson, 'American Cities and the Criminal Classes', Atlanta *Christian Index*, 105 (29 Oct 1925) p. 5; City of New Orleans, Archives Department, *Synopsis of Ordinances, 1841–1937* (typescript, City Archives, New Orleans Public Library) p. 24.

15. Atlanta *Constitution*, (3 Aug 1921).

16. *Atlanta Life*, (3 Aug 1929); Birmingham *Age-Herald* (13 Jan 1927); see also David R. Goldfield and Blaine A. Brownell, *Urban America: From Downtown to No Town* (Boston, Mass. 1979) pp. 340–5.

17. *Tattler*, 1 (25 May 1929): 5; *Nashville This Week*, 1 (28 Sep–5 Oct 1925) p. 3.

18. *Nashville This Week*, (10–17 May 1926) p. 3.

19. Quoted in Franklin Garett, *Atlanta and Environs: A Chronicle of the People and Events* (2 vols, New York, 1954) II, 822.

20. Harland Bartholomew was responsible for numerous city plans throughout the urban South from the 1920s to the 1950s. The evolution of his planning philosophy away from transportation solutions to a more balanced approach is evident in his *A Comprehensive City Plan, Memphis, Tennessee* (Memphis, 1924); *A Comprehensive City Plan, Knoxville,*

Tennessee (Knoxville, 1930); and *A Master Plan for the Physical Development of the City* [Richmond] (Richmond, 1946); see also, Blaine A. Brownell, 'The Commercial–Civic Elite and City Planning in Atlanta, Memphis, and New Orleans in the 1920s'. *Journal of Southern History*, 41 (Aug 1975) pp. 339–67.
21. Quoted in Blaine A. Brownell, 'The Notorious Jitney and the Urban Transportation Crisis in Birmingham in the 1920s', *Alabama Review*, (Apr 1972) p. 114.
22. Atlanta *Independent* (25 Aug 1921); (14 Oct 1926); Arthur Raper, *Preface to Peasantry: A Tale of Two Black Belt Counties* (Chapel Hill, 1936): 174–5.
23. Nashville *Banner* (28 Feb 1922); Birmingham *Age-Herald* (29 July 1926); Atlanta *Constitution* (5 Feb 1923); Nashville *Tennessean* (9 Mar 1927).

7 Some Economic and Social Effects of Motor Vehicles in France since 1890

Patrick Fridenson

Motor manufacturing generates much more employment outside the motor industry than in the motor works themselves. In France in 1984, for instance, only 215 000 people were directly employed in making motor vehicles while perhaps a further 1 900 000 were indirectly involved to a greater or less extent in one way or another. In all, this was approaching 10 per cent of the total French labour force, then 21 483 000. Those outside the motor factories worked in a whole spectrum of occupations stretching at one end from those employed as drivers (766 000, more than three times as many as those in the motor factories), 388 000 in garages and repair work, 315 000 in industries such as metals, rubber, plastics, paint and textiles supplying the motor manufacturers, 155 000 making components for them in outside businesses, 94 000 in insurance and hire-purchase, 80 000 in road-building and repair, round to (at the other end of the scale) 18 000 in driving schools and 7000 working as attendants on autoroutes and in car parks.[1]

In this chapter we shall concentrate upon a few selected aspects of the French economy and society which were directly transformed by motorisation: the relationship between national and foreign technologies; channels of retailing; patterns of consumption; competition between means of transport; social costs of technical progress; and state intervention. Road construction is dealt with elsewhere by Dominique Barjot (Chapter 15).

THE RELATIONSHIP BETWEEN NATIONAL AND FOREIGN TECHNOLOGIES

The beginnings of motorisation in France depended to a very great extent on the import of German technology.[2] So, from the start the question was raised: would motor transport offer opportunities for national innovation? And, if so, what would be the balance of international technology in this important new industry? The answers varied considerably in the course of the twentieth century.

Up to 1914 the growth of the motor industry in France, which held the world leadership to 1905 and the European until 1929, greatly stimulated technical innovation by French entrepreneurs, engineers, workers and scientists. Soon the great majority of patents used in the motor industry were of French origin. Foreign licences – both American and European – continued to be bought, but the export of French patents outweighed them. The overall impact of the motor industry was positive. The electric starter was the only major exception to this French superiority.[3]

The overwhelming American dominance after 1914, however, resulted in a weakening of the French car makers' technical interest and, in consequence, there was a deterioration in the balance of payments for patent licensing. Citroën paved the way in the transfer of American technology to France. Its competitors also bought manufacturing licences or imitated American devices. The emphasis was on the adaptation of foreign inventions and in the research departments of large firms national innovators were no longer held in high esteem. By 1938 only 30 per cent of the patents used in the French motor industry were of French origin.[4] Therefore the shift from small-scale production to mass production clearly meant not only the acceptance of Fordism, but also a decline of the national potential for innovation and long-term specific currency losses.

Prompted by the reflections of some engineers during the Second World War and by France's shortage of foreign currency in the post-war years, there was a renaissance of national innovation in the French motor-car industry, with Renault in the lead for production systems and Citroën ahead for product technology.[6] To be sure, the import of foreign goods and patents continued. For instance, when the U.S. Public Affairs Officer, Lyon, visited the Peugeot motor-car factory at Sochaux in 1949, he remarked, 'American machines were everywhere in evidence. . . . A certain amount of German equipment recently received as reparations was also at work. . . . There were very few French machines to be found, although Mr. Callé was at some pains to point them out when he could.'[7] But the French producers vigorously increased their research activities, especially in the 1950s, and built up 'a strong portfolio of technological know-how'. In the 1970s, for which years we have reliable statistics,[8] the growth of international competition and the energy crisis brought about, as everywhere, another 'upturn in innovative activity',[9] and

Table 7.1   International technological exchanges in the French motor
industry

	Expenditure	Receipts (1000 francs)	Balance	Rate of covering (%)
1970	44 905	58 032	+13 127	129 23
1971	49 302	43 486	− 5 816	88 20
1972	95 661	86 256	− 8 405	91 12
1973	53 792	64 968	+11 176	120 78
1974	43 141	177 709	+134 568	411 93
1975	83 199	158 021	+74 822	189 93
1976	136 379	108 108	−28 271	79 27
1977	183 555	174 396	+ 9 159	95 01
1978	188 622	265 120	+76 498	140 56
1979	222 527	148 695	−73 832	66 82
1980	216 448	109 252	−107 196	50 47
1981	115 647	540 994	+425 347	467 80
1982	134 455	128 066	− 6 389	95 25

hence a more favourable technological balance than in prewar years.[10]

An even more accurate picture can be drawn by considering the impact of the French motor industry on the national machine-tool industry. Some car-makers developed their own machine tools. This does not really apply to the jointly-sponsored company for machine-tools which the French car-maker de Dion founded in 1907, for it disappeared after only six years;[11] But other leading manufacturers, such as Renault and Berliet, created machine-tool departments which invented and produced specific machines. Renault even acquired several machine-tool companies in the 1960s. The latest break-through of the car-makers in this field has been the development of numerical-controlled machines, first at Renault in the 1960s (though with much reluctance by management) and then at Citroën.[12] On the other hand, from the start French car-makers were major importers, mostly of American, German and Swiss machine tools, and thus contributed to that industry's decline, all the more as the branch 'was very flimsy, fragmented into numerous factories and totally inappropriate for meeting the challenge of economic fluctuations'.[13] The high rate of imports for machine tools since the Second World War was paralleled by a large deficit in the technological exchanges in that sector, which the State's intervention from 1981 onward failed to alter. By 1980 the motor industry was buying 20 per cent of machine tools sold in France.[14] On the whole the impact of motor vehicles on France's innovative ability and technological balance of payments is, in the long run, certainly less favourable than industry spokesmen and economists have repeatedly asserted. It shaped a national basis for innovation in modern technologies, but its size and strength seem to have developed in a cyclical way. The impact of the motor car on French retailing and distribution may well have been greater in the long run.

CHANGES IN RETAILING

It took years for motor cars to change the structure of distribution for the process was very gradual. But in the end they profoundly altered the marketing of consumer durable goods, and a new social group was created. Already by 1932 the total number of motor car, motor bicycle and bicycle dealers and agents was estimated to be 15 000.[15]

Initially, marketing of cars was done mostly through existing channels. The early dealers were either connected with horses (black-smiths, keepers of staging-posts, livery stables), with bicycles or with durables (hardware, woodworking machines). All of them, it should be noted, were tradesmen unaccustomed to handling a luxury market.[16] The only newcomers were racers of bicycles, motor cycles or cars, and former employees of the car-makers themselves (mechanics, middle managers).[17] Before 1914, however, car-makers introduced their own marketing hierarchies: regional distributors, local dealers, agents or sub-agents. Simultaneously they opened a limited number of company-owned retailing branches. Meanwhile the dealers felt a need for unity *vis-à-vis* the car-makers. They created voluntary associations for each major make and a national union for the whole trade. All these features, the two distinct channels for marketing, the two types of association, remain today.

The interwar years, with the beginnings of mass production, saw various attempts on the part of manufacturers both to control and to assist the dealers. The most striking was the development of exclusive dealership after the American pattern, a commitment which in France only a grocery company had already developed, and the training of dealers in specialised schools where manufacturers professionalised salesmanship and repairs. The dealers were probably the first trades-men in France to lobby parliament and governments about selling prices and discounts.[18]

Motor showrooms spread rapidly. Garages for service and storage of motor cars have been in existence since the turn of the century, Bordeaux claiming to be where the world's first service station opened, in 1895. But the early motor showrooms which opened before 1914 did not make cars visible from outside or even conveniently available for inspection and testing. Between the wars showrooms were built after the pattern of railway stations or department stores and had large and well-lit premises to display their wares to prospective customers, thus overcoming a traditional weakness of French merchants, i.e. in display and demonstration. The combination of demonstration, credit and repair facilities, in firms under contract filled an important gap in the French commercial system.[19] André Citroën inspired by the American example, went further and inaugurated after-sale services in his whole network. He adopted, for instance, the one-year guarantee for all new cars and their free overhaul after the first 500 km.[20]

The economic influence of dealers in rural areas should not be underestimated. To encourage small farmers to buy their first car or truck, they accepted all kinds of trade-ins: wine, apples, wheat, pianos, agricultural machines. The motor vehicle was also generally the first important claim upon small farmers' accumulated savings and, once bought, it increased their interest in monetary circulation and commercial transactions.[21]

The mass market of the post-Second World War years brought new features. First, the rate of motorisation in France, especially after 1960, caught up again upon other leading European countries, and since 1979 France has ranked first in Europe, with 43.6 vehicles per 100 inhabitants in 1983.[22] This was due in part to the high density of sales outlets: by 1980 4000 dealers (selling on average 500 new vehicles per year) plus 15 000 agents, a total which climbed to 21 000 in 1984.[23]

Second, since the early 1970s the dealer's commercial functions have expanded.[24] He has become, as distinct from the mere garage proprietor, an expert in the second-hand car market. From 1973 onward sales of used cars grew more quickly than those of new ones: on average 4 per cent per year compared to 2.3 per cent earlier. Whereas in 1973 every time one brand-new car was sold, a little more than one used car was sold, the ratio had become in 1982 1 to 2.4: 2 056 490 new cars, 4 808 000 used one.[25] One car usually passed through the hands of 5–6 owners in 1975, and of 3–4 in 1983.[26] Dealers and manufacturers had learnt how to organise the used-car market nationwide, to compete on the prices, to renovate the vehicles and to offer specific guarantees for them: O.R. (Renault), Eurocasion (Citroën), Sécurité.[27] The dealer also increased his spare-part trade. Replacement of parts became often preferable to a repair. As spare parts were very profitable, car-makers controlled a greater proportion of this market: 30 per cent in 1970, 50 per cent in 1983. This favoured dealers *vis-à-vis* independent tradesmen or garage proprietors.[28] The dealer also now had the capacity to present a large range of differentiated financial services, among them leasing, which was meant to attract former customers of the second-hand market, and merchandising.[29]

It is well known that the car-makers from the beginning stimulated their sales by advertising, and that in France motor vehicles were a major force for modern mass advertising, thus influencing advertisement for various kinds of goods.[30] But a simple statistical comparison shows how intimately this effort on advertising relies on the use of the press, and profited the press.

Table 7.2   A breakdown of advertising budgets in 1930[31]

| | Percentage | | | |
	Citroën	Renault	Peugeot	General Motors
Press	45	40	30	45
Catalogues and printed matter	18	25	15	15
Posters	7	7	20	15
Caravans	10	10	15	10
Radio and film	5	3	10	4
Direct advertising	10	10	5	5
Overall fees	5	5	5	5

Table 7.3   A breakdown of Renault's and PSA's advertising budget in 1984[32]

	Renault	PSA
Press	45	44
Radio	23	26
TV	10	14
Posters	19	14
Film	3	2

The share of the press increased in spite of the coming of new media but in the 1980s it came back to the level of 1930.

Car manufacturers professionalised their advertising campaigns by contracts with leading advertising agencies. Peugeot, with La Publicité Française, which it had previously left in 1928!, and Simca, with Havas, initiated that move in the early 1950s; Renault followed suit in 1960 and Citroën in 1968. This shift shows how autonomous *vis-à-vis* the products advertising had become and how important it had become for manufacturers, because of increased purchasing power and of the growth of replacement demand: 'We don't sell the car any more,' a Renault manager said in 1958, 'We sell now the use of the car'.[33] Yet the proportion of turnover devoted to advertising declined, as French people became more accustomed to car purchase, from 2 per cent in the 1930s to 0.6 per cent in 1983. The same year the French detergent and perfume manufacturers spent 14 per cent of their sales income on advertising.[34]

CHANGING PATTERNS OF CONSUMPTION

The motor industry stimulated other branches of business. France was a pioneer of motoring publications, for instance, initially motoring magazines, driving and technical handbooks, then tourist guides and road maps (with Michelin soon in the lead) and books about motoring generally. Special clothes were produced for the early open vehicles. 'Goats skin' coats, dustproof overcoats, large raincoats, caps and women's hats, together with thick goggles. Travel abroad increased: by 1913 5172 Frenchmen owned international driving licences. The same held true for travel in France, and travel firms organised special sections for motorists. Specialised clubs (of which cycling had been the forerunner) became mass membership organisations. By 1913 64 per cent of 125 000 car owners in France were club members and were provided with special services. A great number of goods concerned with motoring also appeared on the market: postcards, gramophone records, sculptures and various personal effects.[35] The middle classes were not so devoted to clubs as the rich, however, and the masses even less so. As motor-car ownership spread down the social scale the proportion belonging to motoring clubs fell dramatically, though the absolute totals continued to rise. In 1979 only 3.5 per cent of French motorists belonged to motoring organisations, though this represented a total of 600 000 drivers.[36]

Motor vehicles also brought about a major expansion of French insurance companies and of family expenditure upon insurance. By 1935 95 per cent of French drivers had insurance cover for their cars. Nevertheless, compulsory insurance arrived relatively late in France in comparison with most western countries. It was made compulsory for public transport vehicles by a decree of 1935, but extended generally to every road vehicle only by a law of 1958. This was due to the resistance of the insurance companies, which, although their profits on this type of risk were very irregular, long refused any control over their premiums. Insurance companies after the Second World War preferred to develop a voluntary association for the improvement of safety on roads and for the education of drivers and passengers, called 'La Prévention Routière', which published a monthly magazine. It had 335 000 members in 1979. This hesitation of the large insurance companies led to the creation of very powerful friendly societies, notably the MAIF, the teachers' road insurance friendly society, founded in 1934 and still based in Niort (Deux-Sèvres), where its sister for tradesmen and industrialists, MACIF, is also located. Relying on the homogeneity of their customers, friendly societies were in a position to offer lower premiums to those more safety-conscious groups. In 1963 the first French assistance company, Europ-Assistance, was created which, during the holiday period, gave subscribers special aid if required, whether medical, technical or legal. Very successful, its example was subsequently followed by insurance companies, friendly societies and automobile clubs. It is now operated worldwide.[37]

A similar extension of existing facilities occurred with instalment plan payments. This is evidenced by the earliest firm specialising in such loans, which I discovered only in 1985. Called 'l'intermédiaire vélocipédique' (the bicycle middleman), it was located in Paris and boasted to be 'the only house selling in monthly instalments bicycles, motorcycles, motor cars and photographic cameras'. This advertisement which associates various symbols of modernity, was issued in 1900.[38] The development of credit sales for motor cars in France has been discussed in my article on French automobile marketing (see note 18). Therefore, I just want to dwell on a few points shown by recent research. The lag between the spread of instalment selling in other industrialised countries and France still exists today, as instalment buying accounts for 50 per cent of total car sales, less than the level reached in the United States in 1920, 60 per cent.[39] In the long run the strength of the second-hand car market provided an alternative to hire-purchase, and the reluctance of small farmers to enter into credit arrangements was not the only obstacle in rural areas. Many rural motor agents did not like the idea of selling on time, and many wealthy farmers were not keen on lending money to peasants for the buying of a car.[40] Nevertheless, motor vehicles have become the largest market in France for credit sales. This makes them a main target for government restrictions on credit buying, as was the case in 1963–5 or in 1974.[41]

What proportion of their incomes do French families devote, on average, to motoring? There has been a dramatic change in the last twenty years. In 1966 a family of upper middle managers spent three times more money on motor cars than a family of workers. In 1979 the difference had fallen to 1.7 times. By 1983 there was virtually no difference. Today every category of the French population spends the same share of its income on the buying, use and maintenance of the motor car, i.e. 13 per cent. This levelling is due to the exceptional effort made by families with lower incomes. They gave higher priority to motoring in their spending, clearly at the expense of housing.[42] But they did not necessarily choose cheaper vehicles: by 1978 two-thirds of the second-hand car-buyers were small farmers, workers, employees and servants.[43]

One further aspect of the impact of motor vehicles on existing services needs to be considered: the shift from the horse cab to the motor taxicab. The transition took many paths at the turn of the century. Existing cab companies bought motor or electric vehicles and some of the old coachmen became the new taxi drivers. New companies were also founded to operate these machines. They were owned by banks, car-makers, car-dealers or mechanics and industrialists. Co-operatives were also created, and there soon were some independent taxi owners too.[44] Concentration developed rapidly. By 1914 the large companies accounted for more than half the taxicabs in Paris and the largest of them still dominates the trade today. The Citroën taxicab company, created in 1924, despite its initial success, did not enjoy a similar life-span; indeed, it did not survive the Second World War.[45] However, in the 1960s the taxicab business changed completely. Because of increasing traffic jams, of too low fares, and of heavier costs of fixed assets, the companies reduced their fleets and sold a large part of their taxicab licences. By 1979, out of 33 000 taxicabs in France, 26 500 were in the hands of owner-drivers. There had been a renaissance of small enterprise.[46]

Motor taxicabs naturally brings us to the question of competition between motor vehicles and other forms of transport.

COMPETITION WITH OTHER MEANS OF TRANSPORT

The horse was the first to be defeated. Yet it took time. In 1899 a mere 1792 motor cars ran alongside 271 000 horse-drawn carriages. In 1909 the numbers were respectively 57 000 and 222 000. The demise of the horse-drawn omnibus was much more rapid and in Paris the last such vehicle ceased to operate in 1913.[47] In 1914 the army was the only domain where the horse continued to reign supreme, largely because the state did not allocate the finance needed for its replacement.[48] Meanwhile, local goods carriers enabled 'the number of horse-drawn vehicles in circulation to grow steadily'.[49] The motoring press repeatedly argued that motor vehicles would be cheaper, quicker, more reliable.[50] Yet, outside towns, the early motor buses and trucks were neither quick, reliable nor cheap. They were as yet no serious challenge either to railways or to the horse-drawn buggies. It was only after 1919 that better and cheaper motor trucks slowly replaced the horse-drawn carts and wagons.[51] From 1925 onwards, motorised trucks and buses began to compete for railway goods and passenger traffic outside towns over shorter distances and on east–west routes. Major railway networks became unprofitable as early as 1928; and the worst was soon to come. Thanks to the introduction of diesel fuel and of lighter chassis, road transport began to bite into the fruitful long-distance traffic of the railways. Even motor manufacturers created large transport companies. The major rail networks responded to this by setting up their own road services. Many small road transport operators were driven out of business.

Then at the request of the state-owned railway network the state decided to 'co-ordinate road and rail transport' by a decree of 19 April 1934. This decree, and the policy it inaugurated, brought about a division of labour, in order 'to allocate long-distance traffic to the railways . . . and short-distance traffic to the roads, thus reducing competition and allowing the railways to close part of the networks to passengers'. In fact there was a share-out of long-distance traffic for goods; truckers' rates were not allowed to be lower than railways'. Most of these measures clearly penalised road transport and arrested its development. Moreover, motor vehicles were subjected to new taxes and no new passenger transport enterprises were allowed to be set up. Road transport competition was now restricted.[52] The war and the immediate post-war years confirmed the railways' privileged position.

From the middle of 1947, however, the rules of the game began to be changed. The state leaned towards the road lobby and took a more liberal view of road–rail co-ordination. One of its salient features was the end of the freeze on the public transport fleet; and competition for the lucrative goods sector was no longer tightly controlled.[53] But if the roads took 'an increasing share of both passenger and goods traffic', there were other reasons besides deregulation. The spread of private motor cars hit rural and suburban train services; and the construction of motorways after 1955 increased road competition over longer distances not only by motor cars and coaches, but especially by trucks. The technology, capacity, comfort and fuel consumption of trucks, buses and coaches were greatly improved, thanks in particular to diesel engines which entirely dominated the range of vehicles between 5 and 15 tons (contrary to the United States). A further source of dynamism lay in the growing specialisation of road transport vehicles according to distances covered and to the type of goods carried, which often enabled them to complement one another.[54]

The post-war years also saw competition between various forms of motor transport. The rise of private cars brought about a sharp decline in rural bus services at a time when school buses grew fivefold.[55] The private car was also one of the reasons for the financial difficulties of city bus services.[56]

By 1982 – the latest statistics available – motor vehicles were responsible for 87.4 per cent of passenger transport. They were less predominant in goods transport, however. Here they accounted for 57.8 per cent of total traffic in tons-kilometres, which was not the highest proportion in the Common Market. Goods traffic by rail had declined from 48.1 per cent in 1965 to 32 per cent in 1982.[57]

The transport law of 1982 recognised these facts. Because of the energy and social costs of the private car, it favoured public transport and relaxed some of the state controls on it while still advocating complementarity between competing transport modes. This was a compromise between the opposite positions of 1934 and 1949.[58]

The challenge which motor vehicles sent to the quasi-monopoly won by the railways back in the 1850s was one of the main issues with regard to which motor vehicles fostered a larger intervention of the state in recent decades.

STATE INTERVENTION

Motor vehicles have made an important contribution to the extension of state power in France.

They reinforced traditional areas of state intervention. Transport policy, which we have just seen, was one. Road traffic regulation was another. The coming of motor vehicles gave rise to many complaints from horse riders and drivers of horse-drawn vehicles as well as from pedestrians, especially in the countryside; and, in the towns, any idea that they would put an end to traffic jams was soon abandoned.[59] A driving licence was introduced for motorists as early as 1891 and was made compulsory in 1893; the Prefects received in 1899 powers to withdraw it after two breaches of traffic regulations.[60] A national highway code was proposed as early as 1902, but the code itself was issued only in 1921–2. The delay was due to the motorists' refusal to accept speed limits or protracted discussions over the powers of municipalities, and to drivers' misgivings about driving on the left.[61] Speed limits on roads and motorways and the compulsory use of safety belts were, indeed, introduced only in 1973 because of a resistance from motorists and their organisations and the motoring press.[62] The number plate was also an extension of earlier registration procedures, directly linked with road accidents. Motorists, their clubs and their press fought against it for several years. The government decided upon it in 1901, but it was actually enforced only after 1909. Once this battle was won, the state was able to organise censuses of motor vehicles, something the French state knew from long experience how to do. Four of them took place in the early 1910s and at least eight in the interwar period.[63] Cab regulation, which had existed since the beginning of the nineteenth century, was extended to taxicabs and considerably refined; taxi drivers themselves came under most careful supervision.[64]

Motor vehicles also opened new areas to state intervention. One was the permanent regulation of sales of durable goods. France was familiar with temporary control of sales during crises and wars, mostly in the form of rationing. But motor vehicles led the state to innovate and regulate the distribution of durables. Instalment sales, second-hand transactions, repair prices and franchising were some of the matters dealt with.[65] The latest development was the compulsory technical control of cars over five years old before a second-hand sale. It was established in 1985, i.e. later and in a more moderate manner than in other European countries such as Sweden. And we have already mentioned compulsory insurance. For the building and the management of motorways, the state entered into partnership with banks and insurance companies to create firms of mixed economy.[66] The social costs of motor vehicles, however, gradually became the main ground for state intervention.

Public awareness of these costs was particularly strong from 1890 to 1914 and from 1960 onwards. The major issues were road accidents and noise throughout, and much later and with less general concern, energy saving, congestion and pollution.[67] The state always reacted with great caution. It imposed few product regulations, often preferring instead to negotiate informal agreements with the carmakers and to wait for the European Economic Community to do the regulations. As Mark Fuller writes,

Although the French had imposed some safety regulations before World War II, they remained more lenient than German or American standards. EEC directives began to supersede national rules in the late 1960s. French opposition to this trend was interpreted by many as an attempt to preserve idiosyncratic requirements as a nontariff barrier to competition. France had no emissions standard until EEC regulations were adopted in the mid-1970s. French representatives were leaders of the 'go-slow' party in emissions deliberations.

The same was true in the negotiations on new European anti-pollution norms since 1982, which led to the Luxemburg agreement of 28 June 1985. 'France imposed no formal fuel efficiency standards. In 1979, the government concluded a voluntary agreement with domestic producers calling for a 10 per cent reduced consumption by 1985. It looks as if the state had clearly followed the domestic producers, to protect a national industry. To be sure, the specific urbanisation of France, the types of vehicles preferred by French customers, and, on another side, the weakness of consumers' movements allowed the French state more latitude than its counterparts to delay action. But it has also been argued that macro-economic developments since the late 1960s left for the French state an increasingly limited margin of freedom and that – with the exception of the years 1981–2 – its industrial policy in general had been progressively defined in narrower terms. The state's typical mode of intervention was rather to encourage research and development by the car manufacturers thanks to important subsidies, and to influence motorists by taxing petrol and vehicles, regulating speed, improving the road network.[68] Not without interesting results. Between the record year 1972 and 1983 road accidents declined by 22 per cent, those killed by 25 per cent. Yet France still had the third highest accident rate in OECD countries. The other social costs were reduced at the same pace.[69]

CONCLUSION

France has had one of the largest working populations dependent in one way or another upon the motor industry. An international comparison of 1980 gave percentages of 10.4 in West Germany, 9.6 in France, 8.9 in Japan, 7.9 in Sweden, 5.8 in the United Kingdom and 4.3 in the United States.[70] These proportions are not just the result of present-day production levels; they are much more deeply rooted in the motor vehicles' acceptance in the various countries. This comparison suggests that motorisation has had a particularly marked effect on the French economy and society. It does not, however, imply that France was always in the vanguard. It lagged in some matters: compulsory insurance, regulation of social costs, instalment sales. It pioneered in others: motoring press, driving licences. It was about average in others; the 13 per cent of families' motoring expenditures in France may be compared with the 12 per cent in the United States or with the 15 per cent in Britain.[71] In spite of the modernity of the motor vehicle itself, these peculiarities may well have to do with some traditional features of French society: the large weighting of rural people and, at the other extreme, of Paris, the strength of the state and, on the contrary, of individualism, the sustained power of the railways and their often bad public image. There has been since the 1910s a strong motoring lobby, but it has never enjoyed so much influence as in other countries. The economic and social effects of motor vehicles in France could not but incorporate these deep-seated national contradictions notwithstanding the growing internationalisation of its economy.

Notes

1. Chambre Syndicale des Constructeurs d'Automobiles, *Statistiques, 1984*, (Paris: GEPAP, 1985) p. 44. For an elaboration of the data of the year

1971, see C. Baudelot, R. Establet, J. Toiser, P. O. Flavigny, *Qui travaille pour qui?*, 2nd ed. (Paris: le livre de poche, 1982).

2. P. Fridenson, 'Les Premiers inventeurs de l'automobile', *L'Histoire* (Dec 1984).

3. C. W. Bishop, *La France et l'automobile. Contribution française au développement économique et technique de l'automobilisme des origines à la deuxième guerre mondiale* (Paris: Librairies techniques, 1971). P. Fridenson, *Histoire des usines Renault*, vol. I (Paris: le Seuil, 1972) pp. 24, 50. J. M. Laux, *In First Gear: The French Automobile Industry to 1914* (Liverpool: Liverpool University Press, 1976) pp. 107, 169, 204.

4. *Les industries de l'automobile et du cycle avant et après la guerre* (Paris: Office professionnel de l'automobile, 1946).

6. J. A. Grégoire, *50 ans d'automobile* (Paris: Flammarion, 1974) vol. 1. J. L. Loubet, 'La Société anonyme André Citroën (1924–1968)', Ph.D. thesis, University Paris X – Nanterre, 1979, pp. 353, 362–3, 419–29. P. Bézier, 'Souvenirs d'un outilleur: après la 4CV, la Dauphine', *De Renault Frères constructeurs d'automobiles à Renault Régie nationale* (June 1985) pp. 272–83.

7. National Archives, Washington, D.C., 851.659/4–449, letter by Horatio Mooers, U.S. Consul to the Secretary of State, 4 Apr 1949, p. 1.

8. J. J. Chanaron, 'L'innovation dans la construction automobile', Ph.D. thesis, University Grenoble II, 1973. F. Picard, *L'épopée de Renault* (Paris: Albin Michel, 1976) pp. 310–11. A. Altshuler *et al.*, *The Future of the Automobile* (London: Allen & Unwin, 1984) pp. 101–04.

9. Archives of the 'Service de la propriété industrielle', Ministry of Industry, yearly statistical reports, 1970–82.

10. General Planning Committee, unpublished note on 'patents in transportation industries' (31 Mar 1983).

11. J. M. Laux, *In First Gear* . . ., op. cit. pp. 123–5.

12. P. Bézier, 'Souvenirs d'un outilleur', *De Renault Frères* . . . (Dec 1975) pp. 176–87. J. P. Poitou, 'Note sur un premier inventaire des archives de Pierre Debos', ibid. (June 1985) pp. 302–3. P. Fontaine, *L'Industrie automobile en France* (Paris: La Documentation Française, 1980) p. 103. P. Bézier, 'Petite histoire d'une idée bizarre', *De Renault Frères* . . . (June 1982) pp. 256–68, and (Dec 1982) pp. 319–31.

13. A. A. Garanger, *Petite histoire d'une grande industrie (histoire de l'industrie de la machine-outil en France)* (Neuilly: Société d' édition pour la mécanique et la machine-outil, 1960). P. Mioche, 'Les difficultés de la modernisation dans le cas de l'industrie française de la machine-outil, 1941–1953' (European University Institute working paper, Florence, 1985).

14. Archives of the 'service de la propriété industrielle', yearly statistical reports, 1970–1980. P. Fontaine, *L'Industrie automobile* . . ., op. cit. p. 193.

15. J. Bouvier, 'L'Automobile et les transports', *Journal de la Société de Statistique de Paris* (Jan 1933) p. 9.

16. C. Harmelle, 'Les Piqués de l'Aigle. Saint Antonin et sa région (1850–1940). Révolutions des transports et changement social', *Recherches* (Nov 1982) pp. 289–90. J. P. Delaperrelle, 'Delaroche, Gaudichet, Turquet', *Cénomane* (Summer 1984) pp. 34–5.

17. G. Hatry, *Louis Renault patron absolu* (Paris: Lafourcade, 1982).
18. P. Fridenson, 'French Automobile Marketing, 1890–1979', in A. Okochi and K. Shimokawa (eds), *Development of Mass Marketing* (Tokyo: Tokyo University Press, 1981).
19. H. Guéné, 'L'Architecture automobile', *Monuments historiques* (Aug–Sep 1984) pp. 33–41. Gérard Monnier, 'Deux chantiers de sculpture monumentale à Marseille en 1925', *Marseille* (Jan 1985). F. Nordemann, *Les Lieux de l'automobile* (Paris: Graphite, 1985). D. Burgess-Wise, *Automobile Archaeology* (Cambridge: Patrick Stephens, 1981) pp. 38–9.
20. J. L. Loubet, *La Société Anonyme André Citroën . . .*, op. cit. pp. 155–8.
21. C. Harmelle, 'Les Piqués de l'Aigle . . .', art. cit. p. 300.
22. Union Routière de France, *La Circulation routiere Faits et chiffres* (Issy: M.C.G., 1984) p. 5.
23. P. Fontaine, *L'Industrie automobile . . .*, op. cit. pp. 78–9. Y. Dupuis, 'Le Devenir de l'industrie française de l'automobile', *Journal Officiel, Avis et rapports du CES* (17 Aug 1984).
24. G. Volpato, 'Innovazione ed evoluzione della struttura de vendita nella commercializzazione dell'automobile', *Commercio*, 18 (1984) pp. 53–5. J. M. Janailhac, *La Vente et la réparation automobile* (Paris: DAFSA, 1979).
25. G. de Bonnafos, J. J. Chanaron, L. de Mautort, *L'Industrie automobile* (Paris: La Découverte, 1983) p. 92. C. Baudclot *et al.*, *Qui travaille pour qui?*, op. cit. p. 145.
26. O. Choquet and H. Valdelièvre, *Acquisition et utilisation de l'automobile* (Paris: INSEE, 1985) pp. 58–60.
27. P. Fontaine, *L'Industrie automobile . . .*, op. cit. pp. 79 and 81. P. Clément, *De la 4CV a la vidéo* (Paris: Communica International) p. 205.
28. J. Baillon, J. P. Ceron, *La Société de l'éphémène* (Grenoble: P.U.G., 1979). G. de Bonnafos *et al.*, *L'Industrie automobile*, op. cit. p. 106. P. Fontaine, *L'Industrie automobile . . .*, op. cit. p.151.
29. P. Fontaine, *'L' Industrie automobile*, op. cit. pp. 69 and 83. G. Wouters, 'La Publicité pour l'automobile: conception et organisation de la publicité. amage sociale de l'automobile de 1937 à 1973', M.A. thesis, University Paris I, 1980, p. 86.
30. P. Fridenson, 'French automobile marketing . . .', art. cit. pp. 131, 139, 143, gives an historical survey of these developments.
31. *Vendre* (June 1930) p. 478.
32. Advertisement data kindly provided by the Sécodip company in Chambourcy.
33. G. Wouters, *La Publicité . . .*, op. cit. pp. 88, 125–6, 163–4. D. Henri, 'La Sociéte anonyme des Automobiles Peugeot de 1918 à 1930', M.A. thesis, University Paris I, 1983, p. 92.
34. C. Zagrodzki, 'L'Automobile et l'image publicitaire' in *L'Automobile et la publicité* (Paris: Musee de la Publicité, 1984) p. 52.
35. P. Gerbod, 'L'Irruption automobile en France (1895–1914)', *L'Information Historique* (Sep–Oct 1983) pp. 189–95. A. Jemain, *Michelin* (Paris: Calmann-Lévy, 1982). D. Burgess-Wise, *Automobile Archaeology*, op. cit. pp. 105–49, 151–2.
36. D. Gaucher, 'La Presse automobile', *Presse Actualité* (Nov 1979) p. 48.

37. I. Bartfeld, *L'Assurance obligatoire dans les Transports publics auto-mobiles de voyageurs et de marchandises* (Paris: Domat-Montchrestien, 1936) pp. 13, 27–30, 75–6, 96–114. G. Neudin, L'Assurance automobile en France aujourd'hui', *Bulletin de liaison et d'information de l'Admi-nistration centrale de l'Économie et des Finances* (Jan–Feb 1973) pp. 48–65. Institut National de la Consommation, *Guide pratique de l'Auto-mobile et des 2 roues* (Paris: Le Seuil, 1974) pp. 79–89, 91–9, 114–15. D. Gaucher, 'La Presse automobile', art. cit. p. 49. Author's interview of Raymond Prothin, secretary-general of the Automobile Club de France, administrator of Mondial Assistance, 20 January 1985.

38. Archives of the Musée de la Publicité, Paris, poster with a drawing by E. Clouet, 1900.

39. P. Fontaine, *L'Industrie automobile* . . ., op. cit. pp. 82–3. P. Clément, *De la 4CV* . . ., op. cit. p. 213. Institut National de la Consummation, *Guide pratique* . . ., op. cit. pp. 19–20. For a detailed breakdown see Choquet and Valdelièvre, op. cit. pp. 68ff.

40. C. Harmelle, 'Les Piqués de l'aigle . . .', art. cit. pp. 300, 302–3.

41. P. Fontaine, *L'Industrie automobile* . . ., op. cit. pp. 69–70. G. de Bonnafos *et al.*, *L'Industrie automobile*, op. cit. p. 101.

42. P. Clément, *De la 4CV* . . ., op. cit. p. 204. For earlier data, C. Baudelot *et al.*, *Qui travaille* . . ., op. cit. pp. 102–4.

43. C. Baudelot *et al.*, *Qui travaille* . . ., op. cit. p. 146.

44. A. Boudou, 'Les Taxis parisiens de la fondation des Usines Renault aux Taxis de la Marne', M.A. thesis, University Paris X – Nanterre, 1982, pp. 53–4, 63–6, 243–4, 259, 274. It supersedes earlier research.

45. A. Boudou, *Les Taxis* . . ., op. cit. pp. 282, 292–302, 325, 330. J. L. Loubet, *La Société anonyme André Citroën* . . ., op. cit. pp. 134–5, 285.

46. P. Laneyrie, *Le Taxi dans la ville* (Paris: Éditions du Champ Urbain, 1979) pp. 31–42, 60–6.

47. P. Gerbod, 'L'irruption . . .', art. cit. p. 190. G. Bouchet, 'La Traction hippomobile dans les Transports publics parisiens (1855–1914)', *Revue historique* (Jan–Mar 1984) p. 134.

48. N. Spinga, 'L'Introduction de l'Automobile dans la société française entre 1900 et 1914, Etude de presse, M.A. thesis, University Paris X – Nanterre, 1973, pp. 153–6.

49. J. Jones, *The Politics of Transport in Twentieth-Century France* (King-ston–Montreal: McGill–Queen's University Press, 1984) p. 22.

50. N. Spinga, 'Comment la France a adopté l'automobile', *De Renault Frères* . . . (June 1974) pp. 38–43.

51. J. Jones, *The Politics* . . ., op. cit. pp. 22–5 and 224. C. Harmelle, 'Les Piqués de l'Aigle . . .', op. cit. pp. 293, 297–9.

52. J. L. Loubet, *La Société anonyme André Citroën* . . ., op. cit. pp. 128–34. C. Harmelle, 'Les Piqués de l'Aigle . . .', op. cit. pp. 304–8. J. Jones, *The Politics* . . ., op. cit. pp. 25–106.

53. J. Jones, *The Politics* . . ., op. cit. pp. 179–193 and 202–3.

54. Chambre Syndicale des Constructeurs d'Automobiles, *L'Automobile en France* (Paris: Riss, 1956) pp. 24–7, 56–7. A. Jardin, P. Fleury, *La Révolution de l'Autoroute* (Paris: Fayard, 1973). P. Fontaine, *L'Indus-trie automobile* . . ., op. cit. pp. 109–10.

55. J. Jones, *The Politics* . . ., op. cit. pp. 198 and 278.
56. J. P. Bardou, J. J. Chanaron, P. Fridenson, J. Laux, *The Automobile Revolution* (Chapel Hill: University of North Carolina Press, 1982) pp. 281–82.
57. Union Routière de France, *La Circulation* . . ., op. cit. pp. 16–17.
58. G. de Bonnafos *et al.*, *L'Industrie automobile*, op. cit. pp. 102–3.
59. V. Breyer, *L'Épopée automobile* (Paris: Société Française des Automobiles Dunlop, 1943) pp. 17–19. N. Spinga, *L'Introduction* . . ., op. cit. pp. 108–30. P. Gerbod, 'L'Irruption . . .', art. cit. pp. 192–3.
60. L. Massenat-Deroche, *L'Automobile aux États-Unis et en Angleterre* (Paris: 1910). V. Breyer, *L' Épopée* . . ., op. cit. pp. 14–15. D. Burgess-Wise, *Automobile Archaeology*, op. cit. pp. 153–4.
61. *Le Gaulois*, illustrated supplement (5 Oct 1921) p. 1. A. Pavie, 'Le Code de la Route – voitures, bêtes et gens', *L'Illustration* (6 Oct 1923) pp. 346–8. N. Spinga, *L'Introduction* . . ., op. cit. pp. 142–6.
62. J. C. Chesnais, *Les Morts violentes en France depuis 1826. Comparaisons internationales* (Paris: P.U.F., 1976) pp. 146 and 151. M. Frybourg, R. Prud'homme, *L'Avenir d'une centenaire: l'automobile* (Lyon; P.U.L., 1984) pp. 56–7.
63. N. Spinga, *L'Introduction* . . ., op. cit. pp. 140–1. C. Harmelle, 'Les Piqués de l'Aigle . . .', art. cit. pp. 290 and 296–9. *L'Automobile et la Publicité*, op. cit. p. 89.
64. A. Boudou, *Les Taxis* . . ., op. cit. pp. 34–9.
65. J. Chardonnet, 'Difficultés actuelles et perspectives d'Expansion de l'Industrie automobile', *Journal Officiel, Avis et rapports du Conseil Économique et Social* (29 Dec 1961). J. P. Jagu-Roche, 'Commerce: le Rouge est mis', *L'Argus*, special issue (Sep–Oct 1984) p. 32. M. B. Fuller, *Note on auto sector policies* (Cambridge, Mass.: Harvard Business School Case Services, 1982) p. 4. J. D. 'Le contrôle à la vente des voitures d'occasion sera obligatoire dès septembre', *Le Monde* (22 Aug 1985) p. 18.
66. C. Rickard, *Les Autoroutes* (Paris: P.U.F., 1984).
67. V. Breyer, *L'Épopée* . . ., op. cit. p. 18.
68. J. L. Flavigny, '1945/1980: Governmental Policies and the French Automobile Industry', M.Sc. thesis, Massachusetts Institute of Technology, 1981, pp. 173–207. M. B. Fuller, *Note* . . ., op. cit., p. 6. Denis Briquet, 'Normes antipollution: un enjeu économique', *Avec* (26 July 1985) p. 3. M. Frybourg, R. Prud'homme, *L'Avenir* . . ., op. cit. pp. 35–58. P. Yonnet, 'La Société automobile', *Le Débat* (Sep 1984) pp. 128–48.
69. Union Routière de France, *La Circulation* . . ., op. cit. pp. 18–20, 28, 30.
70. *Perspectives à long terme de l'Industrie automobile mondiale* (Paris: 1983) p. 11.
71. Ibid. p. 18.

8 Why Did the Pioneer Fall Behind? Motorisation in Germany Between the Wars

Fritz Blaich

As Dr Nübel has pointed out in Chapter 2, Germany's able engineers adapted the gas engine for transport purposes; but Germany failed to drive home its initial advantage. In the present chapter we shall examine the evidence of the interwar period[1] to find reasons for this backwardness and, in doing so, we shall compare the level of motorisation in Germany with that in the three other leading industrial countries of Europe: Great Britain, France and Italy.[2] It is important to note at the outset that the United States, where the social depth of demand was so much deeper than anywhere in Europe, had already forged far ahead. That vast continental country was already in a completely different league. In 1928, for instance, when 20.24 million cars and 2.9 million trucks were registered in the United States, there were still only 351 000 cars and 108 000 trucks in Germany.[3] Ten years later, in 1938, although Germany was catching up a little, the United States' lead was still enormous: 25.5 million cars and 4.24 million trucks to 1.33 million cars and 383 000 trucks in Germany and Austria combined.[4] To compare like with like, therefore, Germany should be compared with its European rivals, not with the United States.

Table 8.1 shows the extent to which Germany lagged behind France and Britain in motor cars per head and, even more remarkable, its meagre lead over Italy until 1934 at a time when only Italy's northern provinces were industrialised.

The same sort of picture emerges when we look at the number of commercial vehicles in use in the four countries, though, since we have chosen to relate the totals to geographical area rather than to population, this weighs more heavily in Britain's favour.

With motor cycles Britain was also in the clear lead until 1931; but here Germany performed much more impressively during the 1930s, outdistancing France in 1930 and Britain in the following year. While in France and Britain motor-cycle ownership per head failed to grow

Table 8.1 Cars in use per 1000 inhabitants in Germany, France, Great Britain and Italy, 1925–1938[5]

Year	Germany	France	Great Britain	Italy
1925	3	11	–	2
1926	3	12	16	2
1927	4	15	20	3
1928	6	16	23	3
1929	7	23	23	5
1930	8	27	27	5
1931	8	31	24	6
1932	8	30	23	6
1933	8	31	25	6
1934	10	34	29	6
1935	12	34	29	7
1936	15	38	35	7
1937	17	40	38	7
1938	19	41	42	8

Table 8.2 Commercial vehicles in use per 1000 square km in Germany, France, Great Britain and Italy, 1925–1938[6]

Year	Germany	France	Great Britain	Italy
1925	17	21	–	9
1926	19	50	98	10
1927	22	51	120	10
1928	26	56	134	11
1929	31	67	137	17
1930	33	63	180	21
1931	34	80	149	24
1932	32	80	155	26
1933	33	79	167	27
1934	41	83	176	30
1935	52	83	176	32
1936	58	84	196	34
1937	68	87	209	34
1938	81	90	215	35

Table 8.3   Motor cycles in use per 1 000 inhabitants in Germany, France, Great Britain and Italy, 1925–1938[7]

Year	Germany	France	Great Britain	Italy
1925	3	3	11	–
1926	4	3	12	2
1927	5	4	16	2
1928	7	6	17	2
1929	10	10	17	2
1930	12	11	17	2
1931	13	12	12	3
1932	13	12	13	3
1933	14	12	12	3
1934	15	13	12	3
1935	16	13	11	3
1936	18	13	11	4
1937	20	13	11	4
1938	23	13	10	4

after that (in Britain it actually fell), in Germany it continued to move ahead and by 1938 it exceeded that in Britain by more than two to one. Germany then had more motor cycles registered per capita than any other country.

DISCOURAGEMENTS TO MOTOR-CAR OWNERSHIP: LOW PURCHASING POWER, HIGH MOTOR TAXATION AND STATE-SUPPORTED RAILWAYS

The high per capita ownership of motor cycles, cheaper to buy and to run than motor cars, suggests that Germany's low ownership of cars was directly related to the country's lower purchasing power. (By contrast, few motor cycles were registered in the United States where purchasing power was far higher and, as we have seen, people could usually afford cars.) In 1927 only 134 679 motor cycles were registered in the United States when Germany already had 339 226; in 1938 the U.S.'s total had fallen to 110 126 whereas the German had risen to 1.58 million.[8] This supposition that Germany's basic problem was lack of purchasing power is strengthened by the fact that the registration in Germany of the lighter and cheaper motor bicycles of under 200 c.c. increased at a faster rate than that of the larger and more expensive ones. In 1928 only 23.7 per cent were under 200 c.c., but by 1936 this proportion had grown to 60.5 per cent and by 1938 to 72 per cent.[9] Such smaller vehicles did not allow sidecars to be fitted or facilities for carrying goods.[10] The light, low-cost motor cycle had become the German People's Vehicle by the 1930s. In the United Kingdom, where real earnings remained higher than in Germany between the wars,[11] demand for motor cycles had reached a peak in 1925 when 71 600 of them were bought. Thereafter, cars came within the means of more middle-class people and the purchase of motor bicycles fell to a minimum of 26 900 in 1933. Moreover, of the motor cycles on British roads in the 1930s, often owned by better-off wage-earners, a higher proportion were of the more expensive and powerful sort, many of them fitted with sidecars.[12]

Germany was also different from her European rivals in having fewer cars registered for private use. This may be related directly to the effects of post-war hyperinflation which deprived professional people and rentiers, whose number, of course, had never been so great as in France, of their savings.[13] Cars registered in Germany were used mainly for commercial purposes. Even by 1934 80 per cent of new-car registrations were declared to be exclusively or mainly for business use.[14] Salaried people, such as better-off civil servants, local government officers, bank officials or school teachers, who could afford cars in Britain and France in the 1930s, evidently could not do so in Germany.

If lack of private purchasing power explains to some extent the low level of car ownership per head in Germany, so, too, does the higher cost of buying and running cars there. Until 1932 motor manufacturers in Germany, which then included subsidiaries of General Motors and Ford, were unable to design a small car of high performance and produce it on a large enough scale to bring down the price low enough to widen the market. Here France and Britain were far ahead. The Fahrzeugfabrik Eisenach, a firm which in 1929 had been taken over by B.M.W., began building the small Austin Seven under licence in 1926. It had a 747-c.c. four-cylinder engine and was marketed with modest success in Germany as the Dixi 3/15.[15] Adam Opel A. G. tried to copy a Citroën design about the same time without a licence; it was taken to law.[16]

Sales of cars in Germany were further hindered by the level of taxation. Engines were taxed on cylinder cubic capacity and there were also heavy import duties on oils and fuels. An investigation by the German government's Statistical Bureau (The Statistisches

Reichsamt) produced the following figures for taxes paid on an average popular type of motor car with a yearly mileage of around 20 000 km. and a fuel consumption of 15 litres per 100 km. in various countries in 1932: in Germany (2500 c.c.) 917 Reichsmarks (RM); in France (18 h.p.) 740 RM (885 RM in Paris itself); in Great Britain (18 h.p.) 589 RM; and in New York (100 kg. net weight) 134 RM.[17]

It must be borne in mind also that the First World War and the years of hyperinflation delayed the improvement of highways, originally built for slow horse-drawn vehicles; and when this much-needed work had eventually started, the financial economies at the beginning of the 1930s led to considerable cut-backs. Expenditure upon a permanently increasing army of unemployed men and women had higher priority.[18] As a result Germany in 1933 could boast only 472 km. of asphalted, well-built roads per 1000 square km., as against 545 km. for Italy, 602 km. for France and 1034 km. for Great Britain.[19] The poor condition of many of Germany's trunk roads with their numerous curves, winding, single-track stretches, and bottlenecks in villages, shared by pedestrians, oxcarts, bicycles and herds of cattle, persuaded many a commercial traveller to put off buying a car and continue to go by train.

In long-distance travel the advantages of the narrow-meshed network of the state-owned railways, the Deutsche Reichsbahn, by far outweighed motor-transport for safety, speed, punctuality and comfort.[20] Moreover the management of the Reichsbahn soon adjusted its fare policies to curb any motor-car or motor-bus competition. Around 1929 it charged only 11.20 RM for a journey of 300 km. A comparable fare for such a distance in Britain was 24.00 RM, in Italy 12.20 RM and in France 9.90 RM.[21] There were, in addition, special reductions for regular commuters and commercial travellers.[22] The Reichsbahn was running a far more extensive passenger service in 1937 than it had done in 1932 and managed to build up its passenger traffic most successfully.

SOME ENCOURAGEMENT AT LAST

Change came during the 1930s when handicaps to the spread of motor cars were removed. Under pressure of the world-wide economic slump, German manufacturers began to develop highly efficient and internationally marketable small cars such as the 'DKW-Meisterklasse', the 'DKW-Reichsklasse', the 'Adler-Trumpf-Junior',

Table 8.4   Rail travel in Germany, France, Great Britain and the United
States in 1932 and 1937[23]
(*000 million passenger-km.*)

	Germany	France	Great Britain	USA
1932	31.48	25.71	28.70	27.35
1937	51.06	27.08	34.31	39.74

Table 8.5   German passenger travel by various transport modes in 1926
and 1937[24]
(*000 million passenger-km.*)

	Railways	Streetcars and suburbans	Bus lines	Air traffic
1926	44.0	15.1	0.9	0.01
1937	51.1	14.6	3.0	0.12

the 'Ford-Köln' or the 'Opel P 4'.[25] Sharp competition between them
cut selling prices.[26] By a law of 10 April 1933 the newly elected
National-Socialist (NS) government removed all taxes upon new cars
and owners of a car which had been licensed prior to 1 April 1933
could disregard the vehicle tax by agreeing to a single payment.[27]

Road improvements became of immediate concern to National-
Socialist traffic planners. On 27 June 1933 the NS-government
founded the enterprise Reichsautobahnen in a major effort to widen
and expand the road system.[28] By 1938 this enterprise had constructed 3065 km. of high-speed, two-lane highways in various parts
of Germany, the so-called *Autobahnen*, which were used exclusively
by motor traffic.[29] These freeways, however, though intended to
allow the driver to travel longer distances quickly and safely, were as
yet not interconnected. Nevertheless, a single autobahn by 1938
stretched all the way from Stettin via Berlin to Munich, a distance of
665 km.[30] Between 1933 and 1938 3.09 billion RM were spent on these
roads.[31]

In 1935 the National-Socialists were already spending on road

Table 8.6   Expenditure upon railway installations and road construction.
(New and replacement investment including funds for maintenance)
(*million RM*)

Year	1926	1927	1928	1929	1930	1931	1932	1933	1934	1935
Rail-ways	1.075	1.261	1.060	1.033	967	691	593	728	814	728
Roads	350	468	424	432	390	269	191	330	610	884

construction and improvements more money than on railway upkeep
and development. This was in sharp contrast to the policy of their
predecessors, the politicians of the democratic Weimar Republic,
who had always preferred investment in the state-owned railways.[32]
But new roads were not sufficient to enable Germany to make up for
lost time as Tables 8.1 and 8.2 make clear. Increased purchasing
power was also needed.

Table 8.7   Average earnings of employees and workers in Germany
during the years 1929 to 1938
(*Index: 1929 = 100*)[33]

	Total salary of an employee per month		Total wages of a worker per week	
Year	in RM	Index	in RM	Index
1929	207	100	31.19	100
1930	208	101	30.57	96
1931	201	97	27.73	89
1932	182	88	22.88	74
1933	174	84	21.88	70
1934	179	87	22.83	73
1935	189	91	24.04	77
1936	199	96	25.25	81
1937	207	100	26.50	85
1938	218	105	27.82	89

From 1936 the German rearmament programme created an increased demand for skilled labour at a time of wage and salary
control. Some employers got round this by offering inducements such
as free light motor cycles or similar incentives.[34] Yet, even by 1938,

the average wage or salary-earner could still not afford to buy or run a small four-seater car. According to an official investigation in the years 1937/8, the DKW-Reichsklasse, equipped with a 585-c.c. dual-action engine, was classified as the lowest-cost vehicle available. Its basic price without any accessories was 1865 RM. With an estimated annual mileage of 15 000 km. and without any garaging, the cost of fuel, oil, maintenance, tyres and insurance premiums amounted to at least 72 RM monthly. Although the Opel P 4, a four-cylinder model with a 1074-c.c. engine was selling at only 1280 RM, its operating and maintenance costs were considerably higher. Its insurance cost 131.88 RM annually as compared to 94.95 RM per annum for the DKW-Reichsklasse.[35] No wonder that the percentage of small cars of under 1000 c.c. only increased from 21.7 to 24.5 per cent of the total between 1928 to 1938.

Cars of medium size, from 1000 and 2000 c.c., on the other hand, enlarged their market share from 36.1 to 59.3 per cent.[36] More better-off salary-earners could now afford cars. This was confirmed when, in August 1937, the Statistical Bureau made an official inquiry into the domestic market and its customers. It found that during the first half of 1937, out of a total of 128 443 newly registered motor cycles, 45.2 per cent had been bought by wage-earners and 13.4 per cent by the salaried of lower or middle rank. In contrast, the share of wage-earners in the new registrations of cars amounted to only 1.4 per cent, whereas salary-earners – excluding managers – reached 12.1 per cent only.[37]

The year 1938 initiated new and promising opportunities for many more Germans to own their own car. Shortly after his seizure of power Adolf Hitler himself had ordered the NS-Party organisation 'Kraft durch Freude' (KdF) to produce at a low price a reliable motor car the costs of operation and maintenance of which would be within the means of the broad majority of the German people. By 1938 this People's Car, later known as the *Volkswagen*, had been developed. Equipped with a fuel-efficient rear-mounted, air-cooled engine the 'KdF-Wagen' offered unique and far-reaching mechanical improvements. Its costs of maintenance proved to be even lower than those of the 'DKW-Reichsklasse'. Anyone who agreed to put down 5 RM a week for a year acquired the right to buy a KdF-Wagen for 990 RM at a future date. How extensive the demand for a small car of this type was, soon became apparent. By March 1939 a total of 170 000 Volkswagen investors had entrusted savings totalling 110 million RM to the NS-government.[38] The number of orders rose to an estimated

253 000 by the end of June 1939. This was on a par with the output of all German car manufacturers in the year 1938.[39] Demand continued to escalate until, by the year 1942, there were over 300 000 Volkswagen investors, 10 per cent of whom classified as wage-earners, 17 per cent as civil servants and 29 per cent as other salaried employees. 69 per cent of the grand total were buying a KdF-Wagen for private, non-commercial use.[40] None of them, alas, received the car as promised. When the Volkswagen works came into production, it was turned over to war production.[41]

FREIGHT TRANSPORT: THE RAILWAYS GAIN MORE TRAFFIC THAN THE ROAD HAULIERS, 1925–1938

An obstacle to Germany's motorisation of freight as with its passenger transport was the keen competition of the railways. Here many small entrepreneurs opposed the monolithic and powerful Deutsche Reichsbahn-Gesellschaft, owned by the German government. Moreover, the horse-drawn cart was still able to hold its position in short-distance transport for an astonishingly long time. In 1933 there were still more than 100 000 drivers of horses busily engaged in commercial road transport alongside approximately 240 000 professional drivers of motor vehicles.[42]

The German motor-vehicle tax in operation from 1922 until 1927 favoured the operation of heavy lorries. Under 2000 kg. unladen weight a tax of 30 RM was levied on each 200 kg. but over 2000 kg., the tax was reduced to 20 RM for each 200 kg.[43] Vehicles under 2000 kg. were usually light delivery vehicles capable of carrying 1.5 tons at best, and were used on short-distance work. They were often used to take goods to and from railway stations and were therefore often welcomed by the directors of the Reichsbahn. Heavier trucks with a dead weight of over 2000 kg., however, were considered to be their serious competitors; provided road conditions were good they could pick up and discharge almost any type of freight from door to door without using the railway at all.[44]

Pressure from the Reichsbahn, as well as fiscal considerations, obliged the state to eliminate the tax advantages enjoyed by these heavier lorries.[45] Moreover, in an effort to counter the impending threat by road hauliers the Reichsbahn speeded up its long-distance goods trains and offered cuts in freight rates. In 1928 the average rate per ton-kilometre on German railways was 4.28 pfennig as against 11 pfennig charged on average by the four British railway companies.[46]

The world-wide economic crisis threatened the profitable and economically sound operation of the Reichsbahn which, during 1931, suffered a deficit of 441.8 million RM.[47] The German government reacted by imposing new regulations on road haulage. Beginning in June 1931 it was subjected to the licensing of journeys over 50 km. At the same time a minimum rate was imposed on all transport of goods by road. This new *Reichskraftwagentarif* was related to the three highest-rate classes charged by the railways; and wherever goods, such as perishable freight, had to be carried in covered, enclosed heavy trucks an even higher minimum rate of 5 per cent had to be charged.[48] At the same time a tax was placed upon trailers and increased tax rates applied on lorries without platform facilities which were used as tractors.[49]

Even the pro-road NS-government hesitated to grant tax benefits to commercial vehicles. Between 1929 and 1932 the Reichsbahn's volume of passenger traffic had declined by 37 per cent, and its freight traffic had fallen by 44 per cent.[50] Not until 1935 were the heavy, long-distance lorries favoured again fiscally; the owner of a truck with an unladen weight up to 2400 kg., which had been issued a permit after 31 March 1935, was held liable for a yearly tax of 30 RM per 200 kg.; but unladen weight over 2400 kg. was taxed at only 10 RM per 200 kg. Motor vehicles already in operation on this target date, however, would continue to be liable to a rate of 30 RM per 200 kg.[51] In 1936 the NS-government reduced these tax rates by one-third on what the regime classified as a cross-country vehicle.[52] The object was to encourage the purchase of heavy lorries required to accomplish large-scale public building operations such as the construction of the Reichsautobahn or, in the longer run, to serve such military purposes as the building of Germany's western fortifications, the *Westwall*.

All these fiscal measures, which considerably increased commercial vehicle ownership, as was seen in Table 8.2, failed to increase the supply of heavy trucking equipment very much.[53] Of an estimated 241 000 licensed trucks of all sizes registered in July 1935, only 5000 or 6000 were regularly engaged in long-distance haulage.[54] On the other hand, almost 46 per cent of a total of 3040 trucks over two tons, which had been issued permits during the first half of the year 1937, had been bought by building contractors and those engaged in the supplying of building materials. Most of these medium and heavy trucks were used in short-distance work, hauling building materials to construction sites which could not be reached by rail.[55]

Table 8.8   German freight transport by rail, water and road in 1925 and
1938
(*million ton-km.*)

Year	Transport by rail	Inland water transport	Motor transport
1925	60.2	13.3	1.6
1938	89.0	18.7	6.7

In the year 1936 45.3 per cent of all trucks registered in Germany
had a carrying capacity of one ton or less, and, two years later, the
percentage was about the same.[56] The high share of light delivery
vehicles helped, rather than hindered, the railway's freight services.
The railway increased its traffic (in ton-kilometres) very much more
than did motor transport between 1925 and 1938.[57]

Table 8.9 shows that the Reichsbahn managed to increase the
distance covered by its goods trains between 1932 and 1937 very
much more than its European competitors and proportionally much
more than the U.S. railways.

Table 8.9   Distances covered by goods trains in Germany, France, Great
Britain and the USA, 1932 and 1937[58]
(*000 million km.*)

	Germany	France	Great Britain	USA
1932	39.31	36.89	22.90	343.55
1937	73.00	37.15	29.33	529.70

The growing dissatisfaction of the NS-leadership with the slow
increase in commercial motor vehicles caused it to turn to imports.[59]
In 1934 Germany imported 501 lorries, in 1935 104, in 1936 63, and in
1937 only 5. In 1938 2614 of them were purchased on world markets,
2112 from the United States.[60]

WAR PREPARATIONS DIVERT RESOURCES

From 1936 no statistics were published of vehicles used by the Wehrmacht.[61] The hasty motorisation of the German armed forces, which implied preparation for a blitzkrieg, diminished the availability of such vehicles, particularly lorries, cars and large-engined motor cycles used for civil purposes. The German General Staff, at the latest by 1934, had decided that motorised and mechanised armed forces would play a key role in its plans.[62]

Since September 1934, when Hjalmar Schacht, Minister of Economics, had proclaimed his New Plan which was to replace free foreign trade by bilateral agreements, sales of foreign-made products on the German domestic market were quickly reduced. An exception was made in favour of Italian motor manufacturers, whose fascist government had concluded a bilateral trade agreement with the Third Reich. Of the 7935 passenger cars which in 1938 were imported into Germany, 7098 were Italian.[63] German exports, on the other hand, were encouraged because of the urgent need to obtain foreign exchange; on world markets only hard currency was able to purchase those raw materials which the regime sorely required for rearmament.[64] In the year 1938 Germany's automobile industry produced 276 663 cars. Of these, 65 069, 23.5 per cent, were exported.[65]

The frequently reappearing bottlenecks in the supply of steel, rubber and various components forced the German motor industry to cut deliveries to private customers. From 1936 all German motor manufacturers had to face the fact that demand for their vehicles grew out of all proportion to productive capacity, which was then limited by a rigid rationing system. The authorities responsible for the rationing of raw and auxiliary materials often failed to allot the necessary goods to the motor works at the right time and in the right quantity.[66]

In the year 1936 the armaments industry's high demands for rubber in particular prevented motor manufacturers from meeting delivery dates. Thus, by July 1936, dealers had to sell 1487 DKW motor cycles delivered to them without rubber tyres. The Zschopauer Motorenwerke, on the other hand, cut back their output to meet the reduced tyre allocation. From 1 May 1937, however, their difficulties were further increased by a complete rationing of iron and steel. They then delivered their model DKW-Reichsklasse with delays of three to four months due to 'material shortages'. In 1938 they estimated their backlog of cars ordered but not delivered at 9050, or approximately a quarter or more of those sold in earlier years. As for the manufacture of motor cycles, there was a shortfall of 1609 units between output target and output rate which was attributable to shortages in small auxiliary steel parts. Delays in delivery rose to six or seven months.[67]

After the breakdown of international trade in 1931 France and Great Britain, too, had considerably curtailed the influx of foreign-made goods in an effort to protect their home industries. But up to the European political crisis of 1938 their governments pursued a peaceful attitude towards the threatening rearmament policy of the Third Reich. Their motor industries did not, therefore, suffer any of the shortages suffered by the German.

CONCLUSION

Germany's poor vehicle ownership per head attributable before 1914 mainly to the country's low purchasing power continued into the interwar period, sustained by the effects upon professional people of hyperinflation in the earlier 1920s and upon the country as a whole by the World Depression of the beginning of the following decade. Most Germans had to be content with motor bicycles – *and* the lower-priced and cheaper sort at that. The state had discouraged motorisa-tion in the 1920s by taxing motor vehicles heavily, spending little on road improvement and, on the other hand, by investing heavily in the railway system, which was able to keep passenger fares low. As far as goods traffic was concerned, the method of taxing road haulage was changed so that the larger, longer-distance vehicles, which competed against the railways, were discriminated against. The NS-government reversed these policies after it came to power. Motor roads were built, the taxation of heavy trucks was (eventually) reduced and orders were issued for the development of a People's Car. After 1936, however, rearmament changed these policies as resources were transferred from civilian production and the People's Car was post-poned. West Germany had to await the 'economic miracle' after the war for the Volkswagen, the further spread of motor vehicles and the overtaking of its other European competitors behind whom it had lagged for so long.

Notes

1. Figures are based on the best available sources. In some cases, however, estimates have been made, and figures are often uncertain. See I. Svennilson, *Growth and Stagnation in the European Economy* (Geneva, 1954) p. 333, table A.43, note.
2. See Svennilson, op. cit. p. 148.
3. Reichsverband der Automobilindustrie E. V. (ed.), *Tatsachen und Zahlen aus der Kraftverkehrswirtschaft*(1931). Auf Grund amtlicher und privater Unterlagen sowie eigener Erhebungen zusammengestellt (Berlin–Friedenau, 1931) p. 112. (Henceforth abbreviated: *Tatsachen und Zahlen.*)
4. *Tatsachen und Zahlen* (1938) p. 124.
5. Registrations refer, as far as possible, to end of calendar year. Only in Germany the deadline of investigation is always on 1 July. All figures are taken from the following sources: *Statistisches Jahrbuch für das Deutsche Reich*, vol. 46 (Berlin, 1927) p. 77*; vol. 47 (1928) p. 75*; vol. 48 (1929) p. 71*; vol. 49 (1930) p. 72*; vol. 50 (1931) p. 78*; vol. 51 (1932) p. 78*; vol. 52 (1933) p. 84*; vol. 53 (1934) p. 100*; vol. 54 (1935) p. 96*; vol. 55 (1936) p. 103*; vol. 56 (1937) p. 106*; vol. 57 (1938) p. 114*; *Tatsachen und Zahlen* (1932) p. 16; (1933) p. 122. Figures concerning Germany refer to the 'Deutsches Reich' within the frontiers of 1937. Fractions as a result of computation have been brought to round figures in order to present complete units of vehicles.
6. Figures are derived from sources as indicated in note 5 and from *Tatsachen und Zahlen* (1936) p. 162.
7. Figures are derived from sources as indicated in note 5 and from *Tatsachen und Zahlen* (1936) p. 166; (1938) p. 127.
8. *Tatsachen und Zahlen* (1931) p. 116; (1938) p. 127.
9. *Tatsachen und Zahlen* (1929) p. 22; (1936) p. 30; (1938) p. 16. Figures given for the year 1938 include motor cycles under 250 c.c.
10. See R. Haller, *Geschäftskraftwagen und ihre wirtschaftliche Verwendung für den Kundendienst* (Berlin, 1929) pp. 12 ff.; R. Burmester, 'Das Kraftfahrzeug im Dienste der Deutschen Reichspost, seine verkehrspolitische und volkswirtschaftliche Bedeutung', Diss. Hamburg, 1940, p. 32.
11. E. H. Phelps Brown; M. H. Browne, *A Century of Pay. The Course of Pay and Production in France, Germany, Sweden, the United Kingdom, and the United States of America, 1860–1960* (London, 1968) pp. 199 ff. See also Svennilson, p. 38.
12. R. Stone; D. A. Rowe, *The Measurement of Consumers' Expenditure and Behaviour in the United Kingdom 1920–1938*, vol. II (Cambridge, 1966) p. 52; D. G. Rhys, *The Motor Industry: An Economic Survey* (London, 1972) pp. 16 ff.
13. For a detailed discussion of this hyperinflation see C. L. Holtfrerich, *Die deutsche Inflation 1914–1923. Ursachen und Folgen in internationaler Perspektive* (Berlin, 1980) pp. 264 ff.
14. According to the list of customers kept by the 'Auto Union' trust, which

can be considered as representative for the German car market. Printed in H. Weinberger, 'Der Inlandsmarkt der deutschen Automobilindustrie', Diss. München, 1936, p. 16.

15. H. Schrader, *BMW Automobile. Vom ersten Dixi bis zum Modell von morgen* (Gerlingen, 1978) p. 35.
16. At last the Citroën firm lost its case in the 'Kammergericht Berlin'. A. Becker, 'Absastzprobleme der deutschen PKW-Industrie 1925–1932', Diss. Regensburg, 1979, p. 192.
17. Statistisches Reichsamt (ed.), *Wirtschaft und Statistik*, vol. 13 (Berlin, 1933) p. 455.
18. R. Adamek; F. Saake (ed.), *Die Straßenkosten und ihre Finanzierung*, main volume (Bielefeld, 1952) pp. 7 ff., 33.
19. Statistisches Reichsamt (ed.), *Das deutsche Straßen- und Wegewesen im öffentlichen Haushalt mit einem Überblick über das Straßen- und Wegewesen im Ausland* (Berlin, 1934) p. 47.
20. See K.-H. Dittebrand, 'Eisenbahn und Kraftwagen in Deutschland', Diss. Göttingen, 1933, pp. 52 ff.
21. O. Krust, 'Eisenbahn und Kraftwagen', Diss. Heidelberg, 1929, p. 74.
22. H. Kröker, Die Entwicklung des Kraftwagenverkehrs hinsichtlich der Gestaltung der Verhältnisse bei der Deutschen Reichsbahn-Gesellschaft, Diss. HH Königsberg, 1933, pp. 114 ff.
23. Svennilson, p. 144.
24. G. Hoffmann, F. Grumbach, H. Hesse, *Das Wachstum der deutschen Wirtschaft seit der Mitte des 19. Jahrhunderts* (Berlin, 1965) p. 400.
25. For a detailed account of the constructional features and of the market data of these models see H. C. Graf von Seherr-Thoss, *Die deutsche Automobilindustrie. Eine Dokumentation von 1886 bis 1979*, 2nd ed. (Stuttgart, 1979) pp. 290, 292, 296,304.
26. W. Dohrn, *Der deutsche Personenkraftwagenmarkt nach der Wirtschaftskrisis*, in *Weltwirtschaftliches Archiv* 44 (1936) pp. 624 ff.; Institut für Konjunkturforschung (ed.), *Haltungskosten von Personenkraftfahrzeugen* (Jena, 1938) p. 93.
27. Reichsgesetzblatt (RGBl.), I (1933) pp. 195 ff.
28. See *Akten der Reichskanzlei. Die Regierung Hitler*, 1. Teil: 1933/4, vol. 1 (Boppard, 1983) pp. 510 ff. See also C. Heinrici, *Die Straßen im deutschen Verkehrssystem* (Berlin, 1938).
29. Statistisches Reichsamt (ed.), *Die Kraftverkehrswirtschaft im Jahre 1938* (Berlin, 1939) (Sonderbeilage zu 'Wirtschaft und Statistik', 19, Nr. 3) p. 32; see also K. Lärmer, *Autobahnbau in Deutschland 1933 bis 1945. Zu den Hintergründen* (Berlin, 1975).
30. Information taken from *Die Straße*, 6. Jg. (1939) pp. 276 ff., and Statistisches Reichsamt (ed.), *Die Kraftverkehrswirtschaft*, p. 32.
31. Adamek, Saake, p. 103.
32. *Wirtschaft und Statistik*, 17, Nr. 9 (1937) p. 332.
33. Länderrat des Amerikanischen Besatzungsgebiets (ed.), *Statistisches Handbuch von Deutschland 1928–33* (Munich, 1949) p. 473.
34. For a detailed account of these employee benefits see R. Hanf, 'Möglichkeiten und Grenzen betrieblicher Lohn-und Gehaltspolitik 1933–1939', Diss. Regensburg, 1975, pp. 134, 184 ff.
35. *Haltungskosten von Personenkraftfahrzeugen*, p. 40.

36. *Tatsachen und Zahlen* (1929) p. 22; (1938) p. 16.
37. *Zur Frage der Motorisierungsmöglichkeiten des Verkehrs. Ein neuer Weg zur Beobachtung und Analyse des Automobilmarktes. Bearbeitet im Auftrage des Reichsverkehrsministers*, 1. Beilage zum Wochenbericht des Instituts für Konjunkturforschung, 10. Jg., Nr. 41 (1937) p. 4.
38. H. Handke, *Zur Rolle der Volkswagenpläne bei der faschistischen Kriegsvorbereitung*, in *Jahrbuch für Wirtschaftsgeschichte 1962*, 1. Teil, p. 54. For an account of the KdF-Wagen savings plan see B. Wiersch, 'Die Vorbereitung des Volkswagen', Diss. Hannover, 1974, pp. 131 ff.
39. See Handke, p. 54.
40. Wiersch, pp. 164 ff.
41. Handke, pp. 55 ff.
42. Institut für Konjunkturforschung (ed.), *Stand und Aussichten des gewerblichen Güterfernverkehrs mit Lastkraftwagen* (Jena, 1937) p. 5.
43. RGBl. I (1922) pp. 396 ff.; RGBl. I (1926) pp. 223 ff.
44. L. v. Héder, *Die Konkurrenzfähigkeit der Kraftwagen und kraftwagenlinien gegen Eisenbahn und Kleinbahn unter besonderer Berücksichtigung der Verhältnisse in Deutschland* (Berlin, 1931) pp. 15 ff., 96, 99, 105.
45. RGBl. I (1927) pp. 509 ff.; see also A. Schmitt, *Verkehrspolitik*, in A. Weber, *Volkswirtschaftslehre*, vol. 4 (Munich, 1933) pp. 254 ff.; A. Lampe, *Zur Problematik der Kraftfahrzeugbesteuerung*, in *Finanz-Archiv*, 47 (1930) pp. 1 ff.
46. Héder, p. 151.
47. P. Wohl; A. Albitreccia, *Eisenbahn und Kraftwagen in vierzig Ländern der Welt. Tatsachensammlung und Bericht* (Berlin, 1935) p. 86.
48. Wohl, Albitreccia, p. 93.
49. RGBl. I (1935) pp. 313 ff.
50. Institut für Konjunkturforschung, p. 36. See also *Akten der Reichskanzlei. Die Regierung Hitler*, 1. Teil, p. 511.
51. RGBl. I (1935) pp. 313 ff.
52. Seherr-Thoss, p. 271.
53. See Lärmer, pp. 110 ff. With the beginning of 1908 military authorities of Imperial Germany paid subsidies to the owners of heavy trucks on condition that these vehicles would be placed at the army's disposal in case of war. G. Horras, 'Die Entwicklung des deutschen Automobilmarktes bis 1914', Diss. Regensburg, 1982, pp. 318 ff.
54. Institut für Konjunkturforschung, p. 3.
55. *Zur Frage der Motorisierungsmöglichkeiten des Verkehrs*, p. 6.
56. *Tatsachen und Zahlen* (1938) p. 16.
57. Hoffmann; Grumbach; Hesse, p. 417.
58. Svennilson, p. 144.
59. See W. Birkenfeld (ed.), *Georg Thomas. Geschichte der deutschen Wehr- und Rüstungswirtschaft (1918–1943/45)* (Boppard, 1966) p. 147, and Lärmer, pp. 105 ff.
60. *Tatsachen und Zahlen* (1938) p. 68.
61. R. Stisser, *Standort und Planung der deutschen Kraftfahrzeugindustrie* (Bremen–Horn, 1950) p. 59.
62. See W. Fischer, *Deutsche Wirtschaftspolitik 1918–1945* (Opladen, 1968) p. 67.
63. *Tatsachen und Zahlen* (1938) p. 68.

64. See J. A. Stölzle, 'Staat und Automobilindustrie in Deutschland', Diss. Freiburg/Br., 1959, pp. 53 ff.
65. *Tatsachen und Zahlen* (1938) p. 72.
66. *Zur Frage der Motorisierungsmöglichkeiten des Verkehrs*, p. 1; Stisser, p. 59; Lärmer, p. 112.
67. Information on Zschopauer Motorenwerke is taken from P. Kirchberg, 'Entwicklungstendenzen der deutschen Kraftfahrzeugindustrie 1929–1939 gezeigt am Beispiel der Auto Union AG, Chemnitz', Diss. Dresden, 1964, pp. 147 ff. On the scarcity of tyres see H. Henning, *Kraftfahrzeugindustrie und Autobahnbau in der Wirtschaftspolitik des Nationalsozialismus 1933 bis 1936*, in *Vierteljahrschrift für Sozial- und Wirtschaftsgeschichte*, 65 (1978) p. 230. The roots of the iron and steel problem lay in the low capacity of that industry and the poorly co-ordinated allocation system. E. L. Homze, *Arming the Luftwaffe. The Reich Air Ministry and the German Aircraft Industry 1919–39* (Lincoln, Nebraska, and London, 1976) pp. 144 ff.

9 Motorisation on the New Frontier: The Case of Saskatchewan, Canada, 1906–34*

G. T. Bloomfield

The 1933 report of a presidential commission examining contemporary changes in U.S. life remarked of the automobile: 'It is probable that no innovation of such far-reaching importance had ever before been disseminated with such rapidity. Its influences ramified throughout the whole of the culture, and the very modes of thought and language have undergone transformation in consequence.'[1]

These consequences of motorisation were most apparent in the countries on the new frontiers of settlement – Canada, Argentina, Australia and New Zealand where agriculture and resource development had stimulated massive investment and immigration. Although lagging behind the United States in the statistical measures of automobile adoption, these hinterland areas were considerably ahead of Europe by the early 1920s and retained their lead until the onset of the Great Depression.[2] These four countries were the principal markets for world motor-vehicle exports; in 1929 almost one-third of all exports were shipped to these areas. Given the similarity of road conditions and social structures, it is not surprising that American vehicles and methods were dominant. Of the 90 000 vehicles Australia imported in 1929, 68 per cent came from the United States;

*An earlier version of this paper appeared as: '"I can see a car in that crop": Motorization in Saskatchewan 1906–1934', *Saskatchewan History*, XXXVII (1) (1984) 3–24.

Part of this work was a contribution to the research effort for the *Historical Atlas of Canada*, volume III, edited by Professor Don Kerr, University of Toronto. I am grateful to the Archives of Saskatchewan and the Legislative Library, Ontario Legislative Assembly, for access to material. Doug Wood and Steve Bellinger assisted with some of the preliminary statistical work.

This paper is one of a sequence of studies examining the earlier phase of motorisation in various parts of Canada.

Canadian manufacturers (all U.S. affiliates) supplied an additional 20 per cent and British makers accounted for only 12 per cent of the market.[3] Argentina was even more dominated by the United States, which supplied 97 per cent of the 1929 imports of 67 000 vehicles.[4]

The countries of the new frontier had broadly similar characteristics of high incomes, very extensive areas of low-density rural settlement and benefited from the export boom of the First World War. Each country motorised very rapidly after 1914. Canadian motor registrations, for example, rose from 74 000 in 1914 to 409 000 in 1920 and 1 232 000 in 1930. All the countries, given their economic focus on primary export commodities and a dependence on outside investment, were devastated by the Depression. As in the United States, the Model T Ford vehicle played a major role in this early phase of motorisation since this 'universal car' was low-priced and could traverse the generally poor roads.[5]

POPULATION, PURCHASING POWER AND MOTOR-VEHICLE NUMBERS IN TOWN AND COUNTRY

Saskatchewan, the central province of the Canadian prairies, typifies the features of frontier motorisation. The new province (created in 1905 from the eastern half of the North-West Territories) covered almost 252 000 square miles, an area larger than France and the Low Countries combined. By every measure of growth, Saskatchewan experienced a massive boom in the first three decades of the century.[6] The total population grew tenfold from 91 000 in 1901 to 922 000 in 1931. Millions of acres of virgin prairie were turned into grain farms all linked by railways, grain elevators and an urban network to the world market. Substantial new cities arose on the grasslands. Regina, the biggest urban place in 1901 with a population of only 2249 had over 53 000 inhabitants thirty years later. Saskatoon, incorporated as a village in 1901 with 113 people, reached a total of 43 000 at the Census of 1931.[7] During this period of extremely rapid growth, Saskatchewan was one of the leading Canadian provinces in the early large-scale adoption of motor vehicles and developed high levels of ownership, especially in rural areas.[8] This paper outlines and explores some of the features of the first phase of motorisation in Saskatchewan when all the elements of ownership and the service infrastructure were developed.[9]

Official recognition of the automobile in Saskatchewan began in 1906 with the first registration of motor vehicles under the new provincial legislation.[10] As in many other parts of Canada the responsibility for maintaining a register of motor-vehicle owners, driver and chauffeurs was given to the Provincial Secretary's Department. This new function was added to the existing duties of issuing licences for marriages, auctioneers, pedlars, company incorporations, theatres, moving-picture exhibitors and dance halls. The issue of annual motor licences grew quickly and within a decade was the largest single function of the Department. By 1919 receipts from motor registrations provided 60 per cent of the revenue of the organisation. As the numbers of motor vehicles expanded other government agencies took on responsibility for administering parts of the growing body of legislative control and rapidly increasing expenditure on this new mode of personal transportation.[11]

Numbers of motor vehicles grew slowly in the first four years of licensing (Table 9.1). Vehicles were not wholly reliable, especially in places distant from service garages, initial purchase costs were high and most residents of the province had higher priorities at that stage of settlement. From 1910 annual registrations began to increase more rapidly. *Gas Power Age* regarded 1911 as the critical year of 'take-off' for automobile sales in Western Canada: 'The farming community are now realizing what a great advantage and pleasure the motor vehicle affords and every new convert is another wheel to the expansion of the automobile industry.'[12] Between 1913 and 1918 the number of vehicles grew tenfold. While a time of international crisis, the period during the First World War was one of extraordinarily rapid transformation in motor-vehicle ownership and usage. Another phase of rapid growth began in 1923; total numbers were doubled by 1929 and trucks increased at a very substantial rate. Trucks as a proportion of total registrations increased from 3.2 per cent in 1923 to 18.6 per cent in 1934. Motor cycles throughout the period were of very limited significance. The effect of the Depression was shown in the rapid decline of numbers of vehicles registered. The 1929 total was not surpassed until 1941, and for passenger cars the 1929 figures was not exceeded until 1948.

How and why did vehicle numbers rise so quickly? Not everyone subscribed to the idea that: 'the heir to the throne of success is the man who owns and uses a good automobile'.[13] But large numbers of households already owned some form of personal transport or aspired to such status. Extensive promotion by merchants, garages, implement dealers, magazines, trade exhibitions and automobile clubs had prepared the way for ready acceptance of vehicles. Schools

168

Table 9.1   Saskatchewan: number of motor vehicles registered, 1906–1934

Year[a]	Total	Passenger cars	Trucks	Livery[b]	Motor cycles[c]
1906	22	...	...	...	...
1907	55	...	...	...	...
1908	74	...	...	...	...
1909	147	...	...	...	...
1910	531	...	...	...	...
1911	1 304	...	...	...	...
1912	2 268	...	...	...	...
1913	4 659	...	...	...	...
1914	8 027	...	...	...	...
1915	10 225	...	...	...	...
1916	15 680	...	...	...	...
1917	33 505	...	...	...	...
1918	55 010	...	...	...	...
1919	56 402	...	...	...	...
1920	60 325	...	...	...	424
1921	60 845	...	...	...	339
1922	60 352	...	...	...	296
1923	64 242	60 931	2 086	1 225	207
1924	69 708	64 666	3 780	1 262	187
1925	77 812	71 205	5 560	1 047	184
1926	95 806	86 105	8 688	1 013	161
1927	104 909	92 640	11 346	923	179
1928	119 682	102 812	16 002[d]	868	174
1929	128 099	108 630	18 671	798	218
1930	126 918	108 161	18 106	651	275
1931	107 565	91 276	15 829	570	306
1932	91 079	75 277	15 318	484	272
1933	84 634	69 540	14 884	210	347
1934	91 434	74 050	17 050	331	358

Notes:
[a] Calendar years from 1919. Fiscal years, ending 31 March, previously.
[b] Includes mostly taxis but also cars for hire.
[c] Not included in the total
[d] 'Public vehicles' (buses) and public freight (licensed carriers) included from 1928.

Sources: Compiled from *Annual Reports* of Provincial Secretary's and High-ways Departments; cross-checked with later Dominion Bureau of Statistics retrospective totals of 1937.

Figure 9.1   Saskatchewan: growth of motor vehicle registrations

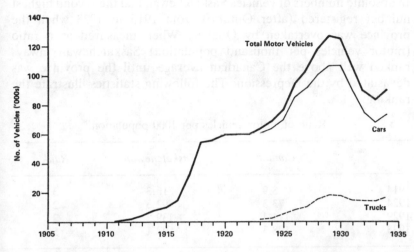

Source: Table 9.1.

for gasoline engine owners and mechanics run by the large gas tractor companies and even local groups such as the Moose Jaw Automobile School provided instruction in driving and maintenance and began to develop an increasing body of trained personnel.[14] As the automobile market expanded and mass production and marketing of standardised vehicles got underway, prices began to fall. A Ford Model T touring car, for example, which retailed at $1025 in Moose Jaw in 1911 was reduced to $800 in 1913, cost only $590 in 1915 and sold for $545 two years later.[15] If public interest had been aroused by promotion and by reliable and cheaper vehicles, the remaining ingredient for large-scale purchase was income. While the precise relationships between farm production, revenue and profitability are still being explored, people at the time tended to perceive their potential income in terms of good harvests.[16] The wheat prospects in the Spring of 1915, which led Mr McNaughton, a Milden minister, to forecast: 'I can see a car in that crop', were reflected in a substantial increase in automobile registrations.[17] The complex linkages between wheat prices, farm incomes and the cycles of automobile buying began during this period. Once the process had begun, the demonstration effect of motor-vehicle ownership was a strong influence in the next cycle of boom.

On a comparative basis Saskatchewan's level of motor-vehicle ownership always had a high ranking among the Canadian provinces. In absolute numbers of vehicles Saskatchewan had the second highest number registered (after Ontario) from 1915 to 1923 when the province was overtaken by Quebec. When measured as a ratio (motor vehicles per thousand population) Saskatchewan always ranked well above the Canadian average until the province was devastated by the Depression. The following statistics illustrate the ranking:

Ratio of motor vehicles per 1000 population[18]

	Canada	Saskatchewan	Rank
1914	8.9	16.3	3
1924	73.3	92.3	2
1929	114.4	139.3	2
1933	108.3	92.1	5

For the earliest period of introduction of the automobile it is possible to show the location of vehicles and owners.[19] The geographical patterns shown in Figure 9.2 illustrate the rapid growth and diffusion of the automobile between 1907 and 1911. The 75 motor vehicles registered in 1907 were located in 18 places. Regina and Moose Jaw, with 19 and 12 vehicles respectively, were the largest centres, followed by Indian Head, Rosthern and Saskatoon with 8 vehicles each. By 1911 there were 821 vehicles in 136 places. Regina, Moose Jaw and Saskatoon were the largest centres (Table 9.2). In the smaller centres there were considerable and unexpected variations in numbers of vehicles in relation to the town's population size. Maple Creek and Rouleau, both places with a population of fewer than 1000 at the 1911 Census, had 24 vehicles registered, thus giving very high ratios of vehicles to population. Prince Albert, North Battleford, Estevan and Melville had, in contrast, low numbers of automobiles in relation to their population. Such variations were not unusual at the time and reflected differences in local economic and urban development.

Details of the location of vehicles were not published after 1911, and therefore it is impossible to identify geographical patterns with any precision. Summary details for eight major cities were published annually from 1923 and these are presented in Table 9.2.[20] One trend evident from the statistics is the declining dominance of the larger

Figure 9.2 Location of registered motor vehicles

(i) 1907

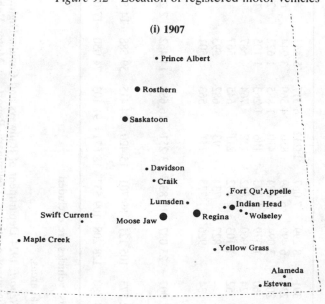

(ii) 1911

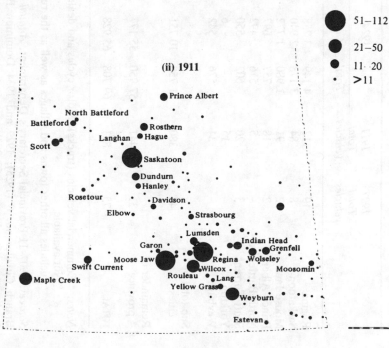

Table 9.2 Location of motor vehicles for selected years

	1907 Total vehicles	1911 Total vehicles	1924 Total	1924 Cars	1924 CVs[a]	1929 Total	1929 Cars	1929 CVs[a]	1934 Total	1934 Cars	1934 CVs[a]
Regina	19	112	4 149	3 758	391	8 299	7 067	1 232	6 317	5 443	874
Saskatoon	8	84	3 178	2 940	238	6 499	5 649	850	4 569	3 944	625
Moose Jaw	12	91	1 980	1 772	208	3 094	2 684	410	1 885	1 622	263
Prince Albert	2	18	757	697	60	1 379	1 193	186	1 263	1 105	158
North Battleford	–	5	516	492	24	960	842	118	764	687	77
Swift Current	3	20	602	550	52	1 103	892	211	645	567	78
Yorkton	–	15	–	6	–	827	734	93	682	593	89
Weyburn	–	34	576	542	34	926	775	151	563	501	62
Sub-total of cities	44	380	11 758	10 751	1 007	23 087	19 836	3 251	16 688	14 462	2 226
Remainder of province[c]	31	441	57 950	55 177	2 773	105 121	89 701	15 420	74 415	59 588	14 827
TOTAL	75	821	69 708	65 928	3 780	128 208	109 537	18 671	91 103	74 050	17 053

Notes: [a] Commercial vehicles (trucks). Motor cycles and dealers' vehicles are excluded.
[b] Not available
[c] Includes all other towns and villages as well as the rural areas.

Sources: 1907–11 Provincial Secretary's Department, Annual Report; 1924 Department of Railways and Canals, Highways Branch Circular no. 6 (1925); 1929 and 1934 Dominion Bureau of Statistics, Preliminary Report on Motor Vehicle Registrations (Catalogue no. 53–204).

centres. Their proportion of all provincial registrations dropped from 58.7 per cent in 1907, to 46.2 per cent in 1911 and to 16.8 per cent in 1924. Although numbers of vehicles increased in the cities, the greatest growth took place in the smaller towns, villages and rural areas. For 1926 it is possible to estimate the proportion classified as 'remainder of the province' in the table.[21] In that year vehicles on farms were enumerated at 55 444 or 57.9 per cent of the total vehicles registered. The seven cities listed in Table 9.2 accounted for 14 949 vehicles or 15.6 per cent of the total. The small towns, villages and hamlets represented the remaining 26.5 per cent. Saskatchewan in 1926, and later, had a high proportion of its total vehicles registered to farmers. In 1931, when a national comparison became possible, 45.8 per cent of Saskatchewan's farms had an automobile the second highest proportion in the country after Ontario, where 60.3 per cent of farms had automobiles.

VEHICLE OWNERSHIP AND USE

Knowledge of the owners, their usage of motor vehicles and the effects of their usage on economic and social life is at best very sketchy. Indeed there is a great need for detailed local history research on this facet of the motor vehicle. Automobiles were initially adopted by two types of individuals: the comparatively wealthy and socially prominent people who enjoyed the recreational and sporting potential of the conspicuous innovation; and the mechanical enthusiasts who were keen on the possibilities of the new machine. Walter E. Seaborn, a barrister in Moose Jaw, who licensed vehicle #2 in 1906, perhaps represented the first type; William Duff, of Regina, who registered a self-made vehicle in 1907 (#28), represents the second type of early enthusiast. Given the comparatively late beginning of automobile registration in Saskatchewan, there were other early adopters who were already well aware of its potential in other parts of the continent.[22] Doctors were among the earliest professional men to buy automobiles, and by 1911 there were at least twenty-six such owners, not only in Regina and the larger centres, but also in the smaller places. Weyburn, for example, had three vehicles licensed to M.D.s at this time. Later adopters were more cost-conscious than the first wave of enthusiasts. Motoring magazines published articles and notes on the operating costs of vehicles. One survey made by a Winnipeg hardware firm showed a 27 per cent saving in delivery costs by using a two-ton truck in place of three wagons.[23] The productivity of a reliable motor vehicle could be very substantial in delivery work.

Business firms were major users of motor vehicles. Possibly half the registered owners in 1911 were engaged in business, with the vehicle serving some economic role during the week and providing for recreation at the weekend. Land and realty companies used motor vehicles to take prospective buyers to view the properties.[24] The Luse Land and Development Company, a St Paul, Minnesota, firm with a local base at Scott, had twelve Reo cars in 1911, the largest fleet of automobiles in the province. Implement dealers were also significant users. International Harvester used its own make of Auto-Buggy at its Regina and Saskatoon branches. The J. I. Case Threshing Machine Co., Tudhope, Anderson & Co. and the American-Abell Engine/Thresher Co. all owned motor vehicles in 1911. The implement firms were probably among the earliest businesses to equip their field representatives with automobiles. Indeed the role of the farm-implement manufacturers in promoting a motor-power era should not be overlooked. Many of the companies were producing gasoline-engined tractor/traction engines, several were manufacturers or agents for automobiles and most organised winter training sessions for operators.[25] Trade exhibitions usually included displays of motor vehicles, a feature which prepared the way for widespread farm adoption of automobiles.[26]

After 1911 the motor vehicle was substituted for many of the horse-drawn vehicles used in urban areas; for coal and general freight delivery, for municipal purposes such as fire and police departments. As the urban middle class adopted motor vehicles, weekend pleasure driving created new patterns of life and business. In rural areas the automobile was widely accepted as a replacement for the horse-drawn buggy, but rather more slowly for the wagon. Haulage from farm to railhead by truck did not occur in a significant way until the late 1920s, when more powerful trucks and tractors became available, thus allowing the horse team to be replaced.[27]

Motor-vehicle regulation by the province was liberal. The early speed limits of 10 m.p.h. within cities and 20 m.p.h. outside, which were specified in the 1906 Act, were modified in 1912 to a speed 'reasonable and proper in the circumstances.'[28] Outsiders commented favourably on the administration of justice in Saskatchewan motoring cases, which, unlike those in Winnipeg, were apparently free from discrimination and unwarranted persecution.[29] Licence fees were $10.00 on initial registration and $3.00 for annual renewal until

1915 when the annual fee was raised to $15.00. Legislators placed few barriers to motor-vehicle ownership and usage until the late 1920s when gasoline taxes were imposed to help in paying for the cost of new road improvements. Problems brought by competition of motor vehicles to railways and urban street railways resulted in the Public Vehicles Act 1928 which allowed the government to regulate trucks and buses through the issue of licences for specific routes and types of vehicles.[30]

Motor clubs were very important in the promotion of early use and adoption through the organisation of displays, races, road tests and summer motorcades for poor children's picnics. Clubs were also active in demanding fair legislation for motorists and promoted the Good Roads Movement.[31] Regina organised a motor club in 1910, Moose Jaw and Saskatoon formed clubs in the following year and Prince Albert in 1912. By 1913 the Saskatoon club was developing a club house with tennis courts and golf links for its members.[32] A provincial league was formed in 1914 and reorganised in 1917.[33] The major emphasis of the provincial group was a campaign for good roads, including the promotion of an interprovincial highway from Winnipeg to Calgary. Local interests which were determined to see taxes devoted to local improvements tended to weaken the role of the provincial body.[34] Motor clubs marked highways and in the 1920s issued route maps, initiated accommodation guides and plans for auto camps, details which were essential components for longer-distance motorised tourism.

Generalisations about motor usage in this period are limited by the paucity of research and surviving materials. The period of initial large-scale adoption up to 1914 is more readily covered than later when the motor vehicle became so commonplace.[35] Business users found new productivity from adoption of automobiles. Isolation of farms was reduced through greater accessibility by farmers, directly through their own vehicles and indirectly through improvements in rural mail delivery.[36] Cottage resorts were developed on a larger scale by city dwellers. Cities had to contend with increased expenditure on road and bridge construction and for pressing demands for parking.[37] While obviously numerous and important, motor vehicles by 1930 were not yet indispensable as shown by the sharp decline in registrations with the onset of the Depression.

Many questions remain to be answered. What were the real changes in accessibility due to the automobile? What effects did motor vehicles have on the commerce of small rural service centres?[38]

When did year-round motoring become possible?[39] How was rural medical practice changed by the automobile?[40] Profound changes did take place in the 1920s, but the details are now obscure.

MAKERS OF VEHICLE, SERVICING AND PETROL SUPPLY

The annual summaries of motor-vehicle registrations from 1915 to 1934 are unusual in specifying all the makes of vehicle in operation. These lists are a kaleidoscope of the successes and failures of North-American motor-vehicle manufacturers in a critical period of mass production and mass marketing. The existence of many small obscure makes in Saskatchewan is a tribute to the energy and persistence of manufacturers and dealers in selling vehicles in a distant market.

The details exhibited in Figure 9.3 confirm the important role of Ford in putting the continent on wheels. Although incomplete, the records for 1911 show that Ford was already the largest make in Saskatchewan and the company retained first position throughout the period. Ford cars were ubiquitous, 'to be seen at every bend in the prairie trail and rows of them, from twelve to twenty, lined up on the single street of a town of a few hundred people . . .'.[41] The Model T Ford became part of rural and family folklore in all parts of Canada as well as the United States.[42] McLaughlin and later Chevrolet, both part of the General Motors empire, were the only other major challengers to the hegemony of Ford. By 1933 all the General Motors makes together were slightly ahead of the total numbers of Fords. Most of the vehicles were produced in Canada by American subsidiaries or affiliates. Only a handful of the expensive makes such as Cadillac and Packard were imported from the United States. The only non-North American vehicles were three British vehicles on the register in 1920, one in 1929 and twenty-one in 1934. The significance of the large prairie market in the wheat boom was appreciated by Ford as early as 1916, when an assembly plant was opened in Winnipeg. General Motors moved more cautiously and, as it turned out, too late, not announcing its plans to build a western assembly plant until 31 May 1928. The Regina plant opened in December 1929, but was forced to close in August 1931 when the regional market demand had collapsed.[43]

Motor vehicles required a complex service system for sales, maintenance, parts, supplies and fuel. The larger motor-vehicle manufacturers established branches in the prairie cities in the same manner as the farm implement companies. Such branches provided display

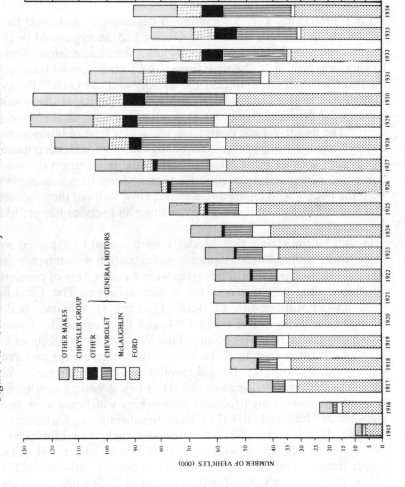

Figure 9.3 Saskatchewan: major makes of motor vehicles 1915–34

space for new automobiles, repairs, stocks of parts and in the early period storage for owners' vehicles. McLaughlin carriages had opened a western branch in Winnipeg as early as 1899 and built automobile branches in Regina (1911) and Saskatoon (1913). Ford opened a branch in Winnipeg in 1910 and one in Saskatoon during 1912. By the following year Ford had fifty subsidiary dealers in the prairies.

Local merchants were also active in developing specialised facilities for the new motor trade. The Garage Ltd, incorporated in 1911 with a capital of $50 000 and newly built premises in Broad Street, Regina, was typical of the new firms. This firm began with agencies for two Canadian automobiles, the McKay and the Galt.[44] By 1913 Regina had twelve auto dealers and was the principal distribution point in the province, handling $1½ million of sales in the previous year.[45] The motor-vehicle dealership was a new type of business with a sales franchise from a large, frequently multinational, corporation. Companies such as Ford set rigid standards, had inspection procedures and demanded a strong commitment from its agencies. Who were the people who entered this trade? How well did they succeed? Much of the history of this type of retailing and service has yet to be uncovered.[46]

In spite of the claims that Model T Fords could be repaired with fence wire, automobiles required sophisticated maintenance and many specialised parts. Rubber tyres were a major item of consumption since they rarely lasted for a high mileage. The Canadian Consolidated Rubber Co. of Berlin (Kitchener), Ontario, built a $40 000 warehouse in Regina in 1914, and the Regina City Council gave a substantial financial bonus to the Western Tire & Rubber Co. to build a plant in the city.[47] In local centres blacksmiths and livery stables added vehicle repairs and gasoline pumps. By the early 1920s the filling station had begun to appear in city streets; a new type of business often with architectural pretensions which are now being recognised.[48] Imperial Oil (the Canadian subsidiary of Standard Oil) was the dominant supplier of petroleum fuels and lubricants in Saskatchewan and developed an extensive provincial network of bulk supply depots. The company had three depots in 1904, 80 in 1912, 138 in 1917; numbers continued to rise until 1929 when there were 497 bulk depots in the province. Demand was so strong that Imperial built a refinery on the outskirts of Regina in 1916. Crude oil from Wyoming and later Montana was processed for the regional market.

Table 9.3   The automotive sector in Saskatchewan's retailing structure
1930

	No. of stores	Percent of total stores	Value of sales $ million	Percent of total sales
Motor-vehicle dealers	421	3.9	18.7	9.9
Filling stations	391	3.6	4.0	2.1
Garages	534	4.9	3.9	2.1
Other establishments	36	0.3	0.4	0.3
Total automotive sector	1 382	12.7	27.0	14.4

Source: Dominion Bureau of Statistics, *Census of Merchandising and Service Establishments*, vol. *x*, Census of Canada 1931.

By 1926 Imperial Oil was supplying about 92 per cent of the province's taxable gasoline sales.[49]

The commercial significance of the automotive sector was clearly apparent by 1930 when the first Census of Merchandising was taken. This sector was found in all parts of the province and automotive sales represented 14 per cent of total retailing (Table 9.3).

IMPROVEMENT OF THE LARGEST ROAD MILEAGE IN CANADA

The provision of improved roads by government was a critical point where the private process of motorisation impinged on public policy. Early motor travellers did much to publicise the inadequacies of the roads. John Mavor's trip across the prairies from Winnipeg to Edmonton in 1912 and Thomas Wilby's 1912 trans-Canada journey by car described the difficulties in detail.[50] Much detailed research needs to be undertaken on the role of the Good Roads Movement and the responses of the provincial government to the problems of the road system.[51]

In some senses the motor vehicle came into Saskatchewan at the wrong time; the process of land settlement was in full swing within the fixed commitments of a rigid land survey system and with railways the dominant form of interurban and long-distance transport. All the early twentieth-century settlement took place within a framework of six-mile by six-mile townships.[52] Road allowances within these survey townships were to be located at one-mile intervals in a north–south direction and at two-mile intervals in an east–west direction (Figure 9.4). If all the roads were developed, a township would have 42 miles of interior roads along with 24 miles around the perimeter in order to serve the possible maximum of 144 farmsteads. Few areas ever developed the maximum allowance of roads. Prairie agriculture developed in very close association with railways, which were expected to build branch lines to serve a close network of grain elevators mostly used in the months following the harvest. Railway-building in Saskatchewan continued on a substantial scale right up to 1929.[53] Such construction activity tended to swallow up most of the capital funds which might have been used for road improvements.

Responsibility for most roads was placed with municipalities following the Rural Municipality Act 1909.[54] The drafters of this legislation had not foreseen the demands of the automobile on the vast mileage of road allowances of the survey system; and Saskatchewan had the greatest mileage in Canada when the first tabulation of roads was made in 1922. With 135 000 miles of road (an unknown proportion being unformed road allowances) Saskatchewan had 35 per cent of the nation's roads. Further settlement in the decade increased this length, and by 1934 the province accounted for a further 3 per cent of the Canadian total.

New legislation in 1912 recognised that the costs of road formation were beyond the financial capabilities of newly settled rural municipalities.[55] A three-member Board of Highway Commissioners was established to plan a highway system and assume part of the costs of municipal roads, while the government was empowered to borrow $5 million for construction and improvement of roads. The Board of Highway Commissioners was replaced by a provincial Department of Highways in 1917, and the road movement received new impetus from the Canada Highway Act 1919 which allocated $1.8 million (of the national total of $20 million) to Saskatchewan.[56] By the early 1920s provincial highways, main market roads, colonisation roads and local roads were defined and some plans were made. The idea of 'main market roads' as a starting-point for highway planning, was promoted as a means of linking farms with market towns and railheads. Most of the early bulletins circulated by the Highways Branch of the Dominion Department of Railways and Canals contained extensive reports of the cost reductions in farm produce haulage

Figure 9.4 Township roads in Saskatchewan

Roads and the Survey System:

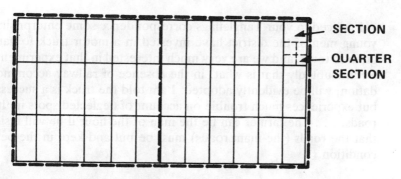

SECTION

**QUARTER
SECTION**

Actual Development: Townships 13, Ranges 16 and 17

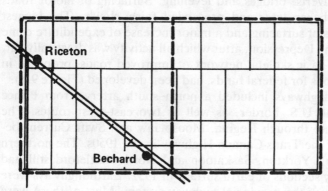

Riceton

Bechard

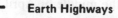

	Earth Highways
	Municipal Earth Roads
+—+	Railway
— — —	Boundary of the Survey Townships

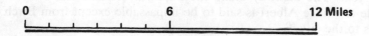

0   6   12 Miles

which came from improved roads.[57] Many letters to the editors of local newspapers supported policies of road development. This one appeared in the *Lloydminster Times* in 1918:

> I noticed in your Tangleflags correspondence some enterprising young men of the district have invested in a motor truck to haul grain to town, and we are very much interested in that experiment, as undoubtedly that is what, in the absence of railway accommodation, will be evidently adopted. I am told the truck is a success, but experiences much trouble on account of neglected spots in the roads. . . . The farmer can be the man of the hour if he will insist that the roads (the main roads) must be put and kept in the best condition.[58]

Highway expenditure began to gather momentum in the early 1920s (Table 9.4, pp. 184–5) with most of the work being concentrated on culverts, bridges and levelling. Surfacing of motor roads with gravel began later and proceeded fairly slowly. The biggest visible effects of surfacing and a major increase of expenditure came just before the Depression, after which all activity was drastically cut. By 1932 the basic skeletal network of improved roads, proposed in the submissions for federal funds, had been developed (Figure 9.5).[59] The gravel highways included a north–south artery, from Prince Albert to the U.S. border, as well as two east–west routes. The southern route through Regina, Moose Jaw and Swift Current became part of the Trans-Canada Highway in the 1950s. The northern route through Yorkton, Saskatoon and North Battleford still had some lengthy sections of earth surface in 1932. Earth highways were the mainstay of the provincial highway system. Most of these were graded and had bridges and culverts over the rivers and gullies. Highway usage was still seasonal and the spring thaw continued to be a major problem for movement. As a correspondent of the *Prince Albert Daily Herald* reported in the middle of April 1935:

> Driven by a woman resident of White Fox, Saskatchewan [70 miles east of Prince Albert], a car thickly covered with mud arrived here last night en route to Saskatoon. Coming by the north road she is probably the first motorist to make the trip in that direction this year. The highway along the Nipawin, Tisdale and Melfort route to Prince Albert is said to be impassable except from Birch Hills to the city.[60]

If most of the larger centres were linked by some type of improved highway during the late 1920s, the hamlets and the farms were dependent on roads of very variable quality. Few farms were located on all-weather roads. In 1931 only 2.7 per cent of Saskatchewan's farms were located on gravel roads; 64.4 per cent were located on 'improved' dirt roads and its remainder were situated on unimproved dirt roads.[61] Weaknesses of the road system persisted throughout the period and beyond. The vast area and enormous length of roads were too large for the public financial resources of a newly settled area.

CONCLUSION

The motorisation of Saskatchewan was accomplished in four major phases. The first phase (c. 1902–9) was the introduction of motor vehicles mostly by enthusiasts for the new sport and technology. Phase two (1910–13) was a critical threshold period of substantial commercial investment in the new mode of transport which prepared all the sales and service foundations for phase three (1914–18) when the mass market first developed. The final, fourth phase (1923–9) was a further extension of ownership and an elaboration of the service facilities. Economic depression after 1929 delayed the process of motorisation for another two decades.

By the late 1920s motor vehicles had become a key element of modernisation and transport in Saskatchewan, their influence permeating virtually every aspect of economic and social life in the province. Automobiles brought a new level of accessibility to rural places, breaking the isolation of farm life, providing easier access to medical services and beginning a more flexible era in the movement of agricultural commodities. In the larger places motor traffic generated new business possibilities and developed new forms of recreation. Motor-vehicle usage had many positive benefits.

There were also penalties. Road improvements and maintenance placed heavy costs on struggling local municipalities especially in rural areas.[62] At the provincial level the costs and benefits of highway expenditure have yet to be explored in detail. Increased road traffic in the late 1920s caused a substantial growth in the death rate from automobile accidents. The total number of fatalities rose from 21 in 1926 to 74 two years later and then declined gradually to 51 in 1930 and 30 in 1934.[63] Motor vehicles were a key factor in the gradual erosion of economic functions in the very small places which formed the base of the extensive prairie community system.[64] The residual

Table 9.4 Saskatchewan: Highways and Rural Roads 1919–1934

	Highways and Rural Roads[a]		Expenditure[b] $ million			Revenue $ million	
	Total mileage (000 miles)	Total surfaced (000 miles)	Total	Construction	Maintenance	Provincial	Federal
1919	..	..	1.3	1.1	0.2	0.7	–
1920	..	..	1.8	1.5	0.3	0.9	–
1921	..	..	2.5	2.3	0.2	0.7	–
1922	135.0	..	2.2	2.0	0.2	0.8	0.2[c]
1923	135.0	..	2.0	1.7	0.3	1.1	0.5[c]
1924	135.0	..	2.1	1.8	0.3	1.2	0.4[c]
1925	152.0	0.1	2.1	1.8	0.3	1.4	0.3[c]
1926	152.0	0.1	2.7	2.3	0.4	1.7	0.1[c]
1927	152.0	0.1	2.5	2.0	0.5	1.9	0.2[c]

1928	152.3	0.4	4.4	3.6	0.8	3.5^d	0.1^c
1929	152.3	0.8	6.7	5.9	0.8	3.8	–
1930	154.6	1.9	10.4	8.4	2.0	3.5	0.5^e
1931	155.6	2.1	8.0	6.9	1.1	2.9	0.8^e
1932	155.6	2.1	0.9	0.3	0.6	2.8	··
1933	155.7	2.2	0.9	0.2	0.7	2.8	··
1934	155.7	2.4^f	1.8	1.0	0.8	3.1	–

Notes: ·· Not available or negligible.

^{a} Urban roads were not counted until 1935.

^{b} Expenditures by province and rural municipalities. Includes highways, rural roads, bridges and ferries.

^{c} Federal contribution under Canada Highways Act 1919

^{d} A gasoline tax was first introduced in 1928. Earlier revenue from vehicle, operator and dealer licences.

^{e} Federal contributions under Unemployment Relief Acts.

^{f} Includes 100 miles of paved road – statistics first collected in 1934.

Source: Compiled from: Canadian Tax Foundation, *Taxes and Traffic*, Canadian Tax Papers No. 8 (Toronto: 1955).

Figure 9.5 Improved roads in Saskatchewan 1932

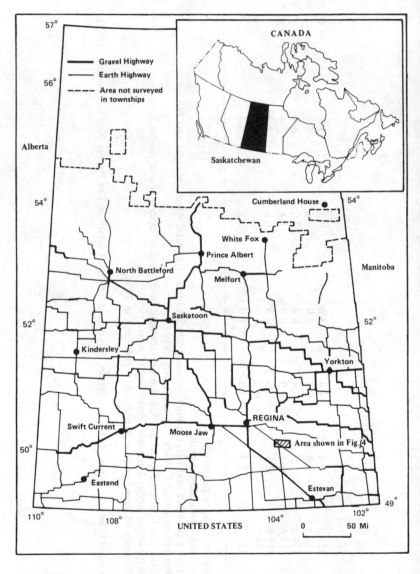

problems created by the original settlement pattern based on railway transport and an obsolete survey continue to plague transport and other planners in the 1980s.

Economic and social changes resulting from the motor vehicle were not the same everywhere. This study of a new rural frontier has emphasised several critical factors which resulted in regional variations. Such factors include the importance of the timing of initial acceptance and large-scale adoption in the context of the development of the general regional economy. The type of vehicles adopted and their potential use were shaped by environmental considerations and the quality of existing roads, as well as by the characteristics of other modes of transport. Although the bulk of the investment was in the form of private vehicles and commercial services, the policies of public authorities towards the automobile were also very significant in shaping the environment for extended use. Appreciation of all these features of the motorisation process may be investigated most fruitfully at the regional and local level.

Notes

1. M. M. Willey and S. A. Rice, *Communication Agencies and Social Life*, President's Research Committee on Recent Social Trends in the United States Monograph (New York: McGraw-Hill, 1933) p. 27.
2. Rates of motor vehicle ownership per 1000 population were calculated as follows:

	1922	*1929*
United States	98.8	218.0
Canada	53.9	114.9
New Zealand	30.8	118.3
Australia	14.9	90.4
Argentina	8.0	32.4
United Kingdom	10.5	31.8
France	6.0	31.4

Source: National Automobile Chamber of Commerce, *Facts and Figures of the Automobile Industry* (New York) and League of Nations, *Statistical Yearbook* (Geneva).

3. Imperial Economic Committee, *A Survey of the Trade in Motor Vehicles* (London: H.M.S.O., 1936) p. 20.
4. Asociacion de Fabricas de Automotores, *Industria Automotriz Argentina* (Buenos Aires: 1973) p. 18.
5. Ford began assembly in Buenos Aires in 1917 and by 1925 had produced

100 000 units in Argentina. C. F. D. Alejandro, *Essays on the Economic History of Argentina* (New Haven, Conn.: Yale University Press, 1970) p. 265.

The dominance of Ford in these frontier markets until the late 1920s is noted in several motoring histories. See for example:

G. W. Taylor, *The Automobile Saga of British Columbia 1864–1914* (Victoria: Morriss Publishing, 1984).

J. Goode, *Smoke, Smell and Clatter, the revolutionary story of motoring in Australia* (Melbourne: Lansdowne Press, 1969).

P. Maclean and B. Joyce, *The Veteran Years of New Zealand Motoring* (Wellington: Reed, 1971).

T. T. N. Coleridge, *Our Motoring Heritage* (Wellington, 1973).

6. The general historical setting of Saskatchewan may be found in:

G. Friesen, *The Canadian Prairies: A History* (Toronto: University of Toronto Press, 1984).

7. A. F. J. Artibise (ed.), *Town and City: Aspects of Western Canadian Urban Development* (Regina: Canadian Plains Research Center, 1981).
8. Some general Canadian context on the adoption and early use of motor vehicles may be found in:

H. Durnford and G. Baechler, *Cars of Canada* (Toronto: McClelland & Stewart, 1973).

9. Recognition in the published literature today is, however, fairly limited. Only one motor-vehicle reference was cited in Ved P. Arora, *The Saskatchewan Bibliography* (Regina, 1980). The *Saskatchewan History Index* includes a few automobile references, mostly covering notes on early auto travel.

There are passing references to the automobile and its significance in the provincial histories and other works as for example:

John H. Archer, *Saskatchewan: A History* (Saskatoon: Western Producer Prairie Books, 1980).

James H. Gray, *The Roar of the Twenties* (Toronto: Macmillan, 1975).

Local community histories through their emphasis on contemporary photographs and family reminiscences frequently provide a strong sense of the importance of motor cars. A view of the town of Lumsford *c*. 1915 shows more automobiles than buildings; see *The Past to the Present: 70 years 1909–1979* (Sceptre – Lumsford Historical Association, 1979).

10. An Act to regulate speed and operation of motor vehicles on highways, *Statutes of Saskatchewan*, 6 Edw. VII, 1906, cap. 44.
11. The Provincial Secretary's Department administered motor registrations from 1906 until May 1930 when this function was transferred to the Department of Highways. The Provincial Tax Commission took over in

late 1937. Other provincial agencies involved were the Department of Public Works, the Department of Highways and the police. All municipal governments were also concerned with provision of services and control of automobiles.

It appears that none of the primary records of ownership have survived. Record-keeping for tens of thousands of automobiles created huge problems for a small clerical staff. In addition to the annual re-registration and growth of new vehicle numbers there were continual changes of ownership and addresses of owners to be recorded. Demands by the police for rapid identification of vehicle ownership also added to the difficulties of maintaining record systems. As the sheer bulk of original annual records made storage unmanageable, most of the records were probably destroyed as they became obsolete.

Summaries published in the Provincial Secretary's Department annual reports contain most of the available details on motor vehicles. The most comprehensive record was that published in 1911 which listed the serial number, name of owner, location and in many cases the make of the vehicle. From 1915 to 1934 only a summary of total numbers was published in the annual reports, although, uniquely in Canada, these were tabulated by make of vehicle.

In the early period of rapid increase in automobiles and close public interest in such growth, the newspapers often published detailed lists of owners and their cars. Regina's paper, *The Morning Leader*, 4 Nov 1911, Second Section, page 1 – in a long piece titled 'Saskatchewan Folks enjoy the Chug-Chug Game' listed owners by place of residence with make of vehicle and licence number.

Apart from the provincial departmental records and newspaper reports there must be many other types of records awaiting rediscovery. Records of court proceedings could shed light on problems of automobile usage; minute books and membership lists of motor clubs would improve our knowledge of the early adopters and promoters of automobiles in specific communities, business records from garages and dealers would help in understanding the ways in which the new transport was introduced and organised.

12. *Gas Power Age*, 4 (5), Nov 1911, p. 10.
13. *Ibid.* 9 (4) Apr 1914.
14. *Ibid.* 7 (4) Apr 1913, p. 27. Students enrolled for a six-week course for a fee of $50.00.
15. The prices quoted include the freight charges from Ford City (Windsor) to Moose Jaw. F.o.b. prices quoted in *Gas Power Age*, 3 (1) (Jan 1911) p. 26, 7 (1) (Jan 1913) p. 55, and the *Ford Times* (Canadian Edition) vol. 3 (Oct 1915) and vol. 4 (May 1917). Freight rates were published in the *Ford Times*.
16. See R. E. Ankli and R. M. Litt, 'The Growth of Prairie Agriculture: Economic Considerations', in D. H. Akinson (ed.), *Canadian Papers in Rural History*, vol. 1 (1978) pp. 35–64.
17. *The History of the Milden Community 1905–1965* (Milden Historical Committee, 1966) p. 110.
18. Compiled from various statistical publications, including *The Motor*

Vehicle, published by the Department of Railways and Canals, Highways Branch to 1928 and thereafter by the Dominion Bureau of Statistics.

Although high by Canadian standards, Saskatchewan's ownership levels were always overshadowed by the neighbouring American states.

	Ratio of motor vehicles per 1000 population		
	Saskatchewan	*North Dakota*	*Montana*
1914	16.3	28.3	22.0
1924	92.3	177.8	145.2
1929	139.3	276.8	264.2
1933	92.1	231.0	208.7

The U.S. ratios were calculated from the statistics in U.S. Department of Transportation, Federal Highway Administration, *Highway Statistics: Summary to 1975* (Washington: U.S. Government Printing Office, 1977).

The substantial regional variations shown in this table are intriguing and are worth further exploration.

19. Provincial Secretary's Department, *Annual Report 1910–11*, pp. 9–22.
20. Department of Railways and Canals, Highways Branch, *The Motor Vehicle 1922*, Circular no. 5 (Ottawa: 1924) pp. 9–10.

The ratio of motor vehicles to population in the city of Regina increased from 111.1 per thousand in 1924 to 156.0 in 1929 and then fell to 118.4 in 1934. These ratios were high in relation to most Canadian cities at the time and were very high in comparison with English cities of a similar size. Carlisle and Worcester county boroughs in 1926 had ratios of 37.1 per thousand and 43.4 respectively. American cities in the same size range (53 000 population) had generally higher ratios in 1930. Durham, North Carolina, had a ratio of 236.8 and Hamilton, Ohio, had 268.8.

21. Census of the Prairie Provinces 1926, and *The Motor Vehicle* (1926).
22. J. J. Flink, *America Adopts the Automobile 1895–1910* (Cambridge, Mass.: M.I.T. Press, 1970).
23. *Gas Power Age* 3 (4) (Apr 1911) p. 65.
24. See W. Pearson, 'Recollections and Reminiscences: Colonization Work in Last Mountain Valley', *Saskatchewan History*, XXI (3) (1978) pp. 111–13.
25. 'Rumely – Oil Pull school of engineering classes for owners and operations in Regina', *Gas Power Age*, 3 (3) (Mar 1911) p. 29.
26. 'The Dominion Exhibition, Regina', *Gas Power Age*, VI (3) (Sep 1911) pp. 31/33.
27. R. E. Ankli, H. D. Hebberg and J. H. Thompson, *The Adoption of the Gasoline Tractor in Western Canada*, Working Paper Series no. 1, (Agricultural History Center, University of California, Davis, 1979) p. 6.
28. *Statutes of Saskatchewan*, 2 Geo. V, 1912, cap. 38, s. 16.
29. *Gas Power Age*, 6 (5) (Nov 1912) p. 38.
30. *Statutes of Saskatchewan*, 18 Geo. V, 1928, cap. 74. Further amendments in the early 1930s limited speeds of loaded and unloaded vehicles, weights and widths of vehicles, qualifications of drivers, etc.

31. 'Saskatchewan first in Good Roads Movement', *Gas Power Age*, 5 (4) (Apr 1912) p. 9.
32. *Gas Power Age* 7 (6) (June 1913) p. 43.
33. *Gas Power Age* 9 (4) (Apr 1914).
34. As late as 1925 a North Saskatchewan Motor League was formed, with fourteen local clubs, to press for road improvements in the region (*Canadian Motorist* (Nov 1925) p. 508).
35. Monthly magazines such as *Gas Power Age*, published Winnipeg 1908–14, provide a comprehensive view of all facets of the motor vehicle and gasoline-powered machines. Library collections of this and other contemporary publications are rare.
36. See J. H. Gray, *The Roar of the Twenties* (1975) p. 47; J. M. Archer, *Saskatchewan: A History* (1980) p. 201–3.
    By 1919 there were 2049 miles of rural mail routes in Saskatchewan. Department of Railways and Canals, *Highways Bulletin*, no. 1 (Ottawa: King's Printer, 1922) p. 15.
37. Were the Saskatchewan cities and towns always ready to accept the motor vehicle? See B. A. Brownell, 'A Symbol of Modernity: Attitudes toward the Automobile in Southern Cities in the 1920s', *American Quarterly*, 24 (1972) pp. 20–44.
    The automobile owners in Regina proposed that the city should pave a section of Victoria Park for car parking. *Gas Power Age*, 7 (6) (June 1913) p. 41.
38. For details of some of the effects of automobile on small towns see:

    P. D. Crouse, *The Automobile and the Village Merchant* (University of Illinois: Bureau of Business Research, 1928).

    J. A. Jakle, *The American Small Towns: Twentieth Century Place Images* (Hamden, Conn.: Archon Books, 1982).

    N. T. Moline, *Mobility and the Small Town, 1900–1930* (Chicago: University of Chicago, Department of Geography, Researh Paper no. 132, 1971).

39. Anti-freeze was being advertised as early as 1912 but did not become fully successful for at least another decade. *Gas Power Age*, 6 (4) (Oct 1912) p. 39.
40. M. L. Berger, 'The Influence of the Automobile on Rural Health Care 1900–1929', *Journal of the History of Medicine and Allied Sciences*, 28 (Oct 1973) pp. 319–35.
41. M. Wilkins and R. E. Hill, *American Business Abroad: Ford on Six Continents* (Detroit, Mich.: 1964) p. 43.
42. R. M. Wik, *Henry Ford and Grass Roots America* (Ann Arbor, Mich.: 1972).
43. The assembly plant reopened in September 1936 and vehicle production continued for another five years before the plant was converted to war-time ordnance work.
44. *Gas Power Age*, 6 (2) (Aug 1911) p. 71.
45. 'Saskatchewan Automobile Trade', *Gas Power Age*, 7 (4) (Apr 1913) p. I. This issue of the magazine included a major section on the motor businesses of Moose Jaw and Regina.

46. For an outline of the evolution of Ford dealers, see H. L. Dominguez, *The Ford Agency: a pictorial history* (Osceola, Wisconsin: 1981).
47. *Gas Power Age*, 9 (3) (Mar 1914).
48. See, for example, Daniel I. Viegra, *Fill 'er up: An Architectural History of America's gas stations* (New York, 1979).
49. Comprehensive details on Imperial Oil may be found in J. S. Ewing, 'The History of Imperial Oil', 4 vols, unpublished manuscript, Business History Foundation, Harvard Business School, Boston, 1951? Located at Imperial Oil Archives, Toronto.
50. See John Mavor, 'Auto Trip across the Prairie', *Alberta History* 30 (2) (1982) pp. 37–38; Thomas W. Wilby, *A Motor Tour through Canada* (London: 1914) also summarised in *Saskatchewan History*. vol. 7 (1950) pp. 23–7.
51. It is likely that substantial volumes of records on highway finance and construction have survived and await the patient research worker. For the general context of road-building see E. C. Guillet, *The Story of Canadian Roads* (Toronto: University of Toronto Press, 1966).
52. D. W. Thomson, *Men and Meridians: the history of surveying and mapping in Canada*, vol. 2 *1867–1917* (Ottawa: Queen's Printer, 1967).
53. The Saskatchewan railway network grew as follows:

1907–2025 miles	1919–6148 miles
1909–2631 miles	1924–6942 miles
1914–5089 miles	1929–7761 miles

See *Canada Year Book*.
54. Province of Saskatchewan, Royal Commission on Agriculture and Rural Life, Report no. 4, *Rural Roads and Local Government* (Regina, 1955) p. 190.
55. *Statutes of Saskatchewan*, 2 Geo. V, 1912, caps 5 and 7. Six other acts passed during this session of the Legislature provided guarantees to the railway companies for the construction of branch lines.
56. The Canadian federal government in highway building and road traffic regulation has always been limited. Roads were clearly designated as a provincial responsibility under the Constitution of 1867. Apart from the financial support 1919–28 and the Trans-Canada Highway Act program (1949–71) federal transport policies have been railway oriented.

The problems of the massive interwar debt burdens of excessive state-supported railway-building were described in L. R. Thomson, *The Canadian Railway Problem* (Toronto: Macmillan, 1938). Clearly the road–rail competition was more than traffic, but was also inextricably linked with financial policies.

General features of the very positive role of the U.S. federal government in highway improvement are outlined in P. J. Hughill, 'Good roads and the automobile 1880–1919', *Geographical Review*, 72 (3) (1982) pp. 327–49.
57. Department of Railways and Canals, Highways Branch, *The Canadian Highway and its Development*, Bulletin no. 7 (Ottawa: King's Printer, 1925).
58. *Lloydminster Times* (14 Nov 1918) p. 7.

59. J. H. Richards and K. I. Fung, *Atlas of Saskatchewan* (Saskatoon, 1969) p. 22.
60. *Prince Albert Daily Herald* (18 Apr 1935) p. 3.
61. Census of Canada 1931, vol. 8, *Agriculture*, Saskatchewan, table 34.
62. Saskatchewan, Royal Commission on Agriculture and Rural Life, *Rural Roads and Local Government*, Report no. 4 (Regina: Queen's Printer, 1955).
63. *Canada Year Book 1936*, p. 693.
   Saskatchewan's ratio of road deaths per 10 000 registered motor vehicles was generally the lowest in Canada. In 1931, the worst year nationally, Saskatchewan had a death rate of 4.61 per 10 000 motor vehicles. The national rate of 10.96 may be compared with Ontario 10.21 and Quebec 19.77.
64. C. C. Zimmerman and G. W. Moneo, *The Prairie Community System* (Ottawa: Agricultural Economics Research Council, 1970).

10 The Internal Combustion Engine and the Revolution in Transport: The Case of Czechoslovakia with Some European Comparisons

Jaroslav Purš

The exploitation of internal combustion engines for powering road vehicles in the Czech Lands began early and was related to earlier traditional industries:

1. The manufacture of carriages and railway carriages in the Koprivnice factory, for instance, later became the renowned firm TATRA (the first car, the 'President' in 1897, later also lorries).
2. The manufacture of bicycles by the firm Laurin & Klement in Mlada Boleslav (trade mark L & K) later developed into the automotive plant SKODA, affiliated to the machine-engineering and armaments concern of the same name (the first motor cycle in 1899, later the well-tried cars L & K, taxis, lorries, buses, ambulances, fire engines, etc.).
3. The machine-engineering company PRAGA, developed in 1909 from the Prague Automobile Factory, founded in 1907.

Before the First World War the manufacture of motor cycles and cars in the Czech Lands played a leading part within the territory of what was then Austria-Hungary, and also had an important place on the European scale. In 1900 there were about 90 cars in the whole of Austria-Hungary. In the spring of 1906 there were 208 cars in Bohemia alone (of which 69 were in Prague) and also 874 motor cycles (of which 248 were in Prague). The number of motor vehicles in Bohemia and Moravia on the eve of the First World War is estimated at about 3000.[1] During this period the motor industry in the Czech Lands exported a considerable amount of its production and was successful at a number of international car shows in Prague

194

and abroad, as well as winning a number of trophies in international races, mostly due to the drivers of L & K motor cycles and racing cars.

Obstacles to the development of motoring, which was considered to be a kind of sport and entertainment in its initial phases rather than a new progressive form of transport, were the poor condition of the roads and the small domestic market, which did not enable the production of accessories, such as electrical installation, tyres, safety glass, etc., to develop; before 1918 these components in aggregate accounted for about 40 per cent of the value of the produced cars and had to be imported. There was also the resistance of the general public, compulsory insurance, unfavourable general economic and fiscal policy of the state, etc. For example, in the early 1920s, a luxury tax was imposed on cars, together with a high import duty on them (65 per cent of the price) and on tyres (67 per cent of the price). Besides this there was another factor adverse to the development of motorism, i.e. the enforcing of the use of an expensive spirit–petrol mixture (20 per cent spirit), the purpose of which was to increase the profits of the large estate owners and of the agrarian bourgeoisie. This put a stop for a long time to the development of motorism, already affected by the First World War (with the exception of the manufacture of military lorries, artillery tractors and other military vehicles, ambulances, etc.). In June 1921 the total power of all cars in Czechoslovakia was estimated at 110 000 h.p., of lorries at about 61 000 h.p., and of the other vehicles at about 17 000 h.p., 200 000 h.p. in all.[2]

In the period between the two World Wars the automobile industry began to concentrate, and companies like PRAGA, TATRA, ASAP (SKODA) assumed leading positions, followed by firms with a lower output like AERO, JAWA, WALTER, WIKOV AND CZ (Cesko-slovenska zbrojovka). Since 1923 the tendency to manufacture and market a 'small car' (like the Volkswagen later in Germany) began to gain ground, when the car changed from a vehicle for sport and leisure into an important means of transport. This trend was further supported by the consequences of the economic crisis in the 1930s.

Motor vehicles were used in passenger and freight transport, for various purposes by the military, and in a number of professions and businesses. Cars were purchased by doctors, for instance, who could then visit patients over a wider area; ambulances were able to transport the sick quickly to hospital and enabled skilled medical care

Table 10.1   Owners of motor vehicles (with the exception of motor cycles and tricycles) in Prague in 1928[3]

Owners of vehicles	Number of vehicles	%
Businessmen and commercial and financial institutions	1 772	24.9
Factory owners and industrial firms	1 405	17.3
Artisans and traders	1 221	15.1
Authorities and public institutions[a]	910	11.2
Officials, officers, etc.	872	10.8
Cab drivers and passenger transport firms	684	8.5
Members of professions	586	7.2
Rentiers and men of means	100	1.2
Estate owners and farmers	38	0.5
Members of various occupations. including 9 students, 8 foremen, 7 chauffeurs, 5 shop managers, 3 shop assistants, 3 small farmers, 2 headwaiters, 3 workers and 1 apprentice	41	0.5
Others	473	5.8
Total	8 102	100.0

[a] Including publicly registered military vehicles (lorries, cars and ambulances) and vehicles belonging to the General Post Office, which also included two tanks in the district of Greater Prague.

to be applied sooner; buses took children to school from remote areas; mobile stores were used in trade, and salesmen were able to search for new customers more easily and maintain contact with them (insurance agents, sale of automobiles, oils, petrol and other goods); fire brigades could act more quickly, using mobile fire engines which took them to the fire or to the site of other disasters in larger numbers and with the necessary equipment; special vehicles were used in removing garbage and to clean the streets in the larger towns.

Table 10.2   Owners of motor cycles in Prague in 1928[4]

Owners of motor cycles	Number of motor cycles	%
Public and private officials	740	28.9
Businessmen and commercial and financial institutions	473	18.6
Artisans and traders	303	11.9
Workers (mostly highly qualified)	279	10.9
Members of professions	170	6.7
Authorities and public institutions	145	5.7
Lower wage-bracket employees	128	4.1
Factory owners and industrial firms	71	2.8
Various (12 men of means, 2 small farmers and 1 large farmer)	15	0.6
Total	2553	100.0

OCCUPATIONS OF MOTOR-CAR, LORRY AND CYCLE OWNERS

Important data on the social and professional composition of more than 8000 owners of cars and lorries and of more than 2500 owners of motor cycles in Prague are available for 1928 (Tables 10.1 and 10.2).

Most of the cars and lorries in Prague in 1928 (57 per cent or more) were obviously owned by those in business; further research evidently needs to be done to decide how many of these were private cars and how many commercial vehicles.[5] Professional people appear well down the list, though some may have had their cars included in the category of officials and officers; and, as Table 10.3 indicates, they formed an increasing proportion of the total. Next to no members of the working classes could afford motor cars in Prague (or anywhere else in Europe), though the inclusion of the 8 foremen, 2 head waiters, 3 'workers' and 1 apprentice is worthy of note. Motor cycles, considerably cheaper to buy and to run, could evidently be afforded not only by the better-off working classes but also by those in the lower wage bracket (whatever that may mean). At the other end of the social scale they were not disdained by some business men and public and private officials.

Table 10.3   Social and professional competition of owners of cars and
lorries in Prague in 1920, 1928 and 1930[6]

Owners of motor vehicles[a]	Percentage		
	1920	1928	1930
Businessmen and commercial and financial institutions	26.1	26.4	23.0
Factory owners and industrial companies	25.6	20.9	17.0
Artisans	16.3	18.2	19.9
Cab drivers and transport companies	20.4	10.2	13.9
Officials	6.4	13.0	12.4
Professions	2.3	8.7	9.5
Men of means	2.3	1.5	1.8
Estate owners and farmers	0.6	0.5	0.7
Others[b]	–	0.6	1.4
Group totals	100	100	100

[a] Excepting vehicles of public authorities, institutions and of the military administration.
[b] Small number of workers, chauffeurs, shop assistants and small farmers.

The development in the social and professional composition of the sample of owners of cars and lorries in Prague in the course of the 1920s (1920, 1928, 1930) is shown in Table 10.3 which supports the conclusions drawn from Table 10.1 and also indicates that, in Prague and in this period, the proportion owned by businessmen and factories decreased.

The proportion owned by cab drivers and transport companies fluctuated, that owned by artisans and shopkeepers and traders increased slightly, and that owned by members of professions and officials increased substantially. This development was apparently due to the special status of Prague as the capital of the Republic (vehicles belonging to shops, authorities, universities, cultural institutions with a considerable percentage of the professions), on the one hand, and to cars being used more and more by the members of the bourgeoisie and petty bourgeoisie.

It is important to compare these data and conclusions with the data in Table 10.4 which give the occupations and percentages of owners of cars and lorries in 1930 separately for Prague, the remainder of Bohemia and for Bohemia as a whole. In Bohemia with the exception

Table 10.4 Social and professional status of a sample of automobile
owners in Bohemia and in Prague, 1930[7]

Owners of motor vehicles	Number of vehicles in the sample			Percentage		
	1	2	3	1	2	3
Businessmen and commercial and financial institutions	264	174	438	23.0	21.3	22.3
Factory owners and industrial companies	195	203	398	17.0	24.9	20.3
Artisans and traders	228	176	404	19.9	21.6	20.6
Cab drivers and transport companies	160	89	249	13.9	10.9	12.7
Officials	142	38	180	12.4	4.7	9.2
Professions	109	56	165	9.5	6.9	8.4
Men of means	21	4	25	1.8	0.5	1.3
Estate owners and farmers	9	52	61	0.7	6.4	3.1
Others	17	23	40	1.4	2.8	2.1
Group totals	1145	815	1960	100.0	100.0	100.0

1 Prague.
2 The remainder of Bohemia.
3 Bohemia as a whole.

of Prague, the percentage of the business groups was slightly lower, of officials considerably lower, which also applies to professions, cab drivers and transport companies. The percentages of factory owners, industrial companies, farmers and estate owners, however, were considerably higher. A slightly higher percentage was due outside Prague to the trades and other occupations, whereas the percentage due to men of means (mainly rentiers and owners of real estate) was slightly lower there. In the figure for Bohemia as a whole these differences as compared to Prague on its own are slightly less pronounced.

Somewhat different conclusions concerning the social and professional composition of owners of cars and lorries follow from the much smaller sample for Slovakia.

According to this evidence, in Slovakia, as opposed to the Czech Lands, the farmers were in first place, followed by artisans and tradesmen and then cab drivers, transport companies and professions

Table 10.5   Social and professional status of a sample automobile owners
in Slovakia, 1930[8]

Owners of motor vehicles	Vehicles	
	number	percentage
Businessmen, commercial and financial institutions	42	12.9
Factory owners and industrial companies	21	6.4
Artisans and tradesmen	50	15.4
Officials	19	5.8
Cab drivers and transport companies	45	13.8
Professions	45	13.8
Men of means	13	2.4
Farmers and estate owners	61	18.8
Other professions (22 small farmers, 10 chauffeurs and 2 waiters)	34	10.7
Total	225	100.0

with exactly the same percentage, only then came commercial activity. This is very probably distorted to some extent because the sample includes only the regions outside Bratislava.[9]

MOTOR-VEHICLE REGISTRATION BY AREA

The difference between Prague and the remainder of Bohemia prompts an investigation into the differences between smaller territorial units, i.e. between the separate vehicle registration districts, on the one hand, and the larger regions and the separate lands (as they were known then) on the other. These differences can be seen in the appendix to this chapter,[10] which relates the intensity of motorism (taking into account cars and lorries) in the smallest administrative units, i.e. vehicle registration districts,[11] to the area of the appropriate territory, the population density and the employment of stationary engine power in industry and the production trades; however, in any further analyses we shall also have to take into account other factors such as the relation between industry and agriculture, the nationality aspect, etc. Table 10.6 gives a very rough account of the intensity of motorism by individual zone, determined with regard to

Table 10.6   Vehicle registration districts according to the degree of motorism in the Czech Lands in 1930

Zone of vehicle · registration districts according to number of inhabitants per 1 vehicle[a]	Number of districts	Number of vehicles in zone	Percentage of vehicles
Up to 100	4	377	0.70
101–200	28	4 804	8.96
201–300	45	13 252	24.73
301–400	35	13 011	24.27
401–500	24	11 538	21.52
500 or more	15	10 627	19.82
Total	151	53 609	100.00

[a] Cars and lorries included only. According to the quoted reference, p. I/30, there was a total of 84 510 of *all* motor vehicles (including automobiles, motor cycles, etc.) in the Czech Lands as at 1 Jan 1930.

the number of inhabitants per vehicle. A more detailed spatial distribution of these values can be seen in the appendix. If we construct a point diagram setting industrial horse-power against the number of motor vehicles in each of the 151 registration districts, there is a clear concentration in districts limited to 20 000 h.p. and about 500 motor vehicles; the points then become rarer in the rectangle up to about 50 000 h.p. and 700 cars per district. The remainder is then distributed in a belt up to 40 000 h.p. and 1500 cars per district, which is only exceeded by the district of Ostrava with a high concentration of power (mining ang iron metallurgy) and by the district of Prague (with a high percentage of cars).[12]

In 1930 what was then the Czech Lands (i.e. Bohemia and Moravia–Silesia) accounted for 56.1 per cent of the area, 72.5 per cent of the inhabitants, 90.3 per cent of the engine power in industry and production trades and for 90.7 per cent of the cars and lorries in the Czechoslovak Republic as a whole. Slovakia accounted for 34.9 per cent of the area, 22.6 per cent of the population and only 9.1 per cent of the engine power and 8.5 per cent of the cars and lorries. It is interesting that the Czech Lands account for nearly the same percentage of engine power in industry (90.3 per cent) and cars and lorries

Table 10.7 Correlation between industrialisation and motorism in the Czech Lands, 1930

Land	Number of registration districts	Correlation coefficient r
Bohemia	107	+ 0.67
Moravia and Silesia	44	+ 0.23
Czech Lands total	151	+ 0.47

(90.7 per cent). This convincingly indicates a close correlation between the degree of industrialisation and the intensity of motorism. This conclusion is supported by the coefficients of correlation between the performance of driving machines in industry in 1930 and the number of cars registered that same year in the 151 registration districts in the Czech Lands (Table 10.7).

The significant positive correlation between these values is conspicuous especially for the industrial, more developed Bohemia ($r = +0.67$); as regards Moravia and Silesia, where the degree of industrialisation was lower and where agriculture was more in evidence, the coefficient was considerably lower ($r = +0.23$).

THE NUMBER OF INHABITANTS PER MOTOR CAR IN VARIOUS EUROPEAN COUNTRIES, 1901–70

Tables 10.8–10 give the ratio of inhabitants per motor car in various European countries between 1901 and 1920, between 1920 and 1938 and between 1945 and 1970. (It is hoped to extend the analysis in due course to include lorries, buses and motor cycles.) Although it is realised that the aim of these statistics may be occasionally influenced by shifts in human birth and death rates, the yardstick of inhabitants per vehicle is a better measure of motorisation than total number of vehicles.

It is interesting to note how little the crisis of the early 1930s affected the growth of motor-car ownership per head: not at all in Britain, France (apart from 1932), Belgium, Switzerland (its set-back was to come in 1935), Sweden (apart from 1933), Italy (apart from 1934). For Austria 1930 was a bad year while Poland, bottom of the league, suffered throughout these years. Czechoslovakia, like Britain

Table 10.8 Development of motorism in Great Britain, France and
Germany, 1901–20

Year	Number of inhabitants per car in		
	Great Britain	France	Germany
1901	. . .	6007	. . .
1902	. . .	4188	. . .
1903	. . .	2969	. . .
1904	4506	2262	. . .
1905	2424	1803	. . .
1906	1678	1646	6055
1907	1211	1243	4219
1908	970	1037	3398
1909	833	872	2590
1910	760	729	2037
1911	566	610	1639
1912	465	510	1328
1913	390	430	1100
1914	313	362	. . .
1915	299	383	. . .
1916	295	386	. . .
1917	380	395	. . .
1918	543	410	. . .
1919	386	417	. . .
1920	228	247	(1497)

and Belgium, enjoyed uninterrupted growth though the Republic
was to suffer a small set-back in 1936.

Table 10.11 makes an overview easier. It shows, *inter alia*, the
differing effects of the Second World War, the very considerable
increase in motorisation (as judged by motor-car ownership) since
then and the narrowing of the gap between the various countries of
western Europe where the range in 1970 was only from 3.5 (Sweden)
to 6.1 (Austria): in 1920 there had been a sixfold difference between
Britain and France, on the one hand, and Italy and Germany, on the
other. In 1970 the gap between western Europe and eastern Europe
had narrowed considerably, too.

We may, if we wish, proceed further and use the method the
author has used to measure the asynchronism in the spreading of the
steam engine in industry in Europe in the nineteenth century in his
report presented at the international colloquium at Pont-à-Mousson,
France, on 30 July 1970,[14] to measure the asynchronism in the
spreading of motorism.[15]

Table 10.9   Development of motorism in some European countries, 1920–38

| | | | | | *Number of inhabitants per car in* | | | | | |
Year	Great Britain	France	Belgium	Czechoslovakia	Switzerland	Sweden	Germany	Italy	Austria	Poland
1920	228	247	(926)	435	(369)	(1497)	1141	4415	1011	14824
1921	174	197	(467)	327	(260)	1012	1068	3695	832	8493
1922	136	161	(314)	260	(199)	767	900	2640	775	6015
1923	112	134	(217)	234	159	631	694	2298	647	4693
1924	92	106	(130)	175	108	483	663	1464	675	3872
1925	75	84	(119)	137	102	369	452	1097	598	3312
1926	64	74	(113)	109	86	316	369	808	541	3088
1927	56	63	(108)	93	75	245	330	689	469	2354
1928	50	54	101	79	65	187	276	550	395	1951
1929	45	44	87	72	62	153	237	433	340	1646
1930	42	37	82	66	59	133	222	343	384	1588
1931	41	32	74	63	58	128	221	289	301	2293
1932	40	34	71	62	61	135	217	229	290	2767
1933	38	30	67	59	63	129	189	190	289	2394
1934	35	28	68	58	61	111	209	171	281	2333
1935	35	27	67	60	57	85	171	162	252	2417
1936	28	26	63	58	53	71	190	169	224	2138
1937	26	24	58	56	47	61	156	157	208	1762
1938	24	23	54	54	40	51	148	148	194	1420

Table 10.10   Development of motorism in some European countries,
1945–70

| | | | | | | Number of inhabitants per car in[1] | | | | | |
| | | | | | | | | | | | |

Year	Great Britain	France	Belgium	Switzerland	Sweden	Federal Republic of Germany	Austria	Italy	German Demo-cratic Republic	Czechoslovakia	Poland
1945	32	41	183	238	133	. . .	. . .	389	. . .	404	1038
1946	27	38	98	70	49	. . .	605	303	. . .	205	1045
1947	25	37	66	55	42	. . .	323	249	. . .	134	1052
1948	25	35	67	43	38	228	217	211	. . .	104	1059
1949	23	34	39	38	36	141	164	174	. . .	101	678
1950	22	31	32	32	28	98	143	137	243	99	623
1951	21	24	29	28	23	75	123	111	219	98	745
1952	20	23	27	26	20	57	111	93	201	97	692
1953	18	21	24	23	17	46	97	78	185	96	694
1954	16	16	20	21	13	36	78	65	171	95	765
1955	14	14	18	19	11	31	50	55	153	94	680
1956	13	13	17	17	10	25	38	48	124	82	622
1957	12	11	15	15	9	21	30	40	105	73	458
1958	11	10	14	14	8	17	25	36	91	66	344
1959	10	9	13	12	7	15	21	30	71	60	279
1960	9.2	8.1	12.1	11.2	6.2	14.8	17.5	25.2	58.2	55.3	254.5
1961	8.6	7.4	11.0	10.0	5.8	10.3	14.8	20.7	45.2	48.1	224.1
1962	7.9	6.6	10.1	8.8	5.3	8.7	12.7	16.8	38.6	43.6	189.1
1963	7.0	5.9	9.0	8.0	4.9	7.6	11.3	13.1	33.7	39.6	160.7
1964	6.3	4.9	8.1	7.3	4.6	6.8	10.1	11.1	29.2	36.8	150.2
1965	5.9	5.0	7.1	6.8	4.3	6.1	9.0	9.5	25.5	34.3	130.7
1966	5.5	4.6	6.3	6.4	4.1	6.1	8.0	8.2	23.3	31.2	112.9
1967	5.1	4.1	6.3	6.1	4.0	5.2	7.3	7.9	20.3	27.5	99.8
1968	4.9	4.2	5.3	5.7	3.8	4.9	6.7	6.5	18.1	24.0	89.6
1969	4.7	4.1	5.0	5.3	3.6	4.6	6.3	5.9	15.9	20.7	80.6
1970	4.6	4.0	4.6	5.0	3.5	4.3	6.1	5.4	14.2	17.9	73.4

Table 10.11 Development of motorism in some European countries, 1920–70

Year	Number of inhabitants per car in									
Year	*Great Britain*	*France*	*Belgium*	*Switzerland*	*Sweden*	*Germany*	*Italy*	*Austria*	*Czechoslovakia*	*Poland*
1920	228	245	(926)	435	(369)	(1497)	1141	1011	4415	14824
1938	24	23	54	54	40	51	148	194	148	1420
1945	32	41	183	238	133	·	389	·	404	4038
						FRG GDR				
1970	4.6	4.0	4.6	5.0	3.5	4.3 14.2	5.4	6.1	17.9	73.4

APPENDIX 10.A Motorism in Czechoslovakia by vehicle registration district in 1930

(a) Absolute numbers

			Horse-power of engines in industry			Number of vehicles		
Registration district	Sq. km.	Inhabi- tants	Primary driving engines	Electric motors	Total	Cars	Lorries	Total
			BOHEMIA					
Aš	142	44 998	10 118	7 979	18 097	131	101	232
Benešov	531	42 611	4 422	1 966	6 388	102	56	158
Bílina	230	39 998	7 824	13 993	21 817	97	54	151
Blatná	681	42 077	1 832	745	2 577	71	28	99
Brandýs n.L.	308	50 636	7 717	4 801	12 518	205	114	319
Broumov	408	50 240	10 316	8 589	18 905	124	55	179

Česká								
Kamenice	182	27 749	9 397	6 430	15 827	128	87	215
Česká Lípa	640	76 524	6 374	7 261	13 635	323	121	444
Český Brod	471	52 331	4 682	3 714	8 396	179	120	299
České								
Budějovice	1015	122 868	22 028	11 274	33 302	341	112	453
Český								
Krumlov	1057	61 870	7 545	25 025[1]	32 570	86	41	127
Čáslav	604	61 280	5 100	2 694	7 794	127	63	190
Děčín	421	98 179	19 797	26 511	43 308	406	162	568
Domažlice	492	46 157	3 983	3 307	7 290	75	24	99
Dubá	407	26 345	778	1 626	2 404	60	35	95
Duchcov	140	50 272	15 566	18 717	34 283	73	65	138
Dvůr Králové								
n.L.	418	73 301	21 547	18 502	40 049	251	139	390
Frýdlant	378	39 957	10 309	11 301	21 610	147	52	199
Havlíčkův								
Brod	590	53 119	4 775	2 562	7 337	109	48	157
Hořice	195	25 620	4 285	2 147	6 432	95	81	176
Hořovice	582	77 719	36 483	36 157	72 640	191	166	357
Horšovský								
Týn	629	48 334	2 614	2 007	4 621	77	48	125
Hradec								
Králové	460	88 018	13 243	14 454	27 697	396	148	544
Jindř. Hradec	711	46 818	3 885	1 743	5 628	113	38	151
Humpolec	311	24 912	6 742	1 268	8 010	63	13	76
Cheb	455	76 979	7 954	8 278	16 232	212	140	352
Chomutov	504	90 228	94 970	40 187	135 157	270	169	439
Chotěboř	539	42 204	2 374	1 114	3 488	50	25	75
Chrudim	706	92 477	9 316	6 648	15 964	224	91	315
Jablonec n.N.	210	101 882	27 385	20 394	47 779	846	206	1 052
Jablonné v								
Podj.	261	28 657	2 547	1 619	4 166	93	44	137
Jáchymov	202	17 997	1 533	3 798	5 331	42	18	60
Jičín	621	64 841	5 853	4 480	10 333	140	59	199
Jilemnice	340	41 693	8 076	3 277	11 353	125	48	173
Jílové	269	22 567	2 765	1 579	4 344	71	47	118
Kadaň	466	42 871	14 532	4 225	18 757	111	65	76
Kamenice								
n.L.	453	31 992	6 206	923	7 129	43	16	59
Kaplice	932	50 284	32 606[2]	4 709	37 315	45	14	59
Karlovy Vary	242	90 655	4 724	11 267	16 991	642	245	887
Kladno	286	84 469	66 152	27 417	93 569	222	178	400
Klatovy	872	75 777	6 157	2 579	8 736	142	47	189
Kolín	489	74 425	20 538	10 283	30 821	388	89	477
Kralovice	658	34 963	3 792	2 518	6 310	56	31	87
Kralupy								
n.Vlt.	217	35 144	9 971	3 747	13 718	157	58	215
Kraslice	172	38 881	4 198	5 010	9 208	104	36	140

Registration district	Sq. km.	Inhabi-tants	Horse-power of engines in industry			Number of vehicles		
			Primary driving engines	Electric motors	Total	Cars	Lorries	Total
Kutná Hora	551	58 672	6 301	4 985	11 286	95	81	176
Lanškroun	457	65 419	10 822	7 079	17 901	132	75	207
Ledeč n.S.	652	44 696	2 930	575	3 505	85	29	114
Liberec	343	135 445	76 619	34 512	111 131	1 027	417	1 444
Litoměřice	628	97 775	15 976	20 072	36 048	301	178	479
Litomyšl	492	46 690	1 974	1 003	2 977	63	54	117
Loket	208	40 795	23 856	9 414	33 270	60	59	119
Louny	358	45 556	6 280	3 451	9 731	163	95	258
Mar. Lázně	322	33 414	7 000	1 504	8 504	207	108	315
Mělník	413	46 618	6 907	10 417	17 324	176	92	268
Milevsko	610	35 496	1 947	641	2 588	58	29	87
Mladá Boleslav	568	86 865	26 706	18 855	45 561	235	137	372
Mnichovo Hradiště	463	40 029	7 910	3 957	11 867	120	55	175
Most	337	108 179	43 246	39 633	82 879	285	116	401
Náchod	232	55 181	23 525	21 704	45 229	189	92	281
Nejdek	242	37 682	10 583	10 211	20 794	84	37	121
Nová Paka	198	29 962	6 915	2 295	9 290	104	65	169
Nové Město n.M.	445	44 633	5 704	2 918	8 622	126	50	176
Nový Bydžov	491	53 582	5 304	4 876	10 180	127	39	166
Pardubice	786	112 669	25 111	12 823	37 934	407	167	574
Pelhřimov	729	47 882	2 040	1 369	3 409	97	34	131
Písek	974	75 371	5 369	4 025	9 394	119	61	180
Planá	561	34 205	2 549	929	3 478	61	39	100
Plzeň	660	184 556	40 565	104 443	145 008	757	324	1 081
Podbořany	580	44 925	1 726	4 011	5 737	123	39	162
Poděbrady	694	85 361	9 849	6 395	16 244	274	109	383
Polička	321	33 070	3 653	2 218	5 871	38	24	62
Praha	893	983 456	108 150	122 322	230 472	10 074	5 611	15 685
Prachatice	1094	69 778	5 878	2 100	7 978	88	58	146
Přeštice	518	44 038	2 044	1 939	3 983	29	18	47
Příbram	708	57 740	4 991	5 325	10 316	162	85	247
Přísečnice	151	28 709	2 270	3 020	5 270	89	54	143
Rakovník	645	54 110	5 136	8 226	13 362	220	93	313
Rokycany	711	59 467	7 603	11 233	18 836	59	38	97
Roudnice	459	55 002	6 134	5 060	11 194	264	91	355

Rumburk	85	28 090	4 401	4 465	8 866	172	49	221
Rychnov n.K	428	52 371	9 215	7 549	16 764	212	84	296
Říčany	234	42 623	1 638	1 266	2 904	87	42	129
Sedlčany	744	48 856	2 400	596	2 996	83	39	122
Semily	314	60 623	24 292	10 910	35 202	121	85	206
Slaný	549	86 084	36 861	23 923	60 784	320	133	453
Sokolov	292	61 629	52 833	23 414	76 247	135	78	213
Strakonice	863	70 821	4 404	2 908	7 313	63	32	95
Stříbro	878	79 103	35 830	15 084	50 914	110	141	251
Sušice	817	48 994	4 580	2 352	6 932	53	34	87
Šluknov	191	53 256	8 037	7 597	15 634	256	87	343
Tábor	978	75 581	4 706	2 698	7 404	169	47	216
Tachov	622	41 159	1 955	719	2 674	59	31	90
Teplá	388	24 757	892	429	1 321	69	42	111
Teplice	197	111 162	17 320	20 589	37 909	431	125	556
Trutnov	516	80 439	79 840	23 012	102 852	158	42	200
Třeboň	893	55 690	3 308	1 547	4 855	76	29	105
Turnov	331	46 795	3 846	3 468	7 314	130	89	219
Týn n.Vlt.	255	16 081	1 522	290	1 812	25	12	37
Ústí n.L.	356	132 977	85 703	54 786	140 489	311	117	428
Varnsdorf	79	38 386	13 071	8 057	21 128	215	68	283
Vlašim	352	23 425	1 481	529	2 010	43	20	63
Vrchlabí	355	42 383	18 114	10 712	28 826	243	65	308
Vysoké Mýto	553	68 785	5 605	3 565	9 170	165	68	233
Žamberk	600	52 921	5 863	3 171	9 034	86	32	118
Žatec	403	50 929	5 317	4 521	9 838	238	61	299
Žluticc	498	27 999	840	1 579	2 419	43	50	93

MORAVIA AND SILESIA

Český Těšín	549	85 334	29 713	68 234	97 947	80	23	103
Bílovec	326	47 675	4 261	4 799	9 060	65	47	112
Boskovice	833	89 915	7 900	5 837	13 737	148	90	238
Brno	774	366 597	77 404	78 981	156 385	1 558	437	1 995
Bruntál	592	49 134	9 066	4 052	13 118	140	41	181
Dačice	809	46 634	3 222	967	4 089	100	36	136
Frýdek	472	115 495	34 604	55 070	89 674	79	38	117
Hlučín	247	48 640	16 239	8 633	25 372	48	29	77
Hodonín	844	108 360	16 627	13 003	29 630	201	101	302
Holešov	809	79 707	12 596	20 061	32 657	83	66	149
Hranice	596	61 652	8 534	3 062	11 596	73	37	110
Hustopeče	648	76 226	7 873	6 162	14 035	69	41	110
Jeseník	736	71 717	12 016	5 274	17 290	198	40	138
Jihlava	572	69 233	9 605	5 824	15 429	197	49	246
Krnov	532	61 995	12 028	6 113	18 141	124	45	169
Kroměříž	464	66 751	8 564	3 276	11 840	152	56	208
Kyjov	462	54 387	4 602	4 455	9 057	75	42	117
Litovel	422	47 709	4 536	2 778	7 314	92	43	135

Registration district	Sq. km.	Inhabi- tants	Horse-power of engines in industry			Number of vehicles		
			Primary driving engines	Electric motors	Total	Cars	Lorries	Total
Mor. Třebová	686	76 716	11 419	7 274	18 693	126	42	168
Mor. Budějovice	660	40 308	2 072	662	2 734	111	38	149
Mor. Beroun	441	27 499	3 113	1 391	4 504	42	19	61
Mor. Krumlov	637	45 010	2 097	1 166	3 263	77	51	128
Mikulov	563	53 563	2 232	919	3 151	87	24	111
Místek	462	58 141	7 745	4 294	12 039	131	61	192
Nové Město na Mor.	817	54 696	2 908	1 010	3 918	73	37	110
Nový Jičín	500	86 242	16 176	18 785	34 961	218	75	293
Olomouc	502	125 089	15 881	14 176	30 057	465	151	616
Opava	748	115 616	15 009	12 838	27 847	404	135	539
Ostrava	356	279 014	328 267	327 456	655 723	962	366	1 328
Prostějov	472	85 731	8 047	8 080	16 127	328	99	427
Prerov	450	84 317	41 113	8 635	49 748	244	79	323
Rýmařov	382	27 584	4 317	2 140	6 457	67	26	93
Šternberk	543	59 060	7 935	5 514	13 449	143	60	203
Šumperk	807	81 183	12 591	8 751	21 342	157	54	211
Tišnov	431	37 818	2 534	2 349	4 883	49	37	86
Třebíč	720	59 991	6 426	3 165	9 591	165	59	224
Uher. Hradiště	872	137 650	9 759	6 350	16 109	344	184	528
Uher. Brod	954	79 309	3 848	1 411	5 259	165	59	224
Val. Meziříčí	452	48 803	3 080	2 556	5 636	89	25	114
Velké Meziříčí	618	39 896	2 184	1 493	3 677	58	45	103
Vsetín	553	42 422	2 752	2 134	4 886	67	26	93
Vyškov	867	100 262	6 841	7 849	14 690	166	124	290
Zábřeh	609	68 804	15 198	10 005	25 203	152	68	220
Znojmo	1 020	103 125	11 817	6 359	18 176	263	131	394

Notes

1. 1900: J. Honc, *Dějiny dopravy na území ČSSR* (History of transport on the territory of the ČSSR) (Bratislava, 1974) p. 7; R. Štechmiler *et al.*, *Naše automobily včera* (Our automobiles of the past) (Praha, 1957) pp. 130–1, etc.
2. R. Štechmiler *et al.*, *Naše automobily* . . . , p. 477.
3. Compiled and computed according to the data in the *Almanach automobilistu 1928*, vol. IX (Praha, 1928) pp. 82–253. Cars and lorries cannot be separated from these data; only lorries with hard rubber wheels are marked separately.
4. Compiled from the same source as the preceding table, pp. 21–81. The table does not include 26 tricycles of which 9 belonged to public institutions, 5 to businessmen, 4 to artisans and shopkeepers, 4 to officials, 3 to factories and 1 to a worker.
5. The scale lorry and luxury car – small car – tricycle – motor cycle (by performance) – bicycle in connection with the social and professional status of the owners at that time evokes an analogous pattern in agriculture: horses and herds of cattle belonging to estate owners and farmers – several cows to the small farmer – a cow for its milk and as traction force to the peasant and a few goats to the cottager.
6. Compiled and calculated from the following references:

   (a) 1920: J. Horký and R. Todt (eds), *Automobilní Československý almanach* (AČA) for 1921, vol. I (Praha, 1921) pp. 59–100;
   (b) *Almanach automobilistu 1928*, vol. IX (Praha, 1928) pp. 82–253;
   (c) R. Todt (ed.), *Almanach československého autoprumyslu a autoobchodu 1930 (AČA)*, vol. IX (Praha, 1930) pp. 9–232.

   All the data for the year 1928 were used if given in the reference given above (see also Table 10.1 in text); where, however, a more extensive sample of automobile owners was taken into account and, therefore, the percentages in the individual categories for 1928 come out slightly differently than in Table 10.3; for 1920 and 1930, a sample including every tenth vehicle in the list was taken into account; where the social and professional status of the owner was not given, data with this characteristic were used instead.
7. Compiled and calculated from the data in R. Todt (ed.), *Almanach československého autoprumyslu a autoobchodu* (1930), cited above, pp. 1–422, 675–794 and 803–18. The motor vehicles of public authorities and institutions and of the military administration are not included. The sample was compiled using the same method as for Table 10.3 in the text (see note 6).
8. The sample consists of 225 owners of motor vehicles (lorries and cars) from 16 vehicle registration districts in Slovakia (excepting Bratislava) whose social and professional status could be determined. See *AČA* . . . *1930*, vol. IX, ibid. part I, pp. 654–64. According to the same reference, a total of 5071 motor vehicles (cars, lorries and buses) were registered with the police directorate in Bratislava, Slovakia (ibid., part I, p. 5).

9. For Ruthenia the sample, drawn from the same sources, recorded 152 of the 560 vehicles registered in the district of Užhorod, of which 25 per cent belonged mostly to firms of the agricultural and forestry industry (distilleries, sawmills, brickworks, breweries, etc.), 17.8 per cent to farmers and estate owner, 17.7 per cent to businessmen, 11.2 per cent to officials, 9.9 per cent to artisans and traders, and the remainder to other groups.

10. Compiled and calculated using the following sources:
    (a) Area and present population, 1930: *Statistický lexikon obcí v zemi České* (Statistical lexicon of communities in Bohemia) (Praha, 1934) pp. xix–xxiv; *Statistický lexikon obcí v zemi Moravskoslezské* (Statistical lexicon of communities in Moravia and Silesia) (Praha, 1935) pp. xvii–xix; *Štatistický lexikon obcí v krajine Slovenskej* (Statistical lexicon of communities in Slovakia) (Praha, 1936) p. xxiii; *Statistický lexikon obcí v zemi Podkarpatoruské* (Statistical lexicon of communities in Ruthenia) (Praha, 1937) p. xvi.
    (b) Motor in industry and trades, 1930: *Československá statistika* (Czechoslovak statistics) vol. 114 (series XVII, no. 1) (Praha, 1935) pp. 158–66.
    (c) Cars and lorries, 1930: R. Todt (ed.), *Almanach československého autoprumyslu a autoobchodu 1930*, vol. IX (Praha, 1930) part I, p. 20, part II, pp. 1–832.

11. These registration districts frequently covered several political districts, and the partial data on the area of the districts, number of inhabitants and motors in industry and trades were derived from the data in the above sources and referred to the appropriate vehicle registration districts. The political district of Zlín (now Gottwaldov) was established only as of 1 Oct 1935 and its individual parts in the jurisdictional districts of Zlín and Vizovice are in this case included in the vehicle registration district of Holešov for 1930.

12. Further statistical information is to be found in the version of this chapter delivered as a paper to the 16th International Congress of the Historical Sciences and published in *History and Society* (Institute of Czechoslovak and World History of the Czechoslovak Academy of Sciences (Prague, 1985).

14. See J. Purš, 'La Diffusion asynchronique de la traction a vapeur dans l'industrie de Europe au XIXe siècle', in *L'Acquisition des techniques par les pays non-initiateurs*, Pont-à-Mousson: Centre Culturel de l'Abbaye des Prémontrés 28 juin–5 juillet 1970, organisé avec la collaboration du International Cooperation in History of Technology Committee (ICOH-TEC), *Colloques Internationaux du Centre National de la Recherche Scientifique*, no. 538 (Paris, 1973) pp. 75–120 and the discussion as far as p. 123). See also *ČSČH*, 18 (1970) pp. 161–94 and *Historica*, 18 (1973) pp. 139–79; *Actes du Cinquième Congrès international d'histoire économique* (Léningrade, 1970) Tome VIII (Moscou, 1977) pp. 51–8 and *Gospodarka przemysłowa i początki cywilizacji technicznej v rolniczych krajach Europy* (Wrocław–Warszawa: Ossolineum, 1977) pp. 75–91 and

also 211–12, 218–19 and 225–6; J. Purš, *Průmyslová revoluce. Vývoj pojmu a koncepce* (The Industrial Revolution. The evolution of the term and concept) (Praha, 1973) chap. XVI, pp. 474–93.

15. J. Purš, 'La Diffusion . . . , op. cit. p. 91.

11 Japan: The Late Starter Who Outpaced All Her Rivals

Koichi Shimokawa

This chapter will set the development of car-ownership and truck transport in post-Second World War Japan into a longer historical perspective. At present over 40 million vehicles are registered and sales on the home market are growing at more than 5 million per year. Eighty per cent of journeys by commercial vehicles, from light pick-ups to heavy trucks, range up to 200 km. An increasing proportion is over 100–200 km. and beyond that. In consequence rail freight traffic, once so important, is declining fast. How are these remarkable changes to be explained and do they conflict?

The first part of the chapter will consider transport in Japan before the Second World War and then outline subsequent changes in its economic and social context and relate it to road-building. We shall be particularly concerned with the rapid growth in car-ownership during the period of high economic growth from the early 1960s and the Japanese motor industry's response to it. The growth in truck traffic and its effect upon the railways will also be outlined. Pollution created by cars and trucks brought society into conflict with the motor-car industry. Its response and that of the policy-makers led to pollution control, safety regulations and traffic management in big cities. The chapter will conclude with consideration of Japan's present-day motor-transport problems seen in their international setting.

1. Slow Beginnings

For many years modern Japan developed its coastal shipping and railways, but not its road services, apart from the electric tramways which ran in the streets of the larger cities. The cheap pedal cycle, however, became a popular means of transport from the 1890s onwards, often being used for the carriage of goods as well as people. At first these vehicles were imported from the Britain and the United

214

States; but domestic production was soon developed, together with the supply of spare parts. Cycle dealers came on the scene and did good business. By 1913 there were already over 400 000 pedal cycles in Japan and the numbers continued to grow steadily until 1938, when over 8 000 000 were in use. Registrations grew at an even faster pace after the war (Table 11.1). Motor vehicles, by contrast, spread remarkably slowly until after the war and, indeed, until 1960; but the breakneck growth in subsequent years can be compared to that in the United States after the coming of the Model T Ford.

In 1912 the U.S. Commercial Agent in Japan reported that, in the previous year there were only 150 automobiles in Tokyo, 'the best market, having many miles of wide and well-made streets'. Yokohama possessed forty motor vehicles, and Kobe a mere four. The whole country had imported only twenty-two of them in 1910 (seven from the United States). The Nippon Automobile Co. of Tokyo was the sole Japanese manufacturer, but it had produced only five cars since its formation four years earlier. Tokyo had four garages and Yokohama two. This American correspondent, while believing that Japan offered a promising market for the sale of cars and trucks, did not see it becoming a large one 'because of the narrowness, poor condition and many steep grades of the roads, the weakness of many of the bridges and the destruction of both bridges and roads by the heavy rainstorms'. The poor infrastructure was not the only problem. Lack of experienced drivers was another. 'They are very hard on the gears,' he noted, 'and among other things recklessly use the foot to bring the car to a sudden stop, thus knocking out the gear teeth.'[1] A photograph of one of Tokyo's wide, lightly trafficked residential streets, taken at about this time, shows a solitary motor taxi cab in the foreground, together with a supporting cast of bicycles, reckshaws, handcarts and pedestrians.[2]

Motor-car registration grew rapidly during the 1920s, from 8000 in 1921 to 56 000 at the end of 1930. (We do not know how many of these were motor taxis.) Totals do not seem to have grown much beyond that figure in the 1930s, though statistics are hard to come by after 1935 because they were no longer officially published. The number of motor cycles (2500 in 1921 and 24 000 by the end of 1930) continued to grow, however, and the total seems to have reached that of motor cars (about 60 000) by 1939.

The numbers of buses and trucks also grew more rapidly than cars. Here the decision of the Tokyo City government to import Ford T trucks after the 1923 earthquake, and to adapt them as motor buses

Table 11.1   Pedal cycle ownership in Japan
(000)

Year	Number
1913	418
1914	608
1915	684
1916	874
1917	1 064
1918	1 258
1919	1 596
1920	2 052
1921	2 318
1922	2 802
1923	3 192
1924	3 648
1925	4 102
1926	4 370
1927	4 751
1928	5 025
1929	5 318
1930	5 779
1931	6 000
1932	6 356
1933	6 524
1934	6 895
1935	7 304
1936	7 722
1937	7 878
1938	8 305
1939	8 311
1940	8 195
1941	8 361
1942	8 618
1943	8 613
1944	8 556
1945	5 686
1946	6 276
1947	6 939
1948	8 013
1949	9 192
1950	10 859
1951	11 693
1952	12 406
1953	13 270
1954	13 667
1955	13 928
1956	15 647

1957	16 005
1958	16 815
1959	18 158
1960	19 559
1961	20 785
1962	21 952
1963	22 931
1964	23 765
1965	24 377
1966	25 430
1967	26 375
1968	27 330
1969	28 241
1970	29 291
1971	30 497

Source: Nihonjitensha Shinkokai (Century of the bicycle) (Tokyo, 1973), p. R518

to take the place of the electric tramways which had been destroyed was of considerable importance. The number of buses and trucks registered increased from under 1000 in 1921 and 2500 at the beginning of 1924 to nearly 7000 in January 1925 and 12 500 at the end of 1926. By the end of 1930 there were 40 000 buses and trucks altogether. We are told that about one-third of this total were used as buses. By then Ford had started assembling these smaller vehicles from completely knocked down (C.K.D.) parts. They were followed by General Motors and Chrysler. Altogether about 20 000 vehicles were assembled annually. They were sold to taxicab proprietors and operators of light truck transport, as well as to private owners.[3]

The Big Three American manufacturers did not assemble heavier vehicles in Japan. Here there was an opening for Japanese enterprise, encouraged by the country's military requirements. The Japanese concentrated upon the manufacture of the $1\frac{1}{2}$-ton truck, taking advantage of the Automobile Manufacturing Business Law of 1936 which provided for the control and exclusion of foreign automobile companies and a licensing system for Japanese makers.[4] Official and unofficial statistics of Japanese motor-vehicle registrations in the 1930s show that the numbers of motor buses and trucks (presumably those in civilian use only) grew from 60 000 in 1930 to perhaps about 100 000 in 1938. Of the latter, perhaps about 34 000 were motor buses.[5] If these figures can be relied upon, while the total of motor cars (no doubt including taxis) remained steady at about 60 000 or a

Table 11.2   Non-motor freight vehicles in use in Japan in 1948 and 1949

	1948	1949
Horse carriage	142 634	121 757
Ox cart	28 318	25 022
Man cart	12 771	11 254
Rear cart	13 236	11 973
Sleigh	48 982	41 262
Sledge	2 511	2 200
Total	248 452	213 448
Horse-drawn	142 797	126 690
Ox-drawn	29 212	25 850

Source: Ministry of Transport, *Rikuun Yoran 1949*, 50 (Tokyo); Tadashi Murao, *Kamotsu Yuso no Jidoshaka*, p. 283.

little more, motor buses tripled in number. Japan was evidently depending increasingly upon motor vehicles for public passenger transport as well as building up its stock of homo-produced $1\frac{1}{2}$-ton trucks.

Although truck production was increased during the Pacific War, ownership when the war was over was little more than it had been at the outbreak; and many non-mechanical forms of transport, vital when there was an energy shortage, were used as may be seen from Tables 11.2 and 11.3. Cycles, too, continued to be increasingly popular (Table 11.1). The latter, however, were to trigger off the rapid increase in ownership of mini or small motor cycles.

2. Post-war Developments: Government Policy, Road Improvement and Freight Transport

Immediately after the war, government policy and attitudes towards road-building were both of importance. Intercity railways were still the preferred transport mode for both passengers and freight; and in the rapidly growing cities, and especially Tokyo and Osaka, private electric railway companies provided a very efficient service for commuters above ground. Road transport's function was to supply short-distance communication to and from railway stations or docks. Investment in trunk roads was, therefore, kept at a low level for many

Table 11.3 Trucks in use in Japan, 1936–55

	Standard truck		Small truck		Triwheeler		Total	
	Business use	Total	Business use	Total	Business use	Total	Business use	Total
1936	42 678	51 338	··	4 272	··	39 891	··	95 501
1946	33 065	75 195	··	11 206	··	33 598	··	119 999
1947	35 432	91 301	··	11 966	··	41 137	··	144 404
1948	36 760	105 376	··	19 139	··	58 806	··	183 321
1949	44 089	115 151	··	24 134	··	81 488	··	220 773
1950	44 752	120 385	1 675	35 101	14 386	120 523	60 813	276 009
1951	44 373	131 802	2 065	41 571	16 342	159 253	62 780	332 626
1952	44 679	142 956	1 751	50 564	21 101	223 462	67 531	416 982
1953	45 085	154 495	2 019	60 091	28 727	310 265	75 831	524 851
1954	46 593	156 988	2 375	75 455	34 500	376 998	83 468	609 441
1955	47 794	159 345	3 022	95 039	40 861	435 415	91 677	689 799

Source: Ministry of Transport, *Unyukeizai Tokeiyoran 1960*; Murao, op. cit. p. 283.

years. It was financed by car registration and petrol taxes, the latter at 53.8 yen per litre, roughly equal to that in Britain and France but nine times that of the United States. This encouraged the manufacture of small, fuel-efficient vehicles which, in turn, brought in more revenue for further road improvement. Between 1960 and 1965, for instance, the total spent on new roads grew from 181 to 605 billion yen (see below, Table 11.8) and in the latter year 365.5 billion yen came from these taxes. Total road expenditure had risen to 4806.6 billion yen by 1980, of which 2077.9 billion yen came from them.

Because of the slow improvement of Japan's narrow, and often mountainous, roads, the government tended to discriminate against motor transport on grounds of road safety. City streets were often dangerous, too. There was strict traffic control, rigorous tests before driving licences were issued and careful inspection of new vehicles, both home manufactured and imported. Periodical inspections were subsequently carried out every two years. A high standard of maintenance was promoted and the manufacture of reliable, safe cars encouraged.

There were significant developments in the reconstruction period

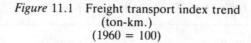

Figure 11.1   Freight transport index trend
(ton-km.)
(1960 = 100)

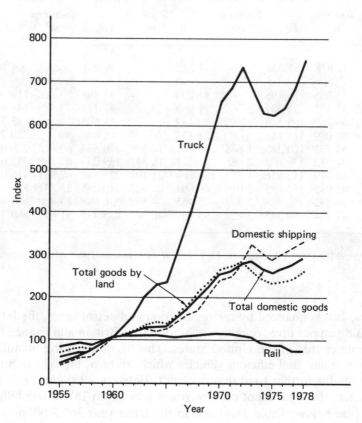

Source: Murao, op. cit. p. 18.

between 1945 and 1954 when motorised freight transport replaced animal haulage over shorter distances.[6] Carriers using tricycles or medium-sized trucks took the initiative and, in the period of more rapid growth between 1956 and 1966 as Figure 11.1 shows, began to win traffic from the railways. Inland freight traffic as a whole multiplied 2.3 times. Truck traffic grew by 6.8 times and coastal shipping enjoyed a threefold increase; but railway traffic (in ton-km.) grew by only 22 per cent, and after 1960 hardly at all. Meanwhile truck traffic went on growing quite spectacularly.

At first truck transport grew over distances up to 200 km. and

Table 11.4 Percentage of allocation of goods volume for distance zone by each measure

Year	Distance zone	Truck	National rail	Shipping	Total
	Under 200 km.	92.3	5.5	2.2	100
	200 km.–400 km.	16.6	49.5	33.8	100
1963	400 km.–600 km.	12.1	45.3	42.6	100
	Over 600 km.	2.8	33.0	64.2	100
	Total	84.8	8.9	6.3	100
	Under 200 km.	93.8	4.0	2.2	100
	200 km.–400 km.	24.2	40.1	35.7	100
1966	400 km.–600 km.	14.0	36.2	49.8	100
	over 600 km.	5.6	27.8	66.6	100
	Total	87.8	6.4	5.8	100
	Under 50 km.	97.5	1.5	1.0	100
1966	50 km.–100 km.	80.8	12.8	6.4	100
Different	100 km.–200 km.	53.5	30.8	15.8	100
classifi-	200 km.–500 km.	21.2	38.3	40.5	100
cation	Over 500 km.	8.0	30.6	61.4	100
	Total	87.8	6.4	5.8	100
	Under 50 km.	98.2	0.7	1.1	100
	50 km.–100 km.	88.7	5.9	5.4	100
1972	100 km.–200 km.	68.1	15.8	16.0	100
	200 km.–500 km.	44.5	19.6	35.9	100
	Over 500 km.	22.6	18.1	59.3	100
	Total	91.9	3.0	5.1	100
	Under 50 km.	97.6	0.7	1.7	100
	50 km.–100 km.	87.0	3.9	9.1	100
1977	100 km.–200 km.	69.2	10.1	20.7	100
	200 km.–500 km.	46.4	11.0	42.7	100
	Over 500 km.	23.3	8.2	68.5	100
	Total	88.9	2.4	8.6	100

Source: Murao, op. cit. p. 292.

coastal shipping up to 400–500 km. From the mid-1960s, however, truck transport gained traffic over longer distances, as may be seen from Table 11.4

It is clear from Table 11.4 that, while coastal shipping retained its dominant position in the longer-distance freight business between

Table 11.5   Profit and loss on the National Railway's freight and passenger traffic, 1966–78
(*hundred million yen*)

Year	Freight Revenue	Freight Profit and loss	Freight Business index	Passenger Profit and loss	Passenger Business index	Total Profit and loss	Total Business index	Contribution rate for the loss Freight	Contribution rate for the loss Passenger	Contribution rate for the loss Total
			%		%		%			
1966	2 184	Δ 535	125	96	98	Δ 439	106	100	–	100
1967	2 344	Δ 743	132	Δ 21	100	Δ 764	109	97.3	2.7	100
1968	2 377	Δ1 108	147	Δ 37	101	Δ1 145	113	96.8	3.2	100
1969	2 414	Δ1 499	162	383	95	Δ1 116	111	100.0	–	100
1970	2 508	Δ1 822	173	497	94	Δ1 325	112	100.0	–	100
1971	2 467	Δ2 153	187	10	100	Δ2 143	119	100.0	–	100
1972	2 370	Δ2 618	210	Δ 564	106	Δ3 182	128	82.3	17.7	100
1973	2 386	Δ3 184	233	Δ1 098	111	Δ4 282	135	74.4	25.6	100
1974	2 399	Δ4 052	269	Δ2 152	119	Δ6 204	145	65.3	34.7	100
1975	2 424	Δ5 141	312	Δ3 551	127	Δ8 692	156	59.1	40.9	100
1976	2 776	Δ5 508	298	Δ3 280	121	Δ8 788	149	62.7	37.3	100
1977	3 072	Δ5 789	288	Δ2 267	112	Δ8 056	138	71.9	28.1	100
1978	3 089	Δ6 076	297	Δ2 588	113	Δ8 664	138	70.1	29.9	100

Note: Δ = minus.
Source: Murao, op. cit. p. 324.

1963 and 1977, truck transport, having grown impressively over medium and longer distances between 1966 and 1972, then ceased to grow for a time because of the first oil shock. The railways, however, having lost traffic to the roads between 1966 and 1972, continued to do so. The National Railway's mounting financial losses, 70 per cent, of which were attributable to freight traffic despite containerisation and other innovations, are set out in Table 11.5.

The success of the road haulier is explained by the building of new toll roads, by larger and heavier vehicles (often travelling at night) and by the larger operators' establishing frequent, regular trunk services.

3. Passenger Travel

In 1983 there were over 42 $\frac{1}{2}$ million motor vehicles running on Japan's roads, just over 26 million of them privately owned passenger

Figure 11.2 Increasing car ownership, 1960–83

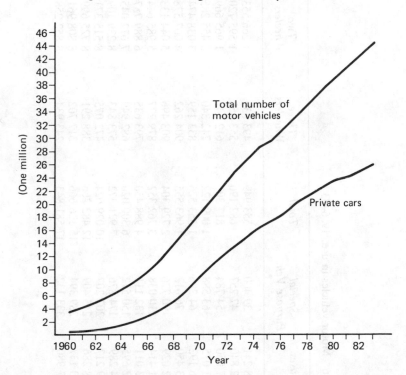

Source: Ministry of Transport, *Transport White Paper* (Tokyo, 1984) p. 63.

cars. About 5.3 million new, and about the same number of second-hand, cars are sold annually on the home market.[7] Figure 11.2 summarises the phenomenal growth since 1960.

The predominance of trucks, which, as we have seen, dated from the 1930s, continued until the later 1960s; but from 1965 motor-car ownership, still relatively tiny in 1960, also soared ahead. The totals displayed in Figure 11.2 are broken down in Table 11.6.

Private cars were responsible for a dramatic increase in personal travel, as may be seen in Table 11.7. Here, however, the National Railway was able to hold its traffic quite well after the passenger-km. peak was reached in 1974. This is explained by intercity modernisation (*shinkansen* and electrified trunk lines). The increase in this traffic, of course, concealed local losses to the motor car. Private railways managed to build up their traffic, despite motorisation, because of the growth of suburbs along their lines. Bus traffic de-

Table 11.6 Motor vehicles in use, 1956-83

End of December	Cars	Trucks	Buses	Special Purpose Veh.	Total	Three Wheelers	Two Wheelers
1956	181 074	295 234	38 241	39 400	553 949	493 839	1 266 553
1957	218 524	373 575	42 840	47 257	682 196	553 958	1 595 720
1958	259 631	455 842	47 050	54 784	817 307	612 342	1 965 669
1959	318 758	575 701	51 075	63 588	1 009 122	742 340	2 455 285
1960	457 333	775 715	56 192	64 286	1 353 526	833 159	3 038 474
1961	663 951	1 159 542	63 450	76 612	1 963 555	904 262	4 067 578
1962	889 032	1 677 467	72 029	90 776	2 729 304	903 496	5 044 133
1963	1 233 651	2 337 249	81 414	110 038	3 762 352	879 277	5 985 644
1964	1 672 359	3 090 969	93 011	132 111	4 988 450	793 635	6 889 757
1965	2 181 275	3 865 478	102 695	150 572	6 300 020	682 949	7 672 045
1966	2 833 246	4 798 961	114 289	174 876	7 921 372	573 843	8 239 109
1967	3 836 409	5 856 191	129 217	207 207	10 029 024	472 075	8 515 371
1968	5 209 319	6 879 252	148 286	245 409	12 482 266	388 291	8 725 699
1969	6 933 732	7 733 403	170 137	289 394	15 126 666	317 702	8 808 961
1970	8 778 972	8 281 759	187 980	333 132	17 581 843	243 934	8 852 258

1971	10 572 122	8 705 716	194 360	385 679	19 857 877	202 906	8 755 466
1972	12 531 149	9 230 385	202 819	444 160	22 408 513	167 671	8 607 560
1973	14 473 630	9 810 306	212 622	502 723	24 999 281	136 465	8 514 140
1974	15 853 548	10 157 905	222 430	547 423	26 781 306	119 659	8 591 688
1975	17 236 321	10 043 853	226 284	584 100	28 090 558	47 998	8 752 980
1976	18 475 565	10 750 017	222 384	621 294	30 069 260	41 406	8 932 404
1977	19 825 712	11 294 549	224 648	663 041	32 007 950	36 229	9 326 721
1978	21 279 689	11 904 903	226 970	709 172	34 120 734	30 978	10 045 622
1979	22 667 297	12 577 139	229 039	757 538	36 231 013	24 298	10 901 116
1980	23 659 520	13 177 479	230 020	789 155	37 856 174	17 724	11 965 547
1981	24 612 270	13 955 848	230 905	821 934	39 620 957	11 453	13 091 427
1982	25 539 061	14 716 570	230 813	849 935	41 336 379	9 277	14 557 879
1983	26 385 444	15 436 908	230 513	878 715	42 931 580	7 848	16 212 645

Note: Truck tractors and bus tractors are included in 'Trucks' and 'Buses', respectively.
Source: Ministry of Transport; Japan Automobile Manufacturers' Association Inc., *The Motor Industry of Japan, 1984*, p. 3.

Table 11.7   Passenger transport in passengers and passenger-kilometres
(*million*)

	Number of passengers							Rate of allocation		
	Automobile		Railway							
	Pas- senger car	Bus	Na- tion- al	Pri- vate	Ship	Air	Total	Auto	Rail	Ship
1960	1 610	6 291	5 124	7 166	99	1.26	20 291	39	61	0.5
1965	4 306	10 557	6 722	9 076	126	5.16	30 792	48	52	0.4
1970	12 221	11 812	6 534	9 850	174	15.46	40 606	59	41	0.4
1973	15 922	11 390	6 871	10 185	193	23.54	44 585	62	38	0.4
1974	16 105	11 206	7 113	10 476	178	25.27	45 103	61	39	0.4
1975	17 681	10 731	7 048	10 540	170	25.47	46 195	62	38	0.4
1976	18 679	10 231	7 180	10 402	164	28.26	46 684	62	38	0.4
1977	19 416	10 189	7 068	10 699	162	32.90	46 195	62	37	0.4
1978	21 446	9 964	6 997	10 763	162	37.12	49 369	64	36	0.3
1979	23 405	9 967	6 931	10 907	166	41.36	51 417	65	35	0.3
1980	23 612	9 903	6 825	11 180	160	40.43	51 720	65	35	0.3
1981	23 673	9 672	6 793	11 425	161	42.10	51 766	64	35	0.3
1982	24 132	9 378	6 742	11 527	156	40.48	51 975	64	35	0.3

(*hundred million*)

	Passenger-kilometre							Rate of allocation		
	Automobile		Railway							
	Pas- senger car	Bus	Na- tion- al	Pri- vate	Ship	Air	Total	Auto	Rail	Ship
1960	115	440	1 240	604	27	7	2 433	23	77	1.1
1965	406	801	1 740	814	34	29	3 824	32	68	0.9
1970	1 813	1 029	1 897	991	48	93	5 871	50	50	0.8
1973	2 257	1 117	2 081	1 048	77	160	6 740	52	48	1.1
1974	2 284	1 158	2 156	1 085	78	176	6 937	52	48	1.1
1975	2 508	1 101	2 153	1 085	69	191	7 107	53	47	1.0

1976	2 645	987	2 107	1 088	67	201	7 095	53	47	0.9
1977	2 640	1 046	1 997	1 126	65	236	7 110	52	44	0.9
1978	2 960	1 070	1 958	1 153	64	269	7 474	54	42	0.9
1979	3 199	1 083	1 947	1 178	64	302	7 773	55	40	0.8
1980	3 213	1 104	1 931	1 214	61	297	7 820	55	40	0.8
1981	3 283	1 088	1 921	1 241	60	310	7 903	55	40	0.8
1982	3 472	1 048	1 908	1 256	59	301	8 044	56	39	0.7

Source: Ministry of Transport, *Rikuun Tokei Yoran*; Nissan Motor Company, *Jidosha Kogyo Handbook* (1984) p. 574.

clined in terms of passengers rather than of passenger-km., showing that the decline was also mainly on the local services. Indeed, it is this loss which has contributed most to the decline in public transport which in itself had encouraged more people to buy cars, especially from the mid-1970s.

There has been a marked switch to the suburbs as the areas with the highest car-ownership per head of population. In 1965 this was to be found in Aichi, Tokyo, Kyoto and Kanagawa, four big city zones. By 1983 the highest ownership per head was to be found in the local prefectures, partly because of two-car families, Tokyo itself had fallen to the bottom of the table. It was then followed by Osaka, Kyoto and Kanagawa.

The huge growth in car-ownership was not foreseen. In 1958, for instance, the Economic Planning Agency forecast a total of only 2 225 000 cars by 1975.[8] This large margin of error was caused by a gross underestimate of income per head, for the high economic growth of the 1960s was never itself foreseen. The forecasters of 1958 assumed a national income per head of $573 in 1975. By 1983 it actually stood at $7821.[9] Another reason for the large underestimation was the failure in 1958 to appreciate that the Japanese motor industry would be capable of mass producing at low cost such high-quality cars. It was never appreciated either by government policy-makers or professional experts that a country with such a small habitable land area and low-price railway travel could expand car-ownership at the rate that western countries had done.[10]

In Japan, as elsewhere, the attractions of the private car were considerable: personal mobility and greater flexibility of travel from door to door in greater comfort and, for the family, perhaps at lower cost. In the earlier stages of higher economic growth additional truck transport still had priority and the Japanese preferred to spend their

Figure 11.3   The Japanese road situation in the early 1980s

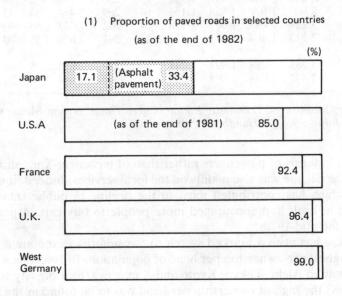

(1)   Proportion of paved roads in selected countries

(as of the end of 1982)

Sources: Japan – *Annual Report on Road Statistics*, 1983 ed. Others – *IRF Statistics*, 1983 ed.

increasing incomes on the purchase of less expensive consumer durables like electric washing-machines, refrigerators, television sets or cameras. Then, with further growth and greater purchasing power per head on the one hand, and, on the other, lower car prices resulting from higher car production, the private car came within the means of an increasing number of people.

4. Problems Arising From the Rapid Spread of Motor Vehicles

The great flood of vehicles on to Japan's roads in the later 1960s and early 1970s called for big road improvements, changes in the city infrastructure, road safety campaigns and anti-pollution measures. A lively public debate arose concerning the social costs of such rapid motorisation, the need to co-ordinate the various transport modes and the possibility of limiting the number of cars in cities so as to reduce traffic congestion. Joint action has been taken by government and people to come to grips with all these problems.

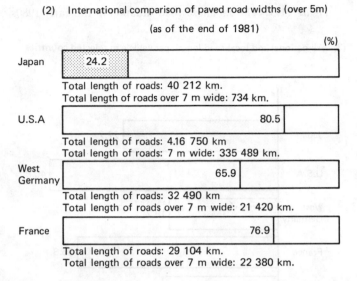

(2)　International comparison of paved road widths (over 5m)

(as of the end of 1981)

(%)

Japan　24.2

Total length of roads: 40 212 km.
Total length of roads over 7 m wide: 734 km.

U.S.A　80.5

Total length of roads: 4.16 750 km
Total length of roads: 7 m wide: 335 489 km.

West Germany　65.9

Total length of roads: 32 490 km
Total length of roads over 7 m wide: 21 420 km.

France　76.9

Total length of roads: 29 104 km.
Total length of roads over 7 m wide: 22 380 km.

Notes: 1. The figures do not include highway and expressway lengths and some are estimates.
2. A 7 m. width for roads and streets is sufficient for two vehicles to pass each other.
Source: *IRF Statistics*, 1983 ed.

Japan's road-modernisation programme, it was felt, then lagged thirty years behind those of western countries. They are still perhaps ten years behind. Japan has fewer surfaced roads, too many narrow roads and not enough motorways. Its inferior position in 1981/2 is shown in Figure 11.3.

Thirty per cent of its main roads are in heavily congested traffic areas.[11] Since the end of the 1960s, however, much has been invested in road construction and maintenance, and investment is now as high as in western countries. The length of road per vehicle is greater than in West Germany or the United Kingdom.

A very effective road-safety campaign has been conducted which has considerably reduced the number of deaths and injuries on the roads despite the vast increase in the number of vehicles. This was achieved by accident-prevention measures, the education of drivers, the building of safer cars and their regular inspection.

Particular efforts have been made to reduce air pollution in congested city centres which developed more rapidly in Japan than in

Figure 11.3   The Japanese road situation in the early 1980s

(3) All national and local road length per vehicle in selected countries

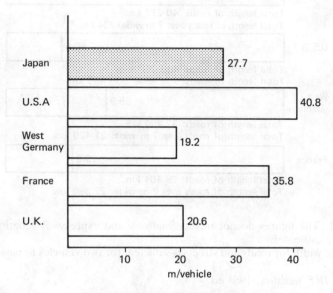

m/vehicle

Notes: Japan – as of the end of 1982. U.S.A. West Germany – as of the end of 1980. Others – as of the end of 1981.
Source: IRF Statistics, 1983 ed.

western countries because of Japan's different geographical and weather conditions. By a law passed in 1973 car-makers were obliged to decrease dangerous exhaust emissions more rigorously than car-makers anywhere else in the world. They reached these target levels within two years, having given priority to the matter in their R & D programmes. (The Americans, following suit, were given ten years to implement their law.) Tokyo and other big cities also have their own anti-pollution regulations, backed by very determined public opinion.

5. Conclusion

On the other side of the account, against the social cost and the social friction it created, must be placed the motor vehicle's undoubted benefits. It contributed to Japan's dynamic growth by providing personal mobility and flexible, low-cost freight transport. Even though

(4)   Length of highways in selected countries
(as of the end of 1982)

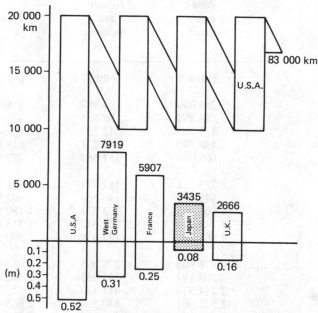

Below line shows highway length per vehicle

Notes: U.S.A. – as of the end of 1981. Japan – as of the end of March 1984.
Sources: Japan – Ministry of Construction. Others – *IRF Statistics*, 1983 ed;
Japan Automobile Manufacturers' Inc., *The Motor Industry of Japan* 1984,
p. 31.

such rapid growth in a limited land area and with intense urbanisation
has raised the National Railway's deficit, an appropriate transport mix
has been achieved by free competition within the existing legislative
framework, which gives reasonable choice for everybody. It is true, as
we have seen, that rapid motorisation did create social problems in the
big cities; but with appropriate legislation and higher investment in the
infrastructure, these problems have been reduced and can be solved in a
step-by-step manner.

When social friction was at its height there were those who saw the
car itself as the chief cause of accidents and pollution. For them, the
solution was its restriction. And, after the first oil shock, there were
advocates of car limitation in order to conserve energy. Others urged

Table 11.8   Road investment in Japan (*million yen*)

Year	New con- struct	Repair and mainte- nance	Total
1960	181 406	59 817	241 223
1961	263 050	91 987	355 037
1962	350 591	98 865	449 456
1963	413 163	112 130	525 446
1964	488 390	130 231	618 621
1965	605 089	152 867	757 956
1966	725 021	167 771	892 792
1967	845 728	146 564	992 292
1968	1 006 903	173 321	1 180 224
1969	1 143 249	202 647	1 345 896
1970	1 355 391	234 979	1 590 370
1971	1 778 905	278 695	2 057 600
1972	2 208 194	333 853	2 542 047
1973	2 214 578	411 140	2 625 718
1974	2 522 961	440 534	2 963 495
1975	2 573 999	484 438	3 058 437
1976	2 758 681	600 694	3 359 375
1977	3 475 482	721 645	4 197 136
1978	4 096 934	863 320	4 960 254
1979	4 439 647	950 842	5 390 489
1980	4 806 618	1 051 523	5 858 141
1981	4 794 067	1 097 469	5 891 536

Source: Ministry of Construction, *Doro Tokei Nenpo*; Nissan Motor Company, op. cit. p. 552.

that motor vehicles should pay the full social cost for the roads they used and the pollution they caused. Higher car taxes from many more vehicles have, to some extent, silenced these critics. On the freight side it was more difficult to attack motorisation, for the advantages of the truck's flexibility and door-to-door service were overwhelming.

In 1972 there was a debate about the co-ordination of transport in which the Ministry of Transport and other interested parties partici-

Table 11.9 Casualties in traffic accidents

Year	1970	1971	1972	1973	1974	1975	1976	1977	1978	1979	1980	1981	1982	1983
Traffic accidents	718 080	700 290	659 283	586 713	490 452	472 938	471 041	460 649	464 037	471 573	476 677	485 578	502 261	525 903
Index	100	98	92	82	68	66	66	64	65	66	66	68	70	73
Deaths	16 765	16 278	15 918	14 574	11 432	10 792	9 734	8 945	8 783	8 466	8 760	8 719	9 073	9 520
Index	100	97	95	87	68	64	58	53	52	50	52	52	54	57
Injuries	981 096	949 689	889 198	789 948	651 420	622 467	613 957	593 211	594 116	596 282	598 719	607 346	626 192	653 620
Index	100	97	91	81	66	63	63	60	61	61	61	62	64	67
Vehicles in use (thousands)	26 581	28 668	31 090	33 727	35 654	37 071	39 437	42 090	45 053	48 221	50 965	54 190	57 558	59 368
Index	100	108	117	127	134	139	148	158	169	181	192	204	211	223
per 10 000 Vehicles Deaths	6.3	5.7	5.1	4.3	3.2	2.9	2.5	2.1	1.9	1.8	1.7	1.6	1.6	1.6
Injuries	369	331	286	234	183	168	156	141	132	124	117	112	111	110

Source: National Police Agency and Ministry of Transport; The Motor Industry of Japan, 1984, p. 32.

pated. This arose from the changing relative importance of railways, motor vehicles, coastal shipping and air transport and was caused mainly by motor vehicles' threat to the railways.[12] The railways hoped that, by marshalling the social friction arguments, then most powerful, they could increase their transport allocation and reduce that going to the roads. They also pleaded that motor vehicles, unlike railways, did not pay their full track costs. This debate did result in more public money being spent on the railways. Total distances run by trucks also declined for a time because of the first oil shock, as may be seen from Figure 11.1. In the longer run, however, the strength of road transport, and the weakness of the railways, became apparent. The National Railway's deficit mounted, to some extent because of its high administrative costs and labour relations problems but also because it was obliged to cut its freight rates, and therefore its profitability given higher overheads, in an effort to meet the road competition.

The fact that trucks have managed to hold their traffic while the increasingly state-subsidised National Railway has not, shows that the former is more efficient and its service preferred by freight consignors. The same may be said of the continued and growing popularity of the private car. In the end transport allocation should depend on free choice by the user. The task of government is to make this free choice possible, not to lay down fixed quotas.

An appropriate transport mix has been reached which meets the public's need in cities great and small, in more sparsely populated areas and in travelling between cities. Many unsolved problems, of course, remain, such as traffic congestion in city centres, the need to reduce road accidents still further (and pollution, too), and the provision of better public transport for those who do not have access to cars. There is clearly scope for improved traffic management: the application of the fruits of the communications revolution by both traffic managers and vehicle drivers may help here. But, when all is said and done, public regulation should result from public consensus. Its aim should be a fair social allocation of available resources.

Notes

1. U.S. Bureau of Foreign and Domestic Commerce, *Special Report No. 53 on Foreign Markets for Motor Vehicles* (1912).

2. 'Sights and Scenes in Fair Japan' (1910), reproduced in Julian Pettifer and Nigel Turner, *Automania* (1984) pp. 56–7.
3. *The Résumé Statistique de L'Empire du Japon* (published annually by the General Statistical Office, Tokyo, between 1912/13 and 1937) does not differentiate various sorts of four-wheeled passenger vehicle. Reports of the U.S. Department of Commerce, the League of Nations (covering 1926–30) and the overseas edition of the *American Automobile* attempt further disaggregation. See Also *Kindaii nihonkeieishi no kisochishiki* (Basic Study of Modern Japanese Business History) (Yuhikaku, Tokyo, 1979) pp. 79, 157–8, 285–6.
4. Ibid. pp. 286–7.
5. *Résumé Statistique* op cit.; *The American Automobile* (Overseas Edition) op cit..
6. Tadashi Murao, *Kamotsu Yuso no Jidoshaka* (The Motorisation of Cargo Transport in Japan) Hakuto-Shobo, Tokyo, 1982) pp. 9–11.
7. Department of Transport, *Unyu Hakusho* (White Paper on Transport) (Tokyo, 1984) p. 63.
8. Genpachiro Konno and Yukihide Okano (eds), *Gendai Jidosha Kotsuron* (Study of Modern Motor Transport) (Tokyo University Press, 1979) p. 39; Economic Planning Agency, *Nihon Kotsu no Genjo to Shorai* (Japanese Transport: Its Present and Future) (Tokyo, 1958) p. 60.
9. Economic Planning Agency, *Kokumin Keizai Tokei Nenpo* (Annual Statistics of National Economy), 1984).
10. Konno and Okano (eds), op. cit. pp. 39–41.
11. Ibid. pp. 79–80.
12. Ibid. pp. 212–20. Department of Transport, *Wagakuni no Sogo Kotsutaikei* (Co-ordinated Transport System in Japan) (Tokyo, 1972) pp. 299–301.

12 Motor Transport in a Developing Area (i) Zaïre, 1903–59

Epanya Sh. Tshund'olela

Historians of Zaïre have until now neglected the study of motor transport, preferring to research further upon the country's railways and waterways and, when it is dealt with at all, to relate motor transport to them.[1] In fact this is seeing twentieth-century transport development from quite the wrong viewpoint. It can be argued that motor transport in Zaïre is now so all pervasive, influencing every aspect of daily life, that rail and water, catering for medium and longer-distance traffic, should be related to it; and, of course, all shorter-distance journeys have always been made by land.

This chapter deals with motor transport in the country from its beginnings to the end of the colonial period. There were three main phases:

(i) the faltering, scattered and certainly very limited beginnings from 1903 to 1920;
(ii) the spread of motor vehicles, though still in a very limited way, between 1921 and 1948, though with more rapid development after 1935;
(iii) the ending of motor transport's dependence upon other forms of transport after 1949 and its more vigorous competition with them.

THE FALTERING BEGINNINGS, 1903–20

Before 1903 the only means of carrying freight overland in Zaïre was on the backs of human beings or of animals, though in the 1890s attempts had been made to improve these age-old methods for the carriage to rail or water of export products such as ivory, rubber or copal (a type of resin for varnish) by the introduction of pack saddles, the training of elephants and the harnessing of cattle to light carts (1000 to 1200 kg.).[2] When none of these experiments succeeded, the

236

authorities turned to mechanical traction which was starting to come into use elsewhere in the world.[3]

The first trials, carried out in 1903/4 over distances up to 10 km., with heavy, $5\frac{1}{2}$-tonne wood-burning *steam* vehicles, were also a failure; but, three years later, in 1906, a state-sponsored scheme, again using wood-fired steam wagons, but this time over relatively long distances with loads of only 800 kg., was more successful. In 1909 there were twelve of them licensed. It was in 1906, too, that the first stretch of road in Zaïre suitable for motors began to be used. Running from Buta to Bambili in the north-east of the country, its construction had been started in 1901; it was not completed over its whole length, 200 km., until 1915/16.[4]

It was also in the north-east of the country (Uele and Ituri), a mining district also rich in ivory and rubber and without any rail and water communication, that motor transport using *petrol* engines was first developed after 1910. Motors were also used near the coast in present-day Bas-Zaïre. These initiatives were, however, halted by the First World War and it was only towards the end of it, in 1917, that the first Lacre lorries came into use with state assistance. In 1918 it experimented with motorised transport on the Buta-Bambili road and in 1919 Ford lorries, with trailers, made their appearance. The results were, however, disappointing. Only 30 tons of exports were carried along the road in 1920, about 500 kg. per week.

Various private ventures by Catholic and Protestant missionaries and business undertakings enjoyed little more success. Motor transport was still of very little importance. This is hardly surprising, for the total length of road was still under 400 km. On the whole, very little had been achieved. The few vehicles which were in use frequently broke down and, when repaired again, ran very slowly. The people of Uele called them *bandakure* (tortoises). Costs were so high as to be almost prohibitive. The white drivers and mechanics had to be highly paid. Motor spirit was scarce and expensive. The backs of sturdy Africans remained much cheaper, sometimes amounting to only a quarter of that charged by the slow and unreliable motor vehicles.[6]

THE SLOW AND UNEVEN SPREAD OF MOTOR TRANSPORT, 1921–35

Progress was perfunctory and, in the absence of adequate capital investment, it depended very largely upon the efforts of the African population. Attention was paid to the construction of main roads as part of the colonial government's efforts to encourage economic growth by the more vigorous exploitation of the country's resources. Much of this development occurred in areas unserved by rail or water. Carriage on men's backs tied up a great part of the available labour force at the expense of agriculture and mining. The resulting shortage of suitable labour slowed down the rate of growth while, at the same time, the need to feed all this transport labour was making additional demands upon food supplies.[7] E. Leplae pointed out in 1918 why the country's transport needs could not continue to be met by the traditional method of carriage:

> The costs that it [human carriage] imposes are the greatest obstacle to commercial and agricultural growth. Since at least three, and often four, porters are needed to carry 100 kg., 40 men are needed to carry a ton of goods. They receive 70 centimes a day and their food. So the carriage of a ton costs 28 francs a day. To walk 100 km. would take these men five days and would, therefore, cost 140 francs per ton, three times the cost from the centre of the Congo to Europe! . . .[8]

If economic growth was to go ahead, there was clearly a strong case for mechanising this expensive and inefficient system of carriage which treated human beings as beasts of burden; but this would involve considerable substitution of capital for labour.

The colonial administration devised a comprehensive road programme and enlisted the support of mining and other business interests in it during the post-war recession of 1920/1 by offering road-building work to their unemployed labour force.[9] As may be seen from Table 12.1 overleaf, this policy really laid the foundations of Zaïre's road network. A mere 2550 km. in 1920 had grown to nearly 12 000 km. by 1925. It exceeded 20 000 km. in 1927, neared 30 000 km. in 1930 and reached almost 50 000 km. by 1934. In pursuing this road-building policy, as the Belgian Minister of Colonies, Louis Franck, told the Chamber of Representatives in Brussels in 1920, the authorities were much influenced by the results of extensive road-building in British East Africa and Uganda.[10]

The Belgian authorities may have perceived the crucial importance of an adequate road network; but they did not appreciate how much capital needed to be substituted for labour. The whole burden of the task, in fact, fell upon an untrained African labour force, much of it press-ganged, which had to meet the huge challenge, working with rudimentary tools for a mere pittance. Not surprisingly, the results were very poor. The work was hurried and skimped, often amounting to no more than the clearing of grass and the removal of tree stumps. Important natural obstacles were by-passed even though this resulted in steep gradients and fierce bends. Bridges were usually makeshift, temporary and limited to narrow crossings; wider rivers had to be crossed by ferry. The heavy rains of the tropics often washed away whole sections of these ill-constructed routes and rendered them quite unusable in the rainy seasons, which lasted for about six months of the year. Indeed, during these months the authorities, well aware of the dangers, had to prohibit all 'heavy' motor lorries (i.e. those carrying loads of over one tonne) from using these newly built roads altogether.[11]

They were built, of course, mainly as feeders to waterways and railways serving the export trade, not for internal purposes apart from the few which served the needs of the colonial administration or of missionaries. In short, the new road layout was determined by international economic considerations. In due course, however, some roads were built parallel to the routes of proposed railways to convey material, labour and material for railway construction. These roads are still in use today.[12]

The roads were pointless without motor vehicles and these spread in Zaïre only slowly. As the *Rapport de la Commission Permanente pour la Protection des Indigènes* put it very bluntly in 1923: 'There are roads but no vehicles. . . . [The Commission] realises that the government has made monetary sacrifices and that the native population has paid with much human effort. Yet it has all been unproductive'.[13] There were many reasons for this. First of all, there was colonial opposition resulting from the high cost and innumerable difficulties encountered during earlier attempts to introduce motor transport. Colonial opinion was hostile to the africanisation of motor drivers and engineers, one obvious way of cutting costs. There were, however, other authorities which took a different point of view, were keen to cut costs and opposed the presence of 'poor whites'. From the early 1920s they encouraged africanisation, especially in the northeast of the country. More conservative colonial opinion was disturbed by this and recommended the establishing of a telegraph connection between the two ends of motor roads!

Another reason for opposition to motorisation was that the vehicles concerned and the sources of oil and petrol were not Belgian,

Table 12.1 Length of motor roads in Zaïre, 1920–59
(km.)

| Year | Grand Total | Public roads | | | Private roads |
		Main roads	Local roads	Total	
1920	2 550	...	...	...	...
1925	11 791	...	...	...	...
1926	16 839	...	...	...	...
1927	20 281	...	...	...	...
1928	24 051	...	...	...	...
1929	26 799	...	...	...	...
1930	29 908	...	...	...	...
1931	35 144	...	...	...	7 975
1932	38 766	...	...	...	9 002
1933	42 756	...	...	...	8 182
1934	49 151	...	...	...	...
1935	54 002	27 391	17 672	45 063	8 939
1936	61 171	...			
1937	68 175	15 002	42 580	57 582	10 493
1938	70 730	14 015	47 883	61 898	8 832
1939	74 039	...			
1940	77 347	...	...		
1941	80 660	...			
1942	83 968	...			
1943	87 277	...	...	...	...
1944	90 586	...			
1945	92 067	...			
1946	94 409	15 207	67 717	82 924	11 485
1947	100 524	15 946	73 459	89 405	11 119
1948	108 498	14 765	82 039	96 804	11 694
1949	111 971				
1950	114 355	15 231	86 185	101 416	12 939
1951	117 436	30 917	73 544	104 461	12 975
1952	119 891	30 983	75 417	106 400	13 491
1953	123 108	31 771	76 856	108 627	14 481
1954	133 443	32 027	86 532	118 559	14 884
1955	135 485	32 987	88 082	121 069	14 416
1956	139 051	33 108	90 338	123 446	15 605
1957	140 676	33 620	89 934	123 554	17 122
1958	145 213	33 787	93 806	127 593	17 620
1959	147 349				

Sources: Royaume de Belgique, Rapports annuels sur l'administration de la Colonie du Congo belge présentés aux chambres législatives, 1933–58; Congo belge, Statistiques relatives a l'année . . . (1956–9); A. Huybrechts, Transports et structures de développement au Congo . . . (Paris–The Hague: Mouton, 1970) p. 383.

and involved purchases in foreign currency and greater economic dependence on other countries. How much better to use railways which burned local wood![14] 'Colonial exploitation', as Colonel Paulis reminded his readers in 1923, 'is intended to produce primary produce for the mother country, not to involve purchases of foreign goods for dollars or pounds.'[15] In order to counter some of these objections, attempts were made to use local sources of energy such as palm oil or cotton seed oil;[16] but they were not a success and, by the mid-1920s, it was realised that there was no alternative to costly imported petroleum products.

From this time onwards it was appreciated that motor vehicles had to have colonial support. Animal haulage disappeared and carriage on men's backs remained the only competitor over long distances because of motor vehicles' continued high fixed and operating costs, their unreliability, short life and slowness, together with the shortage of drivers and qualified maintenance men. The authorities, for their part, believed that profit margins were too high in road haulage. If these could be cut, traffic would grow and even higher returns would accrue to the operators.[17] For the time being, since bulky, low-value freight could not stand the high transport costs to rail or water, economic growth continued to be frustrated.[18] Only when these transport costs fell could African produce be carried from farther afield and more imported products sold in these remoter areas.[19]

When motor transport did arrive the African producers were at a greater disadvantage than the European firms, for it drove down prices in the areas of production; so much so that some native producers preferred to carry their own produce to other places even farther away (50 to 100 km.) where they could obtain three times the price available to them at home.[20] So the older system of porterage continued.[21] As a colonial put it: 'Do they actually think they will make human carriage disappear by substituting for it a form of transport three times more expensive?'[22]

Attempts were made to cut motor transport rates by subsidy from colonial revenues;[23] but without success, for these rates were still not brought down to those of porterage. Only by the colonial administration's direct discrimination against the latter did motor transport slowly overcome the competition. Under a decree of 19 March 1925 provincial governors were empowered to forbid long-distance porterage in regions where modern means of transport were more or less developed.[24] Porterage was abolished in some areas and sharply reduced in others. This raised its price and tipped the scales in favour of motors.[25]

Table 12.2   Increase in the number of motor vehicles by category in Zaïre, 1925–60*

Years	Cars Number	Trucks and vans Total Number	Trucks	Vans	Tractors	Buses	Trailers	Other	General Total
1925	90	140	...	...	45	...	1	...	...
1926	1 410	1 180	...	...	73	...	1	...	...
1927	1 980	1 525	...	...	81	...	1	...	...
1928	2 483	2 497	...	...	174	...	1	...	...
1929	2 996	2 947	...	...	185	...	187	...	...
1930	3 445	3 097	...	...	186	...	171	...	...
1931	2 911	2 959	...	...	183	...	152	...	...
1932	2 692	2 621	...	...	120	...	139	...	...
1933	2 508	2 320	...	...	92	...	132	...	...
1934	2 339	2 188	...	...	103	...	155	...	...
1935	2 662	2 418	...	...	92	...	127	...	...
1936	2 955	2 833	...	...	84	...	152	...	...
1937	3 172	3 480	...	...	92	...	113	...	...
1938	3 677	3 946	...	...	86	...	139	...	...
1939	4 318	4 452	...	...	75	...	146	...	...
1940	4 494	4 381	...	...	72	...	119	...	...
1941	4 672	4 689	...	...	71	...	90	...	...
1942	4 677	5 227	...	...	73	...	129	...	...
1943	4 775	5 253	...	...	75	...	115	...	...
1944	4 930	5 750	...	...	77	...	93	...	...
1945	4 766	4 583	...	...	77	...	101	...	...
1946	4 851	6 684	...	...	103	...	90	...	11 728
1947	5 389	7 733	...	...	167	...	176	...	13 465
1948	6 796	10 089	...	...	206	...	146	...	17 237

The number of motor cars and commercial vehicles increased year by year until 1930, the former only slightly exceeding the latter, as Table 12.2 shows. The grand total, tiny in 1925 (a mere 230), was still only about 6500 in 1930. These were highly concentrated in certain areas: the north-east (Uele, Ituri), the south-east (the mining zones of the present Upper Shaba), in Léopoldville (Kinshasa) and (in smaller numbers) in Kasai. Even within these areas there were heavy concentrations. For instance, in 1928, more than 91 per cent of motor vehicles in Katanga (all of present-day Shoba and some parts of eastern Kasai) were to be found in the Upper Luapula district.[26] There were great disparities in motor transport costs between these

Table 12.2   Increase in the number of motor vehicles by category in Zaïre, 1925–60*

Years	Cars	Trucks and vans			Tractors	Buses	Trailers	Other	General Total
	Number	Total Number	Trucks	Vans					
1949	9 678	11 227	. . .	. . .	294	. . .	177	. . .	21 376
1950	11 137	14 576	. . .	. . .	546	174	438	. . .	26 871
1951	13 540	16 960	. . .	. . .	695	227	562	. . .	31 984
1952	16 602	19 005	. . .	. . .	734	290	685	. . .	37 316
1953	20 049	20 851	. . .	. . .	952	312	671	. . .	. . .
1954	23 048	22 829	. . .	. . .	. . .	387	897	. . .	. . .
1955	24 085	19 179	11 659	7 520	433	328	. . .	36	. . .
1956	29 073	21 001	12 502	8 489	543	388	. . .	75	. . .
1957	33 829	22 285	13 124	9 161	584	402	. . .	92	. . .
1958	35 000	21 858	12 679	9 179	619	489	. . .	58	. . .
1959	38 109	21 252	12 029	9 223	725	473	. . .	58	. . .
1960	34 372	26 483	15 615	10 868	810	638	932	58	71 471

* Only registered motor vehicles which paid tax. Those of governments, state-approved religious institutions and diplomatic corps and ambulances are not included.

Sources: Royaume de Belgique, *Rapports annuels sur l'administration de la Colonie du Congo belge présentés aux chambres législatives, 1925–58*; Congo belge, *Statistiques relatives a l'année* . . . (1956–9); A. Huybrechts, op. cit. p. 383; Mb. Kalongo, op. cit. pp. 19 and 22.

areas where freight rates fell a little and elsewhere where they could be three or four times as high.[27]

Some of these vehicles were operated by the state; others by mining and other businesses; and yet others by the Catholic and Protestant missions. The lorries did not do much work. For instance, on the Buta–Bambili motor road, only one return journey a week was the norm. The carrying capacity was usually limited to two tonnes. In 1932, for example, about 80 per cent of them came into this category and only about 1 per cent carried over four tonnes.[28] (See Table 12.3.)

Haulage at this time was almost exclusively in the hands of Europeans or their associates, usually American. The state widened its interests and a number of professional hauliers appeared, notably (in the north-east) Les Messageries Automobile de la Province Orientale

Table 12.3   Increase in the number of motor trucks by tonnage capacity, 1932–53

Years	Less than 1 tonne		1–2 tonnes		2–3 tonnes		3–4 tonnes		More than 4 tonnes		Total
		%		%		%		%		%	
1932	384	14.7	1 663	63.4	401	18.1	149	5.7	24	0.9	2 621
1933	363		1 449		344		138		26		2 320
1935	421		1 332		490		153		22		2 418
1937	752		1 204		1 148		333		37		3 480
1938	932	23.6	1 122	28.4	1 568	39.7	304	7.7	20	0.5	3 946
1946	1 539		707		2 514		1 670		254		6 684
1947	1 784		659		2 515		2 385		390		7 733
1948	2 668	26.4	1 111	11.0	3 483	34.5	2 078	20.6	749	7.4	10 089
1952	5 492		1 042		3 775		4 781		3 915		19 005
1953	6 171	29.6	944	4.5	4 304	20.6	5 222	25.0	4 710	22.6	20 851

Source: Royaume de Belgique, *Rapports annuels nu l'administration de la Colonie du Congo belge présentés aux chambres legislatives* (1932–53).

(MAPO, established 1925), Les Messageries de l'Ituri Oriental (MIO, 1924) and Les Chemins de Fer Vicineaux du Congo (VICICONGO 1927); and in Shaba and the two areas of Kasai Les Messageries Automobile de Sankuru (MAS 1929).[29] These hauliers tended to monopolise the general haulage business in their areas, especially in the depressed early 1930s when competition intensified and the number of commercial vehicles registered fell from 3000 to just under 2200 (Table 12.2). VICICONGO, which exploited both road and rail transport, fought its road transport rivals by offering inclusive road–rail rates and gave preference to customers who sent all their traffic through their company.[30] The cost of vehicle maintenance also favoured the large, monopolistic concerns. U.S. vehicles were increasingly used in preference to French or Italian. Porterage experienced something of a revival because of the abundance of labour during the slump.

Meanwhile the road system continued to be extended (Table 12.1), also taking advantage of the abundance of labour. African communities were given greater responsibility for building up a secondary road network, only the roads of general interest (RIG), main roads, continuing to be supported by the colonial government itself.[31] The central administration also moved whole villages at this time and resettled them on motor roads, an activity which increased African awareness of the realities of colonial rule.[32]

THE REAL FOUNDATIONS OF PROGRESS, 1935–48

As Tables 12.1–3 show, the years 1935–48 witnessed more rapid development. It is true that the pace of new road-building slowed down (the total length only doubled, from 50 000 km. to just over 100 000 km. in this period); but the network became more coherent.[33] In July 1933 provincial governors had adopted a *Conseil de Gouvernement* which aimed to link the six main provincial towns, and this was completed during the Second World War as part of a policy of economic regional integration.[34] As yet, however, these roads carried little traffic compared with the export routes. Other roads were also built which were not intended to be feeders to rail or water transport, such as those already mentioned, running parallel to these other transport modes. Although this drew protests from the railways and waterways concerned, motor transport as a whole continued to widen their hinterlands and to generate new traffic for them. O. Jadot has noted that, between 1933 and 1937, the hinterland of the B.C.K. (Bas-Congo and Katanga) Railway grew from 50 to 200 km. and provided 200 per cent more traffic.[35] The development of motor transport explains this remarkable growth as well as the cut in water and railway freight rates.

The number of motor lorries running on the older and newer roads grew more than fourfold between 1935 and 1948, from 2400 to about 10 000. Those with a carrying capacity over three tonnes grew faster than the average and, technically improved, could cover greater distances. American vehicles, especially Fords, became much more numerous. Motor cars, it should be noted, continued to be of relatively less importance in Zaïre than in wealthier countries were the number of potential customers for them was far greater. They were, in fact, fewer in number than commercial vehicles, especially in the years after the Second World War. A higher proportion of cars, however, were larger vehicles of 15 h.p. or more.

The standard of some roads was upgraded to some extent during this period by the relocation of some poorly routed stretches; the building of some permanent bridges; and the relaying of some earlier foundations as well as resurfacing. On the whole, however, this work was limited to the major trunk routes. The so-called 'secondary roads' remained much as before[36] and, as traffic increased, so maintenance became more burdensome. The investment in roads was in local labour rather than public capital. Road repair was, inevitably, somewhat superficial and consisted largely of clearing out watercourses or removing accumulations of sand.[37] The cost of maintaining these roads, categorised as 'd'intérêt local', fell increasingly upon the African communities. The colonial government confined its attentions to what were denominated 'routes d'intérêt général' (R.I.G.), the highways which linked the main centres of European residence and the principal industrial areas or those joining the Zaïre road network with that of neighbouring countries.[38]

As may be seen from Table 12.1, the length of highway designated as R.I.G., for which the colonial government was responsible, was halved between 1935 and 1938 and was still at that level, about 15 000 km., in 1950, whereas the length of local roads was almost trebled, to 48 000 km., in 1938 and went on growing, to over 80 000 km., by the end of the 1940s. R.I.G. roads could support vehicles carrying loads of three to four tonnes or even more whereas local roads were usually limited to 1.6 tonnes or less. Roads in the north-east (Uele, Ituri) and in the south-east (present-day Upper Shaba) were the best in the country, partly because of the terrain there. Some people advocated load limitation in order to protect existing roads; but the colonial authorities took the view that roads should be improved to accommodate the heavier loads.[39] Table 12.3 shows that the percentage of heavier loads grew from eight to about twenty-eight between 1938 and 1948. During the war older vehicles had been kept on the road, despite a shortage of spare parts. When they were replaced after 1945 by heavier and more efficient vehicles, the battered old stalwarts could be bought cheaply (mostly by Africans and sometimes on credit) and given another lease of life.

Transport costs per tonne fell as larger consignments of freight became available, as services became more regular and as more efficient vehicles came into use. The training of African drivers and maintenance men also cut costs, as did the better organisation of fuel and spare parts supplies. On the other hand, the shortage of return loads, the lack of competition in some districts and the continued seasonal interruption of services all tended to work in the opposite direction and to push up costs higher than they would otherwise have been. And the lack of complete reliability meant that traders had to maintain higher stocks or to despatch orders sooner than would otherwise have been the case. Peak loads after good harvests also increased costs by demanding larger fleets of vehicles which were under-used at other times.

In Zaïre, as elsewhere, stronger motor competition obliged the railway companies to cut their rates and improve their services.[40] This defensive action did not, however, prevent a significant share of both goods and passenger traffic from being switched to motor vehicles. Some parts of the railway system became quite uneconomic and had to be closed. The Decauville line connecting Charlesville (now Djoko Punda) and Makumbi, which bypassed the waterfalls of Upper Kasai, exploited by Fourminière from 1926, was closed ten years later and its traffic carried by road.[41]

Some farming, industrial or mining concerns bought vehicles to carry their own traffic, and sometimes to carry the goods of others, too. Few Africans, however, were able to operate on this basis for much the same reason that few Africans ran their own cars. But the number of motor transport companies was growing and operating services over longer distances. Competition among them, however, was soon limited by agreement. The state-owned S.T.A. was intended to run passenger as well as goods services in those parts of the colony where private companies did not do so.[42]

All this extension of motorisation stimulated economic growth. It also helped increasingly to determine the location of villages, the development of administrative centres and the movement of population, thereby encouraging the rural exodus.

PROGRESS 1949-59

The spread of motor transport played a vital role in the africanisation process in Zaïre. In the country's first Development Plan 43 per cent of the budget was devoted to transport as a whole, and a third of this went to motor transport.[43] All the previous improvements, to which we have already referred, were taken further and at a more rapid pace. Roads were strengthened so as to be able to carry heavier vehicles, ferries were replaced by bridges and in some cases motor ferries were introduced. It was realised that asphalt surfaces were needed for ten-ton lorries pulling ten-ton trailers. By the end of the Development Plan period, 1545 km. of road had been asphalted – even so, this was a mere 2 per cent of the total – and a further 1100 km. of gravel road had been built, together with 611 bridges and 40 ferries.[44] Manual labour was to some extent replaced by road-making machinery, especially on R.I.G.s. Further attention was paid to drainage and sand removal,[45] and, by a decree of 15 April 1949 which came into force on 1 January 1951, even local roads began to be maintained at state expense. All this was very necessary, for the heavier traffic subjected the roads to greater wear and tear and the ever-present tropical climate made them all the more vulnerable.[46] In these circumstances labour-intensive maintenance continued to be important, and rising labour costs (as well as inflation) contributed to the huge increase in the expense of road upkeep. It grew from 436 francs per km. in 1935 to 12 540 francs in 1951, both at current prices. Such were the escalating costs that in 1954 the whole road programme under the Development Plan had to be scaled down.[47] The 9100 km. of main trunk road and 3135 km. of feeder road due for attention were reduced to 4500 km. and 507 km. respectively. In fact, at the end of the ten-year plan only 2000 km. had been completed altogether.[48] A newspaper reporter described the state of affairs in one particular region in 1957:

I had the impression that only our roads in Lusambo–Panis–Penge were in a bad state . . . Garage men find it hard to stock up with spare parts. Motor vehicles which had to be abandoned are beyond counting. Since the beginning of the rainy season, I have not seen a single M.A.S. lorry operating between Lusambo and Kabunda return from its expedition (and that is the right word for it) undamaged. Every departure . . . is a nightmare for the M.A.S. mechanic at Lusambo. It is unnecessary to record the chapter of accidents to lorries and cars which venture upon these roads: engines seized up; gear boxes and rear axles in fragments; radiators broken.[49]

Yet, despite these disappointments, the ten-year plan had accepted the crucial importance of an interconnected road system as vital to economic growth. Motor transport had come of age even if it had as yet been unable to display its full potential. Without ignoring the feeder roads, top priority was given to the main highways, those going from north to south and from east to west connecting with roads in neighbouring African countries.[50] Within the country these main roads linked with other parts of the road system in such a way as to encourage economic growth in the various regions.[51] These regions became more directly concerned with international trade; the carriage of goods on men's backs over long distances was greatly reduced or even abolished altogether. At the same time the motor lorry penetrated to even the remotest village, thus stimulating rural change.

Table 12.2 shows that the number of motor lorries and vans nearly doubled during the 1950s, and the more detailed returns from 1955 indicate that lorries enjoyed their full share of this increase. And they were getting bigger all the time. By 1953 22.5 per cent of them had a carrying capacity over four tonnes compared to 7.4 per cent in 1948. Traffic, in fact, grew at a faster rate than the road network. Most journeys were over short distances, but all of them became faster, more efficient and reliable; and, with more manufactured goods being imported, back freights became easier to obtain. It is not easy to make any precise statistical analysis because the official returns do not take account of public and other vehicles which were untaxed;[52] but there was clearly much growth in all regions, largely due to the more intensive use of vehicles.[53] The article of 1957 already cited went on to dwell upon the arrival of larger vehicles in Kasai:

> Twenty-five years ago [1932] the small, 500 kg. Ford pick-up made its appearance. Then came the lorries of a ton and a half. Then the two tonners, which were called at that time 'mastadontes', which were looked at with amazement and handled by European drivers. Further evolution followed: $3\frac{1}{2}$ tons, $4\frac{1}{2}$ tons, and $5\frac{1}{2}$ tons; then 7, 8 and 10 tons. . . .[54]

This transition to larger and more efficient vehicles in the more advanced areas of the country – with diesels coming cautiously into use – was delayed elsewhere by the state of the roads. More specialised motor vehicles, such as tankers and ambulances, also made their appearance. All had longer lives than their predecessors because of the greater availability of garages and trained mechanics and improved spare-parts facilities. The motor bus must also be categorised as a specialised vehicle in a country where it was customary for passengers to get a lift by motor lorry. Table 12.2 shows that motor buses increased rapidly in number during the 1950s from a mere 174 to over 470 (and to 638 in 1960). The number of private cars also grew more rapidly during the 1950s and exceeded motor lorries in total after 1953. During these years there was a switch to smaller, less powerful vehicles.[55] Africans owned more motor cars, too – 5817 in 1956 (about one-fifth of the total) and in that year were responsible for nearly 12 per cent of freight carriage by road.[56]

As costs continued to fall, so road transport was able to compete more effectively with rail and water. When an asphalt road was opened between Léopoldville (Kinshasa) and Matadi, most of the competing railway's traffic went to the lorries. The same was true on the resurfaced road between Bukava and Bujumbura via Uvira which drove the competing railway (the Chemin de Fer de Kivu), formed in 1929, out of business. On the Boma-Tshele stretch of railway, road competition quickly won the passenger and fuel traffic, local traffic and a large part of supplies of provisions inwards and produce outwards leaving the Chemin de Fer du Mayumbe with only wood, bananas and oleaginous items. It survived with state subsidy.[57] On the Elisabethville (Lubumbashi)–Jadotville (Likasi) route, road transport took all local passengers and goods, leaving the railway only with mining traffic.

Water transport, which produced poor returns, also felt the keen effects of competition from more efficient road transport. In the dry season, when water levels were low, owners of larger vessels had to rely upon road transport beyond the points to which the waterways were navigable. A large part of this traffic became a source of profit to road-transport operators.[58]

There is no doubt that the economy as a whole benefited from all this competition through cheaper freight rates, lower fares and better service.[59] Motor transport particularly favoured passenger travel and also contributed directly to the growth of imports of petrol, oil, spare parts, etc., for its own particular requirements.

The African road hauliers mentioned earlier took advantage of the ease of entry into the road-haulage business in much the same way as men of modest means managed to do in other countries throughout the world, beginning as owner-drivers, often with second-hand vehicles, and then, if successful, extending their activities by the employment of others. They profited from the increased availability of traffic during the 1950s and managed to break the monopolies of the larger, European-owned concerns. The state, too, changed its attitude, favouring private firms increasingly. The activities of the S.T.A., for instance, were progressively reduced.[60]

Both freight and passenger traffic benefited from the activity of these small operators who organised both urban and interurban

Table 12.4 Motor vehicle penetration in some African countries, 1957–60

Countries	Total road network (km.)	Km. of road for 100 km.² area	Number of motor vehicles for 100 road kilometers	Inhabitants for motor vehicles
Nigeria	67 110	7.26	61	827
Tanzania	45 648	4.87	64	309
Kenya	39 510	6.78	64	105
Zambia	35 536	4.76	117	55
Ghana	30 186	12.66	102	137
Zimbabwe	28 422	7.30	396	22
Ugánda	18 062	7.65		219
Cameroon	15 800	3.66	173	117
Morocco	15 000	3.37	984	74
Ivory Coast	14 000	4.35	164	139
Senegal	11 900	6.03	255	102
Guinea	7 600	3.09	140	250
Zaïre	145 213	6.19	40	235

Source: A. Huybrechts, *Transports et structure de développement* . . ., p. 67 (op. cit. n. 1, below).

services. In the two main cities, Elisabethville (Lubumbashi) and Léopoldville (Kinshasa), however, the administration encouraged the creation of a large concern, Le Société des Transports en Commun du Congo (T.C.C.) which put into service diesel motor buses and gyrobuses.[61] Operating costs and prices fell as the small operators, with lower overheads, were content with smaller returns.[62] The African carriers were prepared to provide door-to-door services, to stop wherever their customers requested and to transport both passengers and goods. They also used African-run garages where charges were lower than those owned by Europeans.

Road accidents increased with the number of vehicles and the government had to enforce stricter traffic regulations. A special traffic police was established in urban centres.

CONCLUSION

By the end of the 1950s Zaïre ranked rather low in the motorisation league of African states. As may be seen from Table 12.4, it came bottom of the list in the number of vehicles per 100 km. of road and near the bottom of that of motor vehicles per inhabitant (only Nigeria, Tanzania and Guinea then had more inhabitants per vehicle). Its lowly position in the ranks of the developing African states makes it particularly worthy of study, for it is clear that even motorisation on such a modest scale could play a significant part in a country's economic development. The townward movement of population was encouraged and, at the same time, the remoter areas were linked to the major markets. Country folk were encouraged to produce for the market and so were drawn into the wider capitalist economy. The different settlements, including new ones along the motor roads, became less confined. They began to be integrated in a more significant way into the regional, interregional, national and international economy. External influences began to play a greater part in the lives of millions of people.

Notes

1. See, among others, A. Huybrechts's studies and particularly *Transports et structures de développement au Congo. Étude du progrès économique de 1900 à 1970* (Paris–The Hague; Mouton, 1970).
2. Paulis (Colonel), 'Voies secondaires pour les Transports interrégionaux au Congo', in *Congo* (1923) I, pp. 592–4; A. Huybrechts, op. cit. pp. 12–13., 23–4; E. Sh. Tshund'olela, 'Politique coloniale, Économie capitaliste et Sous-Développement au Congo belge. Cas du Kasaï (1920–1959)', Ph.D. thesis (History), University of Lubumbashi, 1980, vol. I, p. 416.
3. 'Les Voies de communication au Congo', in *Le Mouvement géographique*, 50 (10 Dec 1899) col.608; 'Les Transports pour automobiles en Afrique', *Le Mouvement géographique*, 51 (17 Dec 1899) col. 616.
4. 'La Route pour automobiles entre le Congo et le Nil', *Le Mouvement géographique*, 4 (24 June 1909) cols. 37–8 and 40; 'La Route pour automobiles dans l'Uele', in *Le Mouvement géographique*, 29 (18 July 1909) col. 343; 'Les Transports automobiles . . .', in *Le Mouvement géographique*, 13 (31 Mar 1907) col. 160.
5. Paulis, op. cit. p. 594.
6. Ibid. pp. 594–5; A. Huybrechts, op. cit. pp. 24 and 35; A. Huybrechts, 'Les Routes et le Trafic routier au Congo', *Economic and Social Papers*, V, 3, Oct 1967, p. 284, and A. Huybrechts, 'La Politique économique du Congo à travers la tarification des transports', in *Études Congolaises*, XII,4 (Oct–Dec 1969) pp. 61–2 and 65.
7. E. Sh. Tshund'olela, op. cit. pp. 419–21; Paulis, op. cit. pp. 593–4.

8. E. Leplae, 'La Question des Transports au Congo belge', *La Tribune congolaise* (11 July 1918) p. 3. See also B. Jewsiewicki, *Agriculture itinérante et économie capitaliste. Histoire des essais de modernisation de l'agriculture africaine au Zaïre à l'époque coloniale* (Lubumbashi, roneo), pp. 95, 119 and 260 and J. P. Peemans, *Diffusion du progrès économique et convergence des prix. Le Cas Congo-Belgique 1900–1960. La Formation du système des prix et des salaires dans une économie dualiste* (Louvain: Nauwelaerts, 1968) p. 59.

9. Royaume de Belgique, *Rapport annuel sur l'administration de la Colonie du Congo belge présenté aux chambres législatives* (1921) pp. 96 and 106–7.

10. 'Moyens de Transport intérieurs et étude des moteurs', in *Congo*, II, 2 (Dec 1920) p. 380 and A. Huybrechts, op. cit. pp. 35–6.

11. A. Huybrechts, op. cit. p. 57; A. Huybrechts, 'La Politique économique . . .', op. cit., p. 61; Paulis, op. cit. p. 594.

12. H. Nicolai and J. Jacques, *La Transformation des Paysages congolais par le Chemin de Fer. L'Exemple du B.C.K.* (Bruxelles: Institut Royal Colonial Belge, 1954) pp. 30–1.

13. L. Guebels, 'Relation complète des Travaux de la C.P.P.I.', *Bulletin du C.E.P.S.I.* 20 (1953) pp. 246–7.

14. Paulis, op. cit. pp. 595–6; M. Mathot, 'Utilisation de l'huile de palme comme Force motrice', *Congo* (1920) II, 2 (Dec 1920) p. 400.

15. Paulis, op. cit. pp. 595–6.

16. M. Mathot, op. cit. p. 395.

17. A. Huybrechts, 'La Politique économique . . .', op. cit. pp. 61–2 and 65; Mb. Kabongo, 'Histoire des transports routiers au Katanga (1920–1959)', M.A. thesis, University of Lubumbashi, 1983, pp. 61, 63 and 88; Royaume de Belgique, *Rapport annuel sur l'administration de la colonie du Congo . . .* (1922) p. 36 and E. Sh. Tshund'olela, op. cit. pp. 386–7.

18. A. Huybrechts, 'La Politique économique . . .', op. cit. pp. 58, 61 and 65; Mb. Kabongo, op. cit. p. 63.

19. R. Engels, 'Du Développement et des possibilités économiques de la province du Congo-Kasaï', *Bulletin de la Société Belge d'Études et d'Expansion*, 65 (avril 1928) p. 187.

20. E. Sh. Tshund'olela, op. cit. pp. 387, 424–5.

21. L. Guebels, op. cit. pp. 246–7.

22. Paulis, op. cit. p. 595.

23. Ibid; A. Huybrechts, op. cit. p. 36; A. Huybrechts, 'Les Routes et le Trafic routier au Congo', op. cit. p. 284.

24. See decrees of 19 March 1925 and of 9 December 1925 *Bulletin Officiel du Congo belge* (1925) p. 179; (1926) p. 7. See also E. Sh. Tshund'olela, op. cit. p. 423.

25. This strategy has been adopted at the end of nineteenth century for railways. See A. Huybrechts, 'La Politique économique . . .', op. cit. p. 62.

26. Royaume de Belgique, *Rapport annuel sur l'administration de la colonie du Congo belge . . .* (1928) p. ; (1931) p. 64.

27. For example in 1931 the cost–price (tonne–kilometre) was:
— 3 francs: Stanleyville (Kisangani) – Buta
— 8 francs: royal Congo–Nile road
— 8/10 francs: network road of the district of Lulua.
(Royaume de Belgique, *Rapport annuel sur l'administration de la Colonie . . .* (1931) p. 64.)
28. Royaume de Belgique, *Rapport annuel sur l'administration . . .* (1932) p. 62.
29. A. Huybrechts, op. cit. pp. 32 and 36; Royaume de Belgique, *Rapport annuel sur l'administration . . .* (1933) p. 350.
30. *Ibid.*; A. Huybrechts, 'La Politique économique . . .', op. cit. p. 68, and E. Sh. Tshund'olela, op. cit. p. 456.
31. See *Bulletin Officiel du Congo belge* (1933) p. 1004.
32. E. Sh. Tshund'olela, op. cit. p. 395; M. S. Inwen, 'La Suppression du Portage au Katanga 1925–1930', M.A. thesis, National University of Zaïre, Campus of Lubumbashi, 1973, p. 31; Mb. Kabongo, op. cit. p. 84.
33. A. Huybrechts, op. cit. p. 42.
34. Ibid. See also Royaume de Belgique, *Rapport annuel sur l'administration . . .* (1935) p. 93.
35. O. Jadot, 'Transports sur Route au Congo belge', *Bulletin des séances de l'I.R.C.B.* (1938) IX, pp. 158–61. See also Mb. Kabongo, op. cit. p. 65.
36. E. Sh. Tshund'olela, op. cit. pp. 388–90.
37. Royaume de Belgique, *Rapport annuel sur l'administration . . .* (1935) p. 94.
38. Ibid. (1937) p. 114; A. Huybrechts, 'Les Routes . . ', op. cit. p. 288.
39. E. Sh. Tshund'olela, op. cit. p. 397.
40. A. Huybrechts, op. cit. pp. 88–90, 113–14; A. Huybrechts, 'La Politique . . .', op. cit. p. 68.
41. E. Sh. Tshund'olela, op. cit. pp. 378–9.
42. B. Kabatantshi, 'Occupation économique de la province du Kasaï 1945–1960, M.A. thesis (History), National University of Zaïre, Campus of Lubumbashi, 1973, p. 164, and Mb. Kabongo, op. cit. pp. 72–3.
43. G. Verhaegen, 'La Planification au Zaïre', *Planification et Développement économique au Zaïre*, J. P. Breitengross (ed.) (Hambourg: Deutsches Institut Für Africa-Forschung, 1974) p. 16; 'Le Plan décennal pour le développement économique et social du Congo belge', *Bulletin Commercial belge*, 7 (25 July 1950) pp. 23 and 43.
44. A. Huybrechts, op. cit. p. 42; 'Le Congo belge', *Bulletin Commercial belge*, 7 (25 July 1950) pp. 43–4; *Congrès Colonial National – VIIè session 1952 (Les transports congolais-La main-d'œuvre) – Rapport et comptes rendus* (Bruxelles, 1952) p. 121; 'Les Investissements réalisés dans le secteur de transport congolais entre 1950 et 1956', *Bulletin de la Banque centrale du Congo belge et du Ruanda-Urundi*, 6–7 (June–July 1958) p. 237.
45. *Plan décennal pour le Développement Économique et Social du Congo belge*, vol. I (Bruxelles: Ed. Vischer, 1949) p. 138; Royaume de Belgique, *Rapport annuel sur l'administration de la Colonie . . .* (1954) p. 309; Mb. Kabongo, op. cit. pp. 58–9; E. Sh. Tshund'olela, op. cit. p. 398.

46. E. Van De Walle, 'Les Transports au Congo 1958–1964', op. cit. p. 405; Royaume de Belgique, *Rapport annuel sur l'administration* (1951) p. 300.
47. Royaume de Belgique, *Rapport annuel* . . . (1951) p. 300; 'Les investissements réalisés . . .', op. cit. 6–7 (June-July 1958) p. 237.
48. 'Les Investissements réalisés . . .', op. cit. p. 237.
49. 'Kabundji, encore ces routes', *Le Commerce colonial* (Luluabourg, Saturday, 2 (1957).
50. *Congrès Colonial national.*, op. cit., . . , p. 12.
51. 'Le Plan décennal pour le développement . . .', *art. cit.*, p. 16 and 44; E. Sh. Tshund'olela, op. cit. pp. 352, 391–2; Mb. Kabongo, op. cit. p. 77.
52. 'Les investissements réalisés dans le secteur . . ., op. cit. pp. 237–8.
53. A. Huybrechts, 'Les Routes et le Trafic routier . . .', op. cit. pp. 311 and 326.
54. 'Kabundji, encore ces routes', op. cit.; *Congrès Colonial national* . . ., p. 25.
55. 'Les Investissements réalisés . . .', op. cit. p. 239.
56. Ibid.
57. A. Huybrechts, op. cit. pp. 41, 89–90, 113–4.
58. E. Sh. Tshund'olela, op. cit. pp. 365 and 392.
59. A. Huybrechts, op. cit. pp. 113–4.
60. E. Sh. Tshund'olela, op. cit. p. 407; Mb. Kabongo, op. cit. p. 72.
61. 'Les Investissements . . .', op. cit. p. 239; A. Huybrechts, op. cit. p. 42.
62. Mb. Kabongo, op. cit. pp. 72–3.

13 Motor Transport in a Developing Area (ii) Soviet Central Asia

M. A. Akhunova, B. A. Tulepbaev and J. S. Borisov

At the beginning of the twentieth century both Kazakhstan and Russian Turkestan (on the territory of which the Uzbek, the Kirghiz, the Tajik and the Turkmen Soviet Socialist Republics of Central Asia have been set up) used to be rather backward colonial outskirts of the Russian Empire, feudal despotic states in a vassal dependence on tsarism. In political, economic and socio-cultural respects they were rather heterogeneous, yet they were united by the very low living standards of their more than 10 million inhabitants.[1]

A feature of this backwardness was the primitive state of transport in the regions which consisted of pack animals, animal-drawn vehicles, small vessels in the Aral and Caspian seas and barges plying the rivers of Amu Darya and Syr Daria. Camel caravans were the most common day-to-day sight in the region. Steam locomotives were an exotic diversion. The short and discontinuous line of the single-track Turkestan railway crossed the south-western corner of that huge area of about 4000 square kilometres. This was far from being an adequate link between Central Russia and the Turkestan Governor-Generalship that vanished into the vast expanses of Siberia. Motor transport was virtually non-existent here in Tsarist times. The occasional motor car or bus was quite an event. By 1917 there were only a few dozen of them in the whole area. As a rule, they belonged either to the top brass of the colonial government or to the local big wigs. For example, in 1917 there were only three motor cars, all of foreign make, in Ashkhabad (now the capital of the Turkmen Soviet Socialist Republic).[2] The narrow and crooked streets of the cities and towns could hardly accommodate pedestrians and traditional pack-animals, let alone motor vehicles. There was no road in the modern sense at that time. The caravan paths and tracks that had been beaten over the centuries were the best that the region could provide.

The October Revolution of 1917 ushered in a new period in the social development of the region. The pace of economic and cultural development quickened. Deep changes in social relations released the forces of economic growth and, as one of the consequences, led to dissatisfaction with the existing forms of transport. It was, however, not easy to overcome the age-old backwardness. Even in the mid-1920s there were still only 310 motor lorries in Central Asia.[3] As the well-known economist, G. N. Cherdantsev, put it in 1928: The ground roads of Central Asia can be divided into cart, caravan, draught and horse-riding roads. . . . Motor transport is in an embryonic state and far from all roads are fit in their present condition for motor communication.[4]

Progress gradually resulted, however, from improved socialist planning, energetic efforts of the Soviet state and help from more developed regions and republics, with that from the Russian Federation to the fore. Motorisation of the former colonial borderlands of the Russian empire became one of the clear demonstrations of the new Soviet national policy. Motor transport became not only part of a more efficient freight transport system but also essential to the whole standard of living in the region in the second half of the 1920s. After the country had recovered from the main consequences of the world and civil wars, economic growth, public education and national culture made exceptionally rapid strides in Kazakhstan and the Central Asian republics, as in the USSR as a whole. By 1984 total industrial output in the country was 5.5 times as great as it had been in 1922. Kazakhstan's industrial production grew 10 times over the same period, Tajikistan's 9 times, Kirghizia's 7.5 times, Uzbekistan's 5 times and Turkmenia's 3 times. Gross agricultural output in the USSR increased 5.5 times on average, compared with 14.5 times in Tajikistan, 13 times in Kazakhstan, 12 times in Uzbekistan and 11 times in Kirghizia and Turkmenia.[5] The planned and rapid advance of industry and large-scale agriculture called for parallel technical development elsewhere, above all in the provision of tractors and motor lorries.

The increase in motor transport, the consequence of economic growth, itself became an important precondition of further economic and social advance. The state made a vast investment in large enterprises for the manufacture and repair of motor vehicles, road-building equipment, road-building itself and the training of personnel for the exploitation of this new technology. All forms of transport

Table 13.1   Bus traffic in the Soviet Union and in Soviet Central Asia in
1940, 1960 and 1983
(million passengers)

	1940	1960	1983		*of which rural areas account for*
USSR as a whole	590.0	11 315.6	44 548.0	(× 75.5)	7 620.1
Uzbek SSR	8.8	346.8	2 217.2	(× 252)	1 101.1
Kazakh SSR	10.3	589.6	3 007.6	(× 292)	352.3
Kirghiz SSR	2.9	141.4	523.0	(× 180)	157.6
Tajik SSR	6.2	89.3	385.0	(× 62)	119.3
Turkmen SSR	17.5	88.5	344.2	(× 20)	40.5

benefited. Since the start of the 1920s railway freight traffic in
ton-kilometres has multiplied 198 times and motor freight transport
has grown an astonishing 4858-fold.[6]

Motor transport has developed particularly fast in Kazakhstan and
Central Asia since the 1950s as a result of new industrial construction
and the opening up of virgin territory. By 1984 in the USSR as a
whole it handled 55 times more freight than in 1940, while in Tajikistan and Kirghizia the growth was about 70-fold, in Turkmenia and
Kazakhstan nearly 80-fold and in Uzbekistan 145-fold.[7] In the latter
Republic motor transport now hauls 15 times more freight than rail,
air and water put together. Indeed, in a number of areas of the region
the motor lorry is the only means of transport. In Kirghizia 97 per
cent of all freight and passenger traffic goes by road.[8] In the Soviet
Union as a whole, the volume of freight carried by road is less than
that carried by railway, pipeline and water;[9] but more passengers
now travel by road than by rail. In 1983 bus traffic accounted for
423 460 million passenger-km. to which the growing motor-car traffic
must be added, while the railways accounted for 361 500 million.[10]
Over longer distances, however, the railways carried more passengers as well as freight.

As can be seen from the Table 13.1 the rate of passenger motorisation in the area since 1940 is very high and exceeded that of the Soviet
Union as a whole. The number of inter-town bus routes in Uzbekistan rose from 591 to 1000 between 1940 and 1985, and, adding urban
bus routes, more than 2500 bus routes covering 119 000 km. now
operate in that Republic. In the last four years 380 new routes were

put into operation. All towns and district centres, and 96 per cent of collective and state farms, are served by buses.[12]

The fleets of taxis in the region have likewise grown far more rapidly than in the Soviet Union as a whole. If by 1984 taxis in the USSR as a whole carried 70 times more passengers than in 1940, in Uzbekistan the increase was 1300-fold.[13] In addition, apart from motor cars used for state purposes, more and more are being sold for private use – and motor cycles, too. In 1965 1700 cars and 19 900 motor cycles and scooters were sold to the population in Uzbekistan and fifteen years later 42 700 and 63 700 respectively. (Note that ownership of private cars, mainly Moskvitch, Zhiguli (Lada), Niva, etc., was growing at a faster rate than that of motor bicycles and scooters.) In 1970 the Uzbek population had 167 200 motor cycles and scooters in personal use and ten years later 351 200. As for private cars, in the early 1980s one family in five had one. In the country as a whole over 20 million cars of different types were in personal use. Their production has grown particularly rapidly since the 1960s, increasing swiftly in the 1970s and 1980s. If in 1940 only 5500 cars were manufactured in the USSR, in the early 1980s the industry was producing upwards of 1.3 million cars a year.[14] Even so the demand, both public and private, far outstrips the supply. Measures are being taken to ensure further expansion of the production of motor vehicles of all sorts.

These developments would have been impossible without a vigorous road-building programme, People's projects were started and spread rapidly in the Republics of the Soviet East in the 1930s. The state financed construction but impetus was given to it by the most active voluntary participation of the local population. The inhabitants of near-by villages, realising that new roads led them from backwardness to progress, appreciated the vital importance of this participation. From 1938 to June 1941 over 5000 km. of regular roads were designed and built in the Soviet Republics of Central Asia. Even the outbreak of war did not stop the process: by the end of it in 1945 another 700 km. of hard-surface motor-roads and about 500 km. of dirt road had been laid.[15]

All this radically changed the transport situation in the region. If by 1922 under 150 km. of hard-surface motor roads had been built in Central Asia and Kazakhstan, in 1984 the figure was 145 000 km., an increase of nearly 1000 times, (out of the total 773 000 km. in the USSR as a whole which had grown only 50-fold in the same period).[16] Motor roads developed much faster than railways, the network of which in the country as a whole more than doubled (reaching some 144 000 km.) during these sixty-two years, while in Central Asia and Kazakhstan it increased 3.5 times bringing it to about 21 000 km.

In the density and quality of motor roads Kirghizia holds one of the first places in the country.[17] In recent years particular attention has been paid to major trunk roads and to the building of motor roads in rural areas, linking by public highway district centres and state farms. Since 1978 responsibility for the building and maintenance of local roads has been transferred from collective and state farms to the state network. This has resulted in more effective road-building and maintenance, for the Ministry of Motor Road Building and Maintenance and other ministries concerned in the republics have more facilities (machinery and skilled personnel) than collective and state farms.[18] The amount of machinery was increased and its quality improved during the 1960s and 1970s. It is supplied on a country-wide basis, but the region has its own enterprises producing it, too. Thus in the 1970s and early 1980s a total of 864 new machines for road-building and upkeep were developed in the whole country, of which twenty-five were developed by designers in Uzbekistan.[19] Better roads have been built using asphalt-concrete and cement-concrete coating. In the second half of the 1970s in Uzbekistan alone the number of asphalt-concrete plants increased from 21 to 35.[20] The new motor roads and the increasing number of vehicles using them greatly stimulated the economic development of the region and brought together its often isolated areas. In a report from Kirghizia *Pravda* wrote: 'Mountain gorges, narrow valleys, the main site of settlements in the republic, owing to the splendid roads, have lost their former quality of god-forsaken areas.'[21]

Motor vehicles always demand their own support systems which are, in fact, separate branches of industry in their own right. The region has gradually built up a network of garages, repair workshops, filling stations and enterprises manufacturing engines, spare parts and, as has been seen, road-making machinery. People previously cattle-breeders or cotton-growers swelled the ranks of those engaged in servicing motor transport. Together with the communication workers, the industry employs about 10 per cent of the total working population (in 1982 the share of Uzbekistan being 6.9 per cent, in Tajikistan – 7 per cent, in Kirghizia – 7.8 per cent, in Turkmenia – 8.8 per cent, and in Kazakhstan – 11.7 per cent), all receiving above-average earnings.[22] The state had created economic incentives in order to attract people to this sphere of activity. The number of drivers in agriculture in Uzbekistan, for instance, grew from 30 000 in 1965 to 60 000 by the beginning of the 1980s.[23] The employees with higher and secondary education account for about 10 per cent of the total number in the transport industry, having grown far more rapidly than the others. In Uzbekistan there were 9000 of the more skilled in the mid-1960s and 33 000 in the early 1980s.[24]

Despite all this economic development with which motor transport has been directly or indirectly involved, a much higher proportion of the population still remains rural in these eastern republics than in the Soviet Union as a whole: 66 per cent in Tajikistan, 61 per cent in Kirghizia, 58 per cent in Uzbekistan, 53 per cent in Turkmenistan and 43 per cent in Kazakhstan (for the USSR as a whole the figure is 35 per cent).[25] At the same time the republics have the highest population growth rates in the USSR and are quite capable of supplying all the region's labour requirements for manufacturing and for the service industries. Motorisation clearly has a continuing part to play in increasing the region's labour mobility as well as in its cultural and technical development.[26]

It is also important for maintaining and improving the health of the region by the distribution of medicines and, especially in the remoter mountainous and desert areas, by the provision of special vehicles which can bring medical aid to the sick and incapacitated and to women in childbirth.

Tourism, which has grown considerably during the past two decades, also depends on an adequate supply of motor vehicles, both private cars and buses. By 1982 there were 520 tour routes in Uzbekistan alone and close on 500 000 people took advantage of them.[27]

Yet, despite all this progress, the population still does not quite appreciate the full possibilities of motor transport or the transport needs of the area.[28] Five-year plans envisage the balanced growth of all transport modes in which motor transport will play a particular part. Uzbekistan, for instance, proposes a 5 per cent increase in freight traffic by 1990, 260 km. more highway and increased sales of motor cars and motor cycles for private use. Similarly high rates of growth are envisaged in the other republics of the area. All this is seen as vital if the dramatic changes which have altered every aspect of people's lives in the course of little more than a single generation are to be sustained.

Notes

1. *The Commemorative Collection of the Central Statistical Department, the Ministry of Interior* (St Petersburg, 1913) p. 41 (in Russian).
2. *The Transport of Turkmenistan – An Historical and Economic Overview* (Ashkhabad, 1974) p. 24 (in Russian).
3. G. N. Cherdantsev, *The Central Asian Republics* (Moscow, 1928) p. 143 (in Russian).
4. Ibid.
5. *The National Economy of the USSR, 1922–1982. A Jubilee Statistical Yearbook* (Moscow, 1982) pp. 74–5 (in Russian); *The USSR National Economy in 1983. Statistical Yearbook* (Moscow, 1984) p. 36 (in Russian).
6. Ibid. p. 325; Ibid. *1983*, p. 314.
7. Ibid. p. 329.
8. *The National Economy of the Uzbek SSR in 1980. A Statistical Yearbook* (Tashkent, 1981) p. 175 (in Russian); A. Zhabarov, L. Akhmetov, *Transport Complex of Uzbekistan in the Period of Developed Socialism* (Tashkent, 1982) pp. 22, 24 (in Russian); *Pravda* (10 Feb 1985) (in Russian).
9. *The National Economy of the USSR in 1983*, p. 314 (in Russian).
10. Ibid. p. 315.
11. Ibid. p. 332.
12. *The National Economy of the Uzbek SSR in 1980*, p. 181 (in Russian); *Pravda* (27 Jan 1985) (in Russian).
13. *The USSR National Economy in 1983*, p. 334 (in Russian); *The National Economy of the Uzbek SSR in 1980*, p. 181 (in Russian); *Pravda* (27 Jan 1985) p. 3.
14. *The USSR National Economy, 1922–1982*, p. 196 (in Russian); *Pravda Vostoka* (4 Dec 1984) (in Russian).
15. L. U. Yusoupov, *Socialism and Creative Activity of the Masses* (Tashkent, 1977) pp. 82, 95, 111 (in Russian).
16. *Transport and Communication in the USSR. A Statistical Collection.* (Moscow, 1980) p. 59 (in Russian); *Pravda* (27 Jan 1985) p. 3 (in Russian).
17. *Pravda* (10 Feb 1980) p. 6.
18. *Materials of the CPSU 26th Congress* (Moscow, 1982) p. 173 (in Russian); A. Zhabarov, L. Akhmetov, op. cit. pp. 82–3.
19. *The USSR National Economy in 1983*, p. 101 (in Russian); *The National Economy of the Uzbek SSR in 1982. A Statistical Yearbook* (Tashkent, 1983) p. 25 (in Russian).
20. A. Zhabarov, L. Akhmetov, op. cit. p. 88.
21. *Pravda* (10 Feb 1985).
22. *The National Economy of the Uzbek SSR in 1982*, p. 187 (in Russian); *The National Economy of the Tajik SSR in 1982. A Statistical Yearbook.* (Dyushambe, 1983) p. 177 (in Russian); *The National Economy of the Kirghiz SSR in 1982. A Statistical Yearbook* (Frunze, 1983) p. 127 (in Russian); *The National Economy of the Turkmen SSR in 1982. A*

Statistical Yearbook (Ashkhabad, 1983) p. 127 (in Russian); *The National Economy of Kazakhstan in 1982. A Statistical Yearbook*) Álmà-Ata, 1983) p. 140 (in Russian).
23. *The National Economy of the Uzbek SSR in 1980*, p. 173 (in Russian).
24. Ibid. p. 211.
25. *The USSR National Economy in 1983*, pp. 12, 16, 17 (in Russian).
26. V. N. Zvanov, *Motorization and Society* (Moscow, 1975) pp. 58–9 (in Russian).
27. *Pravda Vostoka* (4 Dec 1984) (in Russian).
28. *Pravda* (10 Feb 1985) (in Russian).

14 Death on the Roads: Changing National Responses to Motor Accidents*

James Foreman-Peck

> Unwilling marriage, her blood runs with one
> Who bought for a few pounds and pence
> A steel machine able to 'do a ton'
> Not knowing at a ton a straw will pierce a breast:
> No wheel has built-in sense
> Not yet the shiniest and the best

DAVID HOLBROOK: *Unholy Marriage*

The new transport mode killed far more people than had railways and horse-drawn traffic. As motor vehicles spread, the death toll rose. By 1929 in the most thoroughly motorised country, the United States, motor accidents ranked tenth among the principal causes of death.[1] Policy measures in the richer countries eventually restrained the growth of fatalities in road accidents during the 1970s, but by then the contagion had already infected poorer nations. A World Health Organization survey found, for fifteen developing countries in 1972, that road accidents accounted for almost 17 per cent of the total number of deaths studied, a value exceeded only by mortality from enteritis.[2] Here we examine the various policies adopted in industrial states to control the problem and their effectiveness. We pay particular attention to the years between the World Wars in this rather anglocentric discussion because, despite the period's poor reputation, motorisation then proceeded very rapidly and the road-accident

---

* I have been helped by conversations with Michael Jones-Lee and Anthony Ogus, but they are not responsible for the product.

264

problem assumed national importance for the first time. We focus on road deaths because the statistics are more comparable over time and between countries than those on road injuries.

TRENDS IN ROAD DEATHS

At the turn of the century the diffusion of the motor vehicle merely displaced one source of death on the roads by another. Of the 354 fatal traffic accidents between 1896 and 1906 in Manchester, one was due to a motor car and one to a lorry.[3] Even in France, which early took the lead in motor vehicles, cars in 1906 were responsible for only 13 per cent of the total number killed by horses and carriages, and little more than one-fifth of accidental deaths on the railways.[4] Five years later, though, the urban road fatality pattern clearly reflected more intensive motor-vehicle use. Chicago topped the city road death league with 10.16 deaths per 100 000 of population. New York followed with 8.95, and Paris, with 8.28, was close behind.[5] Germany's slow adoption of motor transport was mirrored in Greater Berlin's low fatality rate of 4.81, compared with 7.89 for London, where passengers' transport was being transformed. During the eight years before 1912 the horse-drawn bus had been almost eliminated by motorised competitors. Their improved performance stimulated the demand for transport but brought greater hazards as well. Although buses carried fewer passengers than trams in 1912, they caused more than 3.5 times as many deaths.[6]

On French roads by 1924, cars had overtaken horses and carriages as the major killer and continued to take an increasing toll throughout the interwar years, with slight improvements in 1931 and 1936. Though still far less motorised than France, Germany apparently suffered greater mortality on the roads in 1926 and this type of fatality rose rapidly thereafter.[7] Among the reasons for the heavy incidence of accidents in Germany was the motor cycle, which, as Professor Blaich explains in Chapter 8 above, remained the major form of private passenger transport. As David Holbrook's poem indicates, motor cycles were particularly vulnerable if an accident occurred and especially liable to suffer them. More men between the ages of 15 and 30 were killed in Germany by motor cycles than by cars in 1934. In Britain Motor cycles were also disproportionately represented in road deaths. One-quarter of fatal accident victims in 1928 were motor-cycle drivers or their pillion passengers; drivers and passengers in cars or lorries accounted for only about one-twelfth of the deaths.[8] Pedestrians, however, formed the largest single group of victims, accounting for one-half of the total.

This vulnerability of pedestrians continued to be reflected in British and, to a smaller extent, in continental European, experience. By contrast, in the United States by far the majority of road deaths (three-quarters in 1960 and 1978) are among car-drivers and passengers, because walking is less prevalent. In the early 1960s, Canada and Australia, where vehicle density was also high in relation to size of population, there were very high fatality rates per head.[9]

More surprisingly, Germany's high mortality rates per head continued to approach, and in the 1960s even exceeded, those of the United States. Among European countries British road deaths as a proportion of the population were comparatively low (in 1963 England and Wales had 135 per 100 000 population, Germany 250, France 210, Italy 222).[10] Uniquely in Europe, the absolute number peaked in 1934 and only temporarily reached those levels again in the mid-1960s. The United States also experienced its first downturn in 1938, though subsequently total road deaths continued to rise. If we focus upon deaths per vehicle-mile or per vehicle, both of which have tended to decline since the 1930s, this produces more optimistic results; but they do not solely determine a person's chance of death on the roads when the stock of motor vehicles is rising.

For continental Western Europe the break in trend came in the 1970s. As Table 14.1 shows, by 1980, despite massive increases in motor vehicles, the death toll was lower in Germany, Italy and the United Kingdom than in 1965 and about the same in France. 6000 fewer people were killed on German roads in 1980 than in 1970, even though more than nine million extra cars came on to German roads during that decade.

ACCIDENT POLICY AND THEORY

Although nineteenth-century railway accidents in total were less destructive than those on the road in the twentieth century, they showed a similar rise and fall with the passage of time. Chadwick noted the very poor safety record of mid-nineteenth-century British railways compared with those in France and Germany, a difference that he attributed to greater competition in Britain.[11] Helped by the systematic collection of data the accident rate was eventually brought down. Passenger journeys doubled in the forty years before the First World War, but the number of passengers killed declined and num-

Table 14.1   Road deaths and vehicles in selected European countries, 1965–80

Road deaths in	Year 1965	1970	1980
Germany	15 753	19 193	13 041
France	12 150	15 034	12 384
Italy	8 990	10 208	8 537
United Kingdom	8 147	7 771	6 182
Cars registered in:			
Germany	9 267 423	13 941 079	23 191 616
France	9 600 000	12 900 000	18 400 000
Italy	5 472 591	10 181 192	17 686 236
United Kingdom	9 027 691	11 669 290	15 711 581

Source: *Statistical Yearbook: Transport, Communication, Tourism 1981* (Eurostat, 1983).

bers injured increased by only three-fifths.[12] In the late 1920s similar data were collected about British road acccidents and used to identify danger spots with a view to eliminating them.[13]

The differences from railways were, however, more important than the similarities and created major new problems for policy. Unlike railways, motor vehicles could function without a specially built track (though of course motors were more effective if properly surfaced roads were available). Tracks shared with pedestrians, old and young, as well as with horse-drawn vehicles and cyclists, increased the chance of accidents, a tendency accentuated by the greater speeds that motor vehicles could attain compared with horse-drawn traffic. Moreover motor drivers were not usually servants of large companies, subject to company rules and discipline. At worst, the weekend motorist was a mere amateur – and so were those who were still learning to drive.

By analogy with railways, special motor roads could alleviate the accident problem for pedestrians, cyclists and horse-drawn vehicles, and by separating opposing streams of traffic and eliminating dangerous bends, such roads offered greater protection to motorists. Modifications to the existing system were likely to be cheaper, though. With increasing affluence and expertise, vehicles tended to become safer for their occupants, for example when safety glass and four-wheel braking were introduced during the 1920s in Britain, a trend which was later reinforced by legislation such as that requiring the fitting and wearing of seat belts.[14] Those outside motor vehicles on the old road system could be protected by safer driving, enforced by speed limits, alcohol tests, driving tests and highway patrols. They were helped to save themselves by education campaigns which in the 1920s were beginning to bear fruit in a falling proportion of child road deaths in both Britain and America. For victims of accidents not prevented by these measures, a major problem came to be the establishing of liability and the payment of compensation. Compulsory insurance, the answer, was not needed for many years, because compensation was believed to influence deterrence; the insured might be encouraged to drive more dangerously if they did not have to bear the consequences of an accident (a moral hazard problem).

Underlying this view, appropriately for such an individualistic form of transport, is what might be called 'a rational theory of accidents'.[15] No particular accident can be predicted, but experienced drivers know what behaviour and conditions make accidents more or less probable, and act accordingly. Other things being equal, the evidence suggests that higher speed tends to increase the chances as well as the severity of an accident, but speed reduces the time spent travelling, as well as supplying intrinsic pleasures. The hypothetical rational driver, therefore, chooses his speed and accident chances according to his valuation of the costs he incurs in the event of an accident (including the possibility of death), his assessment of the likelihood of an accident, and other costs and benefits of speed.[16] Reflections such as these have led to the notion of an 'efficient accident rate', that rate at which extra costs of accident avoidance are just balanced by extra benefits of 'the motoring rate' that causes the accident.[17] Over the period with which we are here concerned we need to allow that the amount of travel is also chosen simultaneously with housing location, among other factors, for the motor vehicle radically affected residential patterns.

From this perspective, the justification for regulation of motor vehicles and the accidents they cause stems principally from well-documented systematic errors in drivers' perceptions of speed and of accident probabilities, and from the effects of driving behaviour upon othe people.[18] The speed at which other vehicles are travelling influences each driver's chosen speed and accident probabilities, and these speeds determine pedestrian accident chances in a manner that pedestrians cannot control. Only through the political process can the large numbers of people involved attempt to find a balance among their different interests. When the political system was slow to respond to the spread of the motor vehicle, or reacted inappropriately, death rates were excessive for the reasons mentioned above, even granted the extraordinary benefits in terms of mobility that motorisation conferred.

Accepting that a rational theory of accidents may have something to say about policy does not preclude other theories, or the idea that strong non-rational forces influence behaviour. Aggression arguably is channelled into driving to a greater extent in some countries than in others. During the Second World War Australian road deaths and suicides declined because, it has been contended, aggression temporarily found other outlets.[19] U.S. road deaths by states are significantly correlated with other forms of violent death, with divorce, and with illegitimacy, suggesting that a common factor underlies these deviancies. Institutions at different stages of human life have varying successes in directing or sublimating aggression. The age distribution of those involved in motor accidents and the incidence of motor-cycle deaths in industrial countries is generally consistent with a failure of the relevant institutions to cope with the results of the surfeit of testerone in young adult males. Similar remarks apply to alcohol consumption, generally thought to affect motor accidents. Viewed from a rational theory perspective, these non-rational elements are merely taken as data, with the possible disadvantage that they may be neglected in policy formulation and appraisal.

Nevertheless, rational theory supplies insights not only into the determination of accidents, but also into the policy response. The considerable numbers killed and maimed on the roads early justified radical measures. By 1928 Britain lost more people through road deaths than she had done in the Boer War. Ten years later a deliberately downward-biased calculation of accident costs was 1.3 per cent of Gross National Product (GNP).[20] U.S. road accident costs were variously estimated at between $1 billion and $2½ billion dollars in the early 1930s. By 1978 they had risen to $52.8 billion, 2.5 per cent of GNP.[21] Yet the policy reaction was strangely muted. An explanation may be found in the distribution of the costs and benefits of motor vehicles among the affected population. Those affected by accidents were generally not in a position to influence policy very much and those at risk were subject to such low probabilities that they were easy to ignore. The formation of permanent pressure groups such as the British National Safety First Association in 1923 could only partly overcome these problems.[22] On the other hand, there were powerful pro-motoring pressure groups which had an interest in avoiding constraints on motor vehicles. Legislative adjustment to their social consequences was therefore easiest in countries where no manufacturers existed to protest about the suppression of an infant industry, and where motoring lobbies were weak. The Scandinavian nations, for instance, with few motor vehicles in use by international standards, were among the earliest and most stringent legislators, from 1903 onwards. These states imposed accident liability upon the car-owner; but such laws could only be generally effective with compulsory insurance if road accident victims were to be compensated.[23] In this field, beginning in 1918, the Scandinavians also led.

Where manufacturers existed in a country, they had an interest in influencing policy. The British Society of Motor Manufacturers and Traders' (SMMT) 1938 annual report announced that articles of 'an anti-motoring nature' were answered wherever they appeared. That is not to say, however, that car-makers failed to recognise that they should be making a visible contribution to road safety. During the 1930s the American Automobile Manufacturers' Association established the Automotive Safety Foundation for the provision of financial support to organisations working directly in the interest of safety.[24] Motor clubs and associations, such as the American Automobile Association pushed public safety programmes, including driving and safety courses for schools. Links like this did not always appeal to pedestrians though. In 1954 the Pedestrians' Association was concerned that the new President of the Royal Society for the Prevention of Accidents was the chairman of the Vauxhall Motor Company and that the chairman of the National Road Safety Committee was also head of the legal department of the SMMT.[25]

THE DISAPPOINTING EFFECTS OF ROAD-BUILDING ON ROAD SAFETY BETWEEN THE WARS

Among the policy measures towards road safety, expenditure upon the roads did not divide pressure groups, though financial constraints were often binding. Better roads would reduce congestion, raise speeds and thereby enhance the benefits of motoring. At the same time roads could be designed to reduce contact between motor vehicles and other road-users, and so lower accident rates. That was one of the attractions of special motor roads. Unfortunately they were expensive. Italy tried to get round the financing difficulties with toll roads. In 1925, when there were only 118 000 motor vehicles registered in the country, the first four-lane *autostrada* was begun, connecting Milan with the Lake resorts, a total length of 85 miles.[26] Other autostrada, from Naples to Pompeii, from Florence to the coast, and, in part, from Milan to Turin were built in the 1920s, but the subsequent improvement of the national highways diverted traffic from them. In the following decade Germany began to build its *autobahnen*, a planned network of 4500 miles of double-track road, reserved solely for motor vehicles and free of crossroads.[27] Dominating the proposed network were two motorways running from north to south and four running from east to west. Four years after construction began in 1933, about 1000 miles were open to traffic, though, except perhaps in the Rhine valley, the system was under-utilised. A petrol tax was imposed to repay the construction loans. Even in 1980, Germany and Italy still had larger motorway systems, both in mileage and as a proportion of total network, than France or the United Kingdom.

Both France and the United Kingdom pursued the different strategy of gradually improving existing highways rather than constructing special motor roads. From a safety viewpoint this gradualist policy was likely to be more effective, dealing as it did with the urban traffic problem as well; some critics of the 1930s suggested that motor road advocates had an ulterior motive in facilitating troop movements.[28]

Although at that time there were roughly similar numbers of motor vehicles in use in France as in Britain, French road expenditure was only one-seventh of the British, and French taxes on vehicles and fuel were four times as great as spending on the roads. Nevertheless French accidents were not higher than the British, probably because of the lower population density in France.

Major road projects in Britain were bypasses in the prosperous south, including the construction of the Great West Road out of London. To reduce unemployment the government in the autumn of 1920 allocated £10 million to encourage the construction of new arterial roads and the improvement of existing thoroughfares.[29] Automatic traffic signals crossed the Atlantic in the 1920s, but white lines indicating that main road-users had priority over those joining them were still not standard in Britain even in the early 1930s.[30] Pedestrian refuges, subways and crossings were brought into use especially after the 1934 Road Traffic Act. The government Road

Research Board focused on engineering aspects of road use and construction. They recognised that the reduction of accident ratios involved physiological and psychological effects; but by 1935 they had not initiated any research in those areas.[31] Some of these effects can be analysed in the context of the rational theory of accidents outlined above to suggest why road improvements, as distinguished from specifically road *safety* measures, could have been rather ineffective. If better road surfaces (say) reduce the chances of an accident at a certain speed, then the rational driver will drive faster (and, in the long run, further), partially offsetting his originally lower accident chances. Since his faster driving now increases the likelihood of other road-users being involved in an accident, the total accident rate may not decline at all in response to road improvements.

This is a conclusion that appears to hold for the U.S. road programme of the 1920s. Whereas European road systems were sometimes determined by the exigencies of national defence between the wars, the nature of the American system was dominated by the vast areas of territory to be covered. At the beginning of the 1920s around four-fifths of American roads were earth, and of those which were surfaced, less than 30 per cent were paved or macadam.[32] An Act of 1916 provided federal aid to states with responsible highway departments and, two years later, Pennsylvania began the trend to borrow for highway construction.[33] Surfaced road-building advanced rapidly. In the two years 1923 and 1924, the surfaced road system was expanded by one-quarter. Despite a tradition of greater state involvement and ownership in Europe, public investment in roads was higher in the United States. Earmarked taxes (mainly on petrol) provided a source of funding, but the explanation for the size of the American programme lies in the federal structure which freed state politicians from the wider concerns of national politics.[34] Yet, however much the programme assisted the spread of the motor car, there is no evidence that road deaths were reduced. Controlling for the influence of differences between states in numbers of motor vehicles, in population and in mileage travelled, (proxied by petrol consumption), regression analysis produced no indication that states with higher proportions of surfaced roads in 1929 had significantly fewer road deaths.[35]

More effective, equally uncontroversial and even cheaper than road safety policies, were education campaigns. A 'safety first' campaign and the use of police patrols between 1926 and 1929 enabled Edinburgh to reduce fatal accidents at a time when they were increasing elsewhere.[36] In Salford during the 1930s a drive that included the showing in schools and local cinemas of films and slides giving details of local accidents, together with the introduction of play streets, reduced child road deaths from 11 in 1931 to 6 in 1935 and, between September 1935 and October 1936, to zero.[37] In America during the 1930s, 1101 cities competed for annual awards, based on safety activities, in the National Traffic Safety contests conducted by the National Safety Council.[38] Cities that implemented recommended safety programmes all experienced reductions in road deaths, New York achieving a 16.5 per cent cut. Figures for cities which did not employ safety programmes revealed much higher death-rates.

THE REGULATION OF MOTOR VEHICLE USE AND THE IMPORTANCE OF SPEED LIMITS

Regulations designed to encourage road safety – speed limits, compulsory insurance, driving tests, minimum age restrictions and so on – did not suit the motor lobby, however, for they constrained motor-vehicle users and probably car sales. But some, at least, of these measures were likely to be more effective than road improvements, according to the rational theory of accidents. Speed limits, when credible and enforceable, directly attacked a major cause of accidents and would not generate the offsetting behaviour that could be expected to result from road improvements. Unfortunately speed limits in the first third of the century lacked credibility and enforceability. The speed restrictions of the British 1903 Motor Car Act (20 m.p.h. on the open road and much lower in towns) were abolished in 1930 because they were so widely ignored. Even the road safety lobby (rightly) did not envisage adverse consequences from the abolition because, four years earlier, there had been none when the speed limit in Northern Ireland had been removed.[39] In America the autonomous states had a rich and bewildering variety of choice as Table 14.2 shows, by 1931–32 had maximum speeds for private cars in open country mostly between 35 and 45 m.p.h., seven states required 'reasonable and proper' speeds without stating what these were, six states specified speeds, from 25 to 40 m.p.h., beyond which higher speed was deemed prima facie evidence of improper driving, and three states had no limit at all.

By the standards of later periods these limits seem low; but more primitive vehicles running on winding roads were often incapable of much more. When speed observations began in Rhode Island in

Table 14.2   U.S. State legislation on certain aspects of motor vehicle regulation in 1931–2

	Operator licence	Minimum age for operators	Examination for operator licence	State patrol force or inspectors	Maximum speed permitted private passenger cars in open country
		Years			Miles per hour
Alabama		16			45
Arizona		16	l	x	35
Arkansas			ldv	x	45
California	x	16	ldv	x	45
Colorado	x	15	ldv		rp
Connecticut	x	16	ldv	x	rp
Delaware	x	16	ldv	x	40
Florida		14			45
Georgia		16			40
Idaho				x	35
Illinois		15			45
Indiana	x	16		x	rp
Iowa	x	15	l	x	rp
Kansas	x	16	ld	x	rp
Kentucky		16			rp (40)
Louisiana				x	45
Maine	x	15		x	rp (35)
Maryland	x	16	ldv	x	40
Massachusetts	x	16	ldv	x	rp (30)
Michigan		14		x	rp
Minnesota					45
Mississippi					30
Missouri		16			rp (25)
Montana				x	rp
Nebraska	x	16			45
Nevada	x	15			rp
New Hampshire	x	16	ldv	x	35
New Jersey	x	17	ldv	x	40
New Mexico		14			45

New York	x	18	dv	x	40
North Carolina		16			45
North Dakota				x	50
Ohio					45
Oklahoma					45
Oregon	x	16	ldv	x	45
Pennsylvania	x	16	ldv	x	40
Rhode Island	x	16	lv	x	rp (35)
South Carolina	x	12	ld	x	45
South Dakota		15			40
Tennessee				x	ns
Texas					45
Utah		16			45
Vermont	x	18	ldv	x	ns
Virginia[+]			(+)	x	45
Washington	x	15		x	40
West Virginia	x	15	ldv	x	45
Wisconsin	x	16		x	ns
Wyoming		15			rp (35)
District of Columbia	x	16	ldv		30 [*]
Total	26		30		

[+] Operator licence law adopted in Virginia in 1932, examination of new applicants (minimum age, 16 years) obligatory after 1 July 1933.

[*] On certain streets in the District of Columbia only; speed limit elsewhere, 22 miles per hour.

Note – The following abbreviations are used, either singly or in combination:

    x – Legislative and/or administrative action
    d – Driving test
    l – Written or oral test on motor-vehicle laws
    v – Test of vision
    rp – Not greater than is 'reasonable and proper' figure in parentheses indicates limit beyond which higher speed is deemed prima facie evidence of improper driving
    ns – Not specified.

Source: 'Digest of Regulations Governing the Operation of Motor Vehicles throughout the United States and Provinces of Ontario and Quebec, Canada, 1931–1932', compiled by E. Austin Baughman, Commissioner of Motor Vehicles, State of Maryland, Baltimore, Md, cited in *Public Roads* (Dec 1932) p. 16.

1925, the average vehicle speed was only 25.6 m.p.h. and on East Massachusetts highways in the summer of 1934 the average was 33 m.p.h.[40] By 1973, with many miles of specially built motorways and other highway improvements, average vehicle speeds in the whole of the United States reached 65 m.p.h. before falling again under the influence of lower limits and the need to save petrol. Yet, though vehicle speeds in the 1920s and early 1930s were generally low, state speed limits do not seem to have reduced fatalities according to a regression analysis similar to that undertaken to find the effects of the road programme.[41] Similarly, minimum ages for drivers, which ranged from 12 in South Carolina to 18 in New York and Vermont, did not appear to have any effect upon deaths. Nor did the sixteen states with driving tests experience lower accident rates, other things being equal. Only the existence of a state patrol force or inspectors showed some tendency to reduce state road fatalities, and then not significantly so.

Why did some states adopt these ineffective policies? An examination of some of the more obvious variables – fatalities per head, number of cars per head or the absolute number of cars or trucks, the mileage driven in the state (as measured by petrol consumption), the absolute number of road deaths and the year of statehood – did not reveal any single one that could significantly discriminate between states with speed limits and those without, or between those with and without highway patrols.[42] The year of statehood was included among the variables in an attempt to indicate the strength of political forces bent upon restrictive legislation, for a state's 'political age' has been found a significant determinant of certain socio-economic rigidities, in which regulation might be included. Pressure groups aiming at measures likely to redistribute income in their direction are believed to be more likely to emerge the longer the undisturbed political life of the state.[43] Year of statehood appeared as the sole significant discriminating variable for driving tests; knowing a state was long-established allowed it to be classified as one likely to require driving tests. States with and without minimum-age restrictions could be distinguished by fatalities per head. Those with high road-accident rates imposed the restriction.

Before the 1930s, then, the regulatory reaction was almost as arbitrary as it was ineffective, but as experience accumulated, speed limits had greater effect. In March 1935 Great Britain introduced a 30 m.p.h. limit in all built-up areas, but, as William Plowden has remarked, it produced no dramatic fall in total casualty figures.[44]

They remained stable for a while and then began to rise again. Both advocates and enemies of the limit could claim their point of view was vindicated, but Plowden neglects the crucial question of whose point of view was right. The number of *fatal accidents* fell by 15 per cent in the following year despite a normal increase in motor-vehicle registration. Personal injury accidents fell by 6 per cent. As has been seen, numbers killed did not reach 1934 levels again until the mid-1960s despite a quadrupling of motor vehicles in use. Other changes at the same time included the introduction of driving tests and pedestrian crossings.

The evidence favours the speed limit as the main reason for casualty reductions. It is true that fatal and all personal injury accidents were higher in January and February 1935 than in the corresponding months of 1934; but the reverse was true for March–December in the two years. Moreover the ratio of accidents in the towns to those in the country showed a downward shift in 1935. Using the ratios of 1934 to 1936 implies that the introduction of the speed limit was followed by a reduction in fatal accidents of 15 per cent (a result broadly confirmed by regression analysis) and in personal injury accidents, a fall of 3 per cent in town police districts compared with county districts. Because roads in built-up areas with speed limits did not exactly coincide with the urban boundaries, the apparent effect of the 30 m.p.h. speed limit was less than the actual outcome.[45] A similar effect was observed in 1957 when West Germany introduced a 31 m.p.h. limit in built-up areas, following abolition in 1953: a 30 per cent reduction in fatalities. Switzerland's 37 m.p.h. built-up limit of 1959 seems to have reduced deaths by 21 per cent and reported accidents by 2 per cent. In almost every case the introduction was followed by a reduction in pedal-cyclist and motor-cyclist fatalities. The effects on other classes of road-users was more varied.

If the market and legal system operated ideally, then accurate cost-benefit analyses of the general speed limits, increasingly reimposed from the 1960s, should have yielded negative figures, the greater time costs incurred at the lower speeds required by the new limits outweighing the deaths and injuries forgone, for instance. A crucial value for such a calculation, but one hard to assess is the value of life and limb, as a British study of the effects of the 70 m.p.h. general speed limit of 22 December 1965 shows. The Road Research Laboratory calculated that there were 44 fewer fatal motorway accidents in 1966 as a consequence of the restriction. These accounted for more than half the benefits when a life saved was valued at £5000. The Laboratory concluded the accident costs saved could not be said to outweigh the time costs minus running costs (£0.93 million) incurred with any degree of certainty.[46] However a higher value of life, such as is obtained from questionnaire approaches, would undoubtedly have yielded a net benefit for the speed-limit introduction on the motorway.

Such a conclusion is more consistent with the effects of speed limits imposed in the summer months of 1968 and 1969 on trunk roads in Sweden outside towns. The 90-k.p.h. limit showed a reduction in accident costs greater than the increase in transport costs. A similar result was obtained in the German Federal Republic where, from 1 October 1972 a 100 k.p.h. speed limit on a trial basis was applied to all roads that did not have at least two traffic lanes in each direction outside built-up areas. Assuming DM 200 000 for one fatality avoided, in the first year of operation, 1972/3, the minimum value of personal death and injury avoided was DM 100 million compared with a time cost of DM 76.5 million.

The energy crisis of 1973/4 precipitated the reduction of the general speed limit in the United States from 65 to 55 m.p.h. outside built up areas. Fatal accidents on state roads fell 21 per cent and on controlled roads 32 per cent. A study of California attributes 40 per cent of the fatality reduction to the speed restrictions. Applying this figure across the whole United States, and using $205 000 per fatal casualty, savings in 1973 were $920 million. Increased travelling time was 600 million hours, but reduced travel costs were $440 million–$675 million. So if the value of travel time was less than $2.27–2.66 per hour, the speed limit was beneficial, providing no value is placed on the intrinsic benefits of speed, and the costs of enforcement and loss of freedom are neglected. Noise and pollution costs have also been ignored.

ACCIDENT LIABILITY AND COMPULSORY INSURANCE

Although these cost-benefit assessments of the general speed limits of the 1960s and 1970s suggest that accidents were above the 'efficient rate', and therefore implies that the market and the legal frameworks were not previously functioning in an ideal manner, a number of steps had been taken in the interwar years to change, clarify and enforce the obligations of motor-vehicle owners, which tended to push the accident rate downwards. The French tried to limit road accidents by providing detailed traffic rules in the 1922 Code de la

Route; but these proved inadequate and the judicial system by 1930 responded to the accident-compensation problem by effectively reversing the fault liability rule of the 1806 Civil Code, and introducing liability consequent upon control.[47] Vehicle drivers had more control over the likelihood of accidents. Making them liable tended to encourage safety and ease the task of the courts.

We have already noted that legal liability only made sense when drivers could meet their financial obligations. For that, compulsory insurance was required. Despite an enormous volume of litigation in the U.S. courts during the 1920s and large numbers of uncompensated victims, the United States had no *general* insurance requirement. Among U.S. states in this period Massachusetts was unique in requiring vehicles be insured in an attempt to provide the possibility of compensation for road-accident victims. It is true that, beginning with Connecticut in 1925, by 1932 some eighteen states carried on their statute books financial responsibility laws designed for the most part to require those owners and drivers shown to be at fault either to pay damages or to insure themselves in the future; but these were generally regarded as ineffective.[48] Although the Massachusetts Compulsory Automobile Liability Security Act was intended to promote distributive justice rather than safety, or 'an efficient accident rate', the legislation may in fact, have reduced accidents. Some accident costs that were borne by victims under the former regime, (before 1 January 1927) and in all other states, were now met by the vehicle owners who paid insurance premiums at rates fixed by the state Commissioner of Insurance. Compulsory insurance of state-registered vehicles was restricted to bodily injury caused, subject to an upper limit of $5000 for any one person and $10 000 for any one accident.[49] Because drivers did not have to be fully insured, the moral hazard problem alluded to earlier should not have caused rational motorists to drive less carefully; some of the costs of accidents would have been borne by the drivers and therefore they still had an incentive to avoid them. On the contrary, compulsory partial insurance provided some incentives to drive *more* carefully.

In a system where accident proneness is penalised by high insurance premiums, drivers have financial reasons to change their behaviour compared to a regime where they can avoid any financial responsibility for accidents by going bankrupt. Under the Massachusetts legislation there was no system of merit-rating, rewarding careful drivers and penalising careless drivers, and premiums became matters of political controversy, the insurers claiming that they were too low, and vehicle owners maintaining that they were too high. For the first two years of the Act there was no basis for the calculation of rates. From 1929 premiums began to creep up at the same time as prices fell; but in nominal terms the 1926 levels were not exceeded until 1931.

Though by 1936 rates had risen by one-half over the 1927 levels, more than the national average, in California the increase was almost 80 per cent. There is little doubt that the Act increased claim frequency but the average cost per claim fell as expected if the Act was achieving its intended purpose.[50] Since the premiums did discriminate between types of car, drivers could reduce their payments by switching to lower-rated cars with better accident records, and in so doing lower the death-rate. We would, therefore, expect that the Act caused drivers in Massachusetts to use less powerful cars with lower petrol consumption. Other things being equal, a given quantity of petrol consumed represented a greater mileage of lower horse-power cars in Massachusetts. The greater mileage increased accident risk but the change in vehicle composition reduced it. We have attempted to identify the net effects of the Massachusetts legislation relative to other states between 1929 and 1931 by a regression model similar to that employed earlier. The results suggest a less than 2 per cent chance that the legislation had no effect upon fatalities, assuming the model has identified all other influences upon road deaths that distinguished Massachusetts from other states. The best estimate of the reduction in road fatalities is that they were cut by the legislation to around 73 per cent of the level they would otherwise have been. Since there were around 800 deaths a year in Massachusetts, the implication is that as many as 290 lives a year might have been saved by the Act between 1929 and 1931.[51]

Similar consequences must have flowed from the taxation of petrol. Our estimates suggest a less than 4 per cent chance that state petrol taxes had no effect on state petrol consumption, and therefore on mileages as well as vehicle types. The best estimate is that an increase from 4 to 5 cents a gallon reduced mileage by almost 2.5 per cent, and deaths by much the same proportion.[52]

A second opportunity to assess the consequences of early compulsory insurance legislation is provided by the British 1930 Road Traffic Act. This required all motor-vehicle users to have insurance against third-party personal injury. In contrast to the earlier Massachusetts legislation, there was neither rate- nor premium-fixing. Though fatal road accidents fell the following year, so also did the number of vehicles registered. The fall in registration was, however, restricted to motor cycles; that of cars increased. Moreover, whereas the decline in motor-cycle registrations might possibly be attributed to the greater impact of unemployment on lower-income motor cyclists, the decline continued throughout the 1930s, during the recovery period. Apart from the tendency for the burdens of compulsory insurance to eliminate impecunious motor cyclists, the fall could have been influenced by the shift in the relative price of cars and motor cycles between 1930 and 1931, thanks to the introduction of smaller cars. The ratio of Stone and Rowe's car-price index to motor-cycle index fell from 4.6 in 1930 to 3.9 in 1931.[53]

A test of the influences on the fatal accident rate of the interwar years was to use time-series regression analysis to identify the role of changes in the composition of vehicles in use (horse-drawn vehicles numbered 207 000 in 1923 but only 12 000 in 1937),[54] and dummy variables to isolate the impact of the 1930 and 1934 legislation. Then we tried to identify the forces making for a decline in motor-cycle usage, distinguishing the contribution of income and relative price changes from that of legislation. The 1930 Act might be expected to have a stronger effect than the Massachusetts legislation because the insurance companies were free to act in a more market orientated fashion. But in the 1970s the ability of insurance companies to deter costly driving behaviour by loading premiums has been questioned. The administrative costs of increasing the number of insurance categories, so it has been maintained, is too high, and outside life assurance, statistical methods play only a small part in fixing premiums, it is asserted.[55] In the less-sophisticated interwar years this was even truer. During the 1920s claims for damages to cheap and low-powered vehicles were almost as costly as those to expensive cars, paying higher premiums.[56] Yet, while the British Royal Exchange Assurance companies complained that there was no profit in the small-car business, premiums were not raised. Eight years were to pass between the R. E. A. beginning to investigate varying rates according to district and the introduction of district rating in 1935.

A second potential source of deviation of insurance premiums from those that provide ideal deterrence are court valuations of compensation for death and injury. Court awards are reflected in premium levels in actuarially fair payments for insurance against third-party harm. British interwar awards were distinctly erratic. Compensation for the death of children under 15 was based upon the cost of rearing a replacement.[57] An Act of 1935 permitted the award of damages to the deceased's estate, thus opening a Pandora's Box for the courts in ensuing years.[58]

Premiums may still deter even though they are not ideal. That, apparently, was the effect of the 1930 British legislation, as it was for the Massachusetts law from 1927. The regression results imply that motor cycles were around $1\frac{1}{3}$ times as likely to be involved in a fatal accident as other motor vehicles.[59] The 1930 Act had no significant effect on fatalities, independent of changes induced in the vehicle stock; but the 1934 Act (with the urban speed limit) reduced fatal accident by around 20 per cent. Reducing the proportion or the numbers of motor cyclists was, therefore, an effective means of lowering road deaths. This the 1930 Act seems to have achieved according to regression analysis, cutting the ratio of motor cycles to other motor vehicles by 40 per cent in the following year by raising the financial costs of motor cycling.[60] Putting together these two results, and assuming that no cars were priced off the road by compulsory insurance, the best estimate is that fatal accidents were about one-fifth (or more precisely 22 per cent) lower than they would have been otherwise.[61]

MOTOR VEHICLE SAFETY LEGISLATION

Liability rules and insurance, like speed limits, could be justified in terms of the damage that motorists did to others. Much vehicle-safety legislation presupposed that motorists did not know their own interests. In this category come compulsory crash helmets for motor cyclists and seat belts for car drivers and passengers. If motorists wanted to be safer, would they not wear seat-belts and buy cars incorporating safety devices? After all, motor vehicles have become progressively safer, generally, since they were introduced, without legislation. Admittedly the way in which some safety features became incorporated was idiosyncratic. The enterprising maker of Triplex safety glass during the 1920s read that Mr Ford had been cut in a motor accident and telegraphed to him to fit safety glass, thereby ensuring that all Fords acquired Triplex windscreens.[62] Here perhaps lies a clue to the desirability of vehicle safety legislation: manufacturers have more information about the properties of their vehicles than do the individual buyers. If the makers were made liable for accidents caused by their vehicles, then vehicles would be considerably safer and more expensive, so long as the courts were prepared to award appropriate damages. Vehicle safety legislation removes that particular problem from the court's domain but achieves the same object.[63]

In the United States an important influence in the passing of the 1966 National Traffic and Motor Vehicle Safety Act, requiring seat belts in cars, was Nader's critique of motor vehicles in his *Unsafe at Any Speed*. Peltzman employed a version of the rational theory of accidents in an attempt to show that the legislation actually increased deaths: by reducing the costs to the driver the Act encouraged faster driving and greater risks to pedestrians.[64] The ratio of pedestrian to vehicle-occupant deaths certainly rose slightly after the legislation, but it rose even more before.[65] Moreover Peltzman's regression equations, which he uses to predict fatalities in the absence of the legislation, is inconsistent with his model.[66]

Evidence provided by the accident experience of the VW Rabbit in the late 1970s suggests seatbelt-wearing did indeed make a great difference and that the decision not to wear a seat belt was not a rational one.[67] Most probably individuals have difficulty making rational decisions about behaviour which involves very low probabilities of highly adverse events. Between 1975 and 1979 about 180 000 VWs were equipped with a passive seat-belt system (a shoulder-belt attached to the door so that when the door is closed it works as an automatic restraint). The utilisation of belts on VW Rabbits with manual belt systems was less than half that of the passive system and the fatality rate for drivers of the first group was double that of the second. The benefit-cost rate from passive belts was computed as ranging from 2.2 to 11.7. Insurance companies offered lower rates on the passive belt cars, suggesting that they had not detected the fully offsetting behaviour allegedly identified by Peltzman. That is not to deny that *some* offsetting behaviour occurred. An updated and modified version of Peltzman's study concluded that car safety regulation in the United States had indeed reduced road deaths between 1966 and 1980, but pedestrian deaths had been increased by regulation.[68] Vehicle regulation including compulsory seat-belts needs to be supplemented by enforced speed limits and better road safety designs to reduce pedestrians' risks.

CONCLUSION

In the richer countries the human cost of the motor vehicle has been reduced by a variety of measures, among which legislative, adminis-trative and education policies ultimately have been at least as effective as engineering solutions. Society's problem was, and is, to decide how much death and injury the enormous powers conferred by motor vehicles are worth. Low individual probabilities of death allowed policy-makers generally to pay little attention to the large and rising death toll for much of the century, despite the substantial proportions of GNP that this tragedy wasted. Early legislation was therefore ineffective. Thresholds of tolerance were reached in Britain in the 1930s and 1960s, in the United States in the 1930s and 1970s and, helped by higher petrol prices, in other major industrial countries in the early 1970s. At these points policy measures began to reduce road deaths despite the continuing spread of the motor vehicle.

These measures were socially desirable because without them biases in drivers' perceptions and interaction effects with other road-users resulted in a greater than socially efficient accident rate, broadly interpreted. For non-industrial countries whose road-accident rates are still extraordinarily high, the question remains how quickly and how effectively their political systems will react.

Notes

1. *Statistical Abstract of the United States* (1931) p. 81.
2. G. D. Jacobs and P. Hutchinson, 'A Study of Accident Rates in Developing Countries', Dept of Environment, Transport and Road Research Laboratory Report (*TRRL*) *Report* LR 546; G. D. Jacobs and P. R. Fouracre, 'Further Research on Road Accident Rates in Developing Countries', *TRRL*, SR 270.
3. *Royal Commission on Motor Cars* (1906) P.P. p. 233.
4. *Annuaire Statistique de la France* (1908) p. 82.
5. *Report of the Select Committee on London Traffic* (1913) P.P. app C61, p. 1037.
6. Ibid., p. 1140.
7. *Statistische Jahrbuch das Reich* (1926, 1934). 'Apparently' because the French statistics are enigmatic. In addition to roads deaths caused by cars, horses and bicycles, there is also a category 'mort subitement sur la voie publique'. A comparison of the 1938 deaths in the 1940–5 *Annuaire*, using this classification, with figures for the same year in the 1952 *Annuaire* yield very different results. The French data are discussed in J. C. Chesnais, *Les Morts violentes en France depuis 1826: Comparaisons international* (Paris: P.U.F., 1976).
8. J. A. A. Pickard, 'Accident Prevention', *Journal of the Institute of Transport* 12 (1930) pp. 73–82.

9. *Annuaire Statistique: Retrospectif* (Partie Internationale, 1966) p. 37*.
10. *Annuaire Statistique: Retrospectif*, p. 37*.
11. E. Chadwick 'Results of Different Principles of Legislation and Administration in Europe', *Journal of the Statistical Society*, 22 (1859) pp. 381–420.
12. *Report on Railway Accidents* (1929) U.K. P.P., Cmd 3379, table 3, p. 5.
13. Royal Society for the Prevention of Accidents, *Fifty Years of Service* (London, 1966); J. E. Holmstrom, 'Prevention of Road Accidents', *Fortnightly Review*, 133 (1933) pp. 619–29.
14. C. F. Caunter, *The Light Car: A Technical History* (London: HMSO, (1970) pp. 87, 99–100, 164–5.
15. Elements of this theory are to be found in D. Ghosh, D. Lees and W. Seal, 'Optimal Motorway Speed and Some Valuations of Time and Life', *Manchester School* (1975) pp. 134–43, and S. Peltzman, 'The Effects of Automobile Safety Regulations', *Journal of Political Economy*, 83 (1975) pp. 677–725.
16. Those who find it hard to believe people could ever behave rationally in the face of perceived risks of death may be convinced otherwise by Robert Graves's autobiographical account of trench warfare in the First World War:

    'Like everyone else I had a carefully worked out formula for taking risks. . . . In a hurry we would take a one-in-two-hundred risk, when dead tired a one-in-fifty risk. . . .'
    R. Graves, *Good-bye to All That* (1979) (1st ed. 1929), pp. 116–17.

    With lower and less frequent risks of death in civilian life there is obviously less incentive to articulate the bases of behaviour, but the types of judgements might well be similar.
17. G. Calabresi, *The Costs of Accidents: A Legal and Economic Analysis* (New Haven, Conn.: Yale University Press, 1970).
18. European Conference of Ministers of Transport, *Costs and Benefits of General Speed Limits* (Paris: OECD, 1978), R. J. Arnaud and H. Grabowski, 'Auto Regulation: An Analysis of Market Failure', *Bell Journal of Economics*, 12 (1981) pp. 27–48.
19. F. A. Whitlock, *Death on the Road: A Study in Social Violence* (London: Tavistock Publications, 1971) pp. 90, 102, 105, 134.
20. J. H. Jones, *Road Accidents: Report* (London: HMSO, 1946).
21. H. H. Kelly, 'The Problem of Motor Vehicle Regulations', *Public Roads*, 13 (1932) pp. 153–68.
22. Royal Society for the Prevention of Accidents (ROSPA), *Fifty Years of Service*.
23. A. Tunc, 'Traffic Accident Compensation: Law and Proposals', in *International Encyclopaedia of Comparative Law*, 11, ch. 14.
24. S. J. Williams, 'Accidents on the Road', *Public Roads* (1938) pp. 77–82.
25. W. Plowden, *The Motor Car and Politics in Britain*, (Harmondsworth, Middx: Penguin Books, 1973) p. 344.
26. H. H. Kelly, 'A Study of the History and Present Status of Toll Roads', *Public Roads*, 12 (1931) pp. 2–3.

27. J. Cracroft Haller, 'Highways for Transport', *Journal of the Institute of Transport*, 19 (1938) pp. 108–11.
28. F. J. Wymer, 'Transport in France', *Journal of the Institute of Transport*, 19 (1937) pp. 65–74.
29. *10th Annual Report of the Administration of the Road Improvement Fund* (1921) p. xvii.
30. Holmstrom, op. cit. n. 13 above.
31. *Report of the Road Research Board 1935* (D.S.I.R., HMSO).
32. *Statistical Abstract of the United States* (1924).
33. F. L. Paxson, 'The Highway Movement 1916–1935', *American Historical Review*, vol. 51 (1946) pp. 236–53.
34. J. A. Dunn jr, 'The Importance of Being Earmarked: Transport Policy and Highway Finance in Great Britain and the United States', *Comparative Studies in Society and History*, 20 (1978) pp. 29–53.
35. The regression equations for a cross-section of forty-three states in 1929 with road deaths as the dependent variable are to be found in Appendix 14.A.1.
36. Pickard, op. cit. n. 8 above.
37. B. Preston, *Focus on Road Accidents* (London: Public Affairs News, 1954).
38. Williams, op. cit. n. 24 above.
39. Pickard, op. cit. n. 8 above.
40. W. G. Eliot 'The Rising Accident Rate', *Public Roads*, 16 (1935) pp. 7–11.
41. The regression equations below are estimated from pooled cross-section/time-series data for 1929–31 for 138 observations on U.S. states. While time-series coefficients are generally taken to measure short-run responses and cross-section coefficients, long-run responses, there was generally not a great difference in coefficient sizes among the annual strata or with larger subsets of the data. Only the precision of the estimates was increased by pooling. (See Appendix 14.A.1.)
42. The procedure followed was to estimate the discriminant function, the linear function of a subset of the variables that distinguished best between states which did and which did not adopt the different regulations. On discriminant analysis see, for example, R. A. Cooper and A. J. Weekes, *Data, Models and Statistical Analysis* (Deddington: Phillip Allan, 1983). The variable selection routine involved computing the significance level of each variable in turn. If a variable did not attain a 10 per cent significance level it was excluded. Fatalities per head had a F statistic of 4.9310 (sig. level 0.0316) in the minimum age classification, the functions being:

$$-2.8162 + 0.02745 \text{ (min. age)} \quad \text{and} \quad -4.9324 + 0.036329 \text{ (min. age)}$$

<table>
<tr><td>11 obs.</td><td>35 obs.</td></tr>
<tr><td>(without min. age)</td><td>(with min. age)</td></tr>
</table>

For driving tests, year of statehood was significant at 1.5 per cent level with an F of 6.3972, the functions being:

−209.05 + 0.49622 (year)   and   −193.10 + 0.47691 (year)
31 obs.   15 obs.
(without tests)   (with tests)

South Dakota is excluded from the sample. Logit regression analysis produced broadly similar results, although when all variables were included in some cases variables were statistically significant, when on their own they were not.

43. M. Olsen, *The Rise and Decline of Nations* (New Haven, Conn., and London Yale University Press, 1982).
44. Plowden, op. cit. p. 287 (n. 25 above).
45. Road Research Laboratory, *Research in Road Safety*, pp. 155–7.
46. European Conference of Ministers of Transport, *Costs and Benefits*.
47. F. Deak, 'Automobile Accidents: A Comparative Study of the Law', *University of Pennsylvania Law Review*, 79 (1930–1) pp. 271–305.
48. Tunc, op. cit. n. 23 above.
49. R. H. Blanchard, 'Compulsory Motor Vehicle Liability in Massachusetts', *Law and Contemporary Problems*, 3 (1936) pp. 533–53.
50. A. D. Little Inc., *The State of the Art of Traffic Safety: A Comprehensive Review of Existing Information* (New York: Praeger, 1970). In 1964 the Massachusetts claims frequency was almost double that of the second highest state.
51. See the variable Mass. × mileage in Appendix 14.A.1.
52. The pooled sample of American state for 1929–31 was employed. The resu'ting regression equation was:

$$2n. \left\{ \begin{array}{l} \text{mileage} \\ \text{or petrol} \\ \text{consumption} \end{array} \right\} = 3.4433 - \underset{(16.289)}{0.091101n.} \text{ tax} + \underset{(2.0767)}{0.160952n.} \text{ population} \underset{(4.7317)}{}$$

$$+ \quad \underset{(8.6153)}{0.33097 \ 2n.} \text{ trucks} \quad + \quad \underset{(12.588)}{0.50241 \ 2n.} \text{ cars}$$

$$R^2 = 0.9807 \qquad \qquad F = 1689$$

t ratios in parentheses

53. R. Stone and D. Rowe, *The Measurement of Consumers Expenditure and Behaviour in the United Kingdom 1920–38*, vol. II (Cambridge: Cambridge University Press, 1966).
54. *U.K. Annual Abstract of Statistics 1924–37*.
55. P. S. Atiyah, *Accident Compensation and the Law* (London: Weidenfeld & Nicolson, 3rd ed. 1980) ch. 24.
56. B. Supple, *The Royal Exchange Assurance: A History of British Insurance 1720–1970* (Cambridge: Cambridge University Press, 1970) ch. 18.
57. Jones, op. cit. n. 20 above.
58. W. A. Dinsdale, *A History of Accident Insurance in Great Britain* (London: Stone, 1954) pp. 213–14.
59. See Appendix 14.A.2. Because horse-drawn vehicles showed a similar pattern in decline to the rise of motor vehicles exclusive of motor cycles, their separate influence are hard to disentangle (their correlation

coefficient is 0.978). The regression equations show the influence of the multicollinearity in the extraordinary change in the vehicle coefficients when horse-drawn vehicles are included. The effect of motor cycles on fatal accidents though positive in both cases is doubled when horse-drawn vehicles are included, and the coefficient on motor vehicles changes from a very significant +0.74 elasticity to and statistically insignificant −0.4476. The former result has more *a priori* plausibility. A 1 per cent increase in motor cycles causes a smaller percentage increase in accidents than does an increase in motor vehicles because they were a smaller proportion of the total vehicle stock, but each one was much more likely to be involved in a fatal accident. The coefficient is two-thirds that of other motor vehicles, but during the 1920s motor cycles accounted for little more than one-third of total vehicles and therefore for one-half of other motor vehicles. The unit response changes with the proportion of motor cycles, but is approximately $(\frac{2}{3})/(\frac{1}{2})$. Thus motor cycles were one-third more likely to be involved in a fatal accident.

60. log. (m. cycles/other m. vehicles) = 18.67 − 0.502 (1930 Act)
    (6.23)   (4.609) Dummy

    − 2.692 log *GDP* − 0.06 log (price m. cycle/car price)
    (7.2572)   (0.1963)

    R^2 = 0.9686   *F* = 123.37   *DW* = 1.7592

    t ratios in parentheses. The *DW* statistic is in the indeterminate range.

    The calculations underlying the result in the text are: logs are to the base *e*, therefore the Act reduced (motor cycles/motor vehicles) to exp (−0.502) = 0.605, 60.5 per cent of the former value. Collinearity required the specification adopted here.

61. Assuming no change in other motor-vehicle numbers, the reduction in fatal accidents is the antilog of the product of the 1930 Act coefficient in the equation in the preceding note and the motor-cycle coefficient in the first equation of note 59, Appx 2: 0.498 × −0.502 = −0.25, exp (−0.25) = 0.778.

62. G. S. Davison, *At the Wheel* (London: Industrial Transport Publications, 1931) pp. 50–1.

63. The argument is analogous to that advanced by Atiyah for the desirability and effectiveness of the Workmen's Compensation Act reducing the number of industrial accidents (see n. 55 above).

64. Peltzman, op. cit. n. 15 above.

65. The ratios of pedestrian to motor-vehicles' occupant deaths were 0.197 (1960), 0.217 (1965), 0.219 (1970). *Statistical Abstract of the United States* (1978) p. 647.

66. Speed erroneously appears as an independent variable forecasting counterfactual fatalities, in the absence of the 1966 legislation, whereas it is properly a dependent variable in Peltzman's account (see n. 15 above).

67. Arnaud and Grabowski, op. cit. n. 18 above.

68. T. J. Zlatoper, 'Motor Vehicle Deaths in the United States', *Journal of Transport Economics and Policy* (Sep 1984) pp. 263–74.

Appendix 14.A.1 OLS regressions: fatal road accidents in the United States 1929–31

Dep. variable road fatalities	Constant	log state surfaced/ earth road	log rural surfaced/ earth road	log cars	log mileage	log population	log. mileage surfaced road	log mileage earth road (state)	log rural surfaced road	log rural earth road	R^2	F stat.
	-0.4188 (0.3814)	—	—	-0.6558+ (2.9921)	1.2794+ (5.4999)	0.3295 (3.2516)	-0.0334 (0.5094)	-0.0347 (1.3221)	0.0634 (1.3477)	0.0242 (0.5126)	0.974	189
	-0.0309 (0.0415)	0.0207 (0.8888)	0.0199 (0.7678)	-0.5539+ (2.8566)	1.1922+ (5.6477)	0.3354+ (3.6201)	—	—	—	—	0.973	268

Notes: t-statistics in parentheses; + indicates significance at 5 per cent level.
All data from *Statistical Abstracts of the United States*.

Dep. variable fatalities	Constant	log trucks	log cars	log miles	log population	Mass. dummy	Min. age	Driving test	State patrol	Speed limit	Mass. × mileage	R^2	F
	2.1500 (4.3921)	0.0300 (.4142)	-0.3120 (3.465)	0.6681 (4.940)	0.6292 (12.130)	-0.3043 (2.354)	0.1322 (2.6188)	0.0531 (1.0926)	-0.0573 (1.0926)	0.0485 (1.2039)		0.968	390.26
	1.6546 (3.3409)	-0.0387 (.5356)	-0.4050 (4.535)	0.8568 (6.6099)	0.6047 (11.811)	-0.3129 (2.444)	—	—	—	—	—	0.961	650.08
	1.6544 (3.4405)	-0.0387 (0.5359)	-0.4051 (4.535)	0.8586 (6.610)	0.6047 (11.811)	—	—	—	—	—	-0.0367 (2.445)	0.961	650.11

Appendix 14.A.2 OLS regressions with dependent variable fatal road accidents in Great Britain 1923–38

Constant	log m. vehicles exc. m/c	log m/cycles	1930 Act Dummy	1934 Act Dummy	log horse-drawn	D.W. stat.	F. stat.	R²
-0.0884 (0.1611)	0.7426+ (8.4643)+	0.4980+ (3.7563)	0.0564 (1.0677)	-0.2013+ (3.4252)	–	1.6212	130.99	0.979
7.5925+ (3.9526)	-0.4476 (1.4687)	1.0086+ (6.5136)	-0.0601 (1.3262)	-0.1604+ (4.4039)	-0.5175+ (3.9764)	2.8450	249.07	0.992

Notes: t stats. in parentheses, + = significant at 1 per cent level, DWs are in the indeterminate range at 5 per cent level.

Sources: Annual Abstract of Statistics (various), Royal Commission on Transport(1929) Cmd 3365, London and Cambridge Economic Service, The British Economy: Key Statistics 1900–1970 (London: Times Newspapers, ?1971).

15 Advances in Road Construction Technology in France

Dominique Barjot

Branded as the foremost social evil by some, hailed as indispensable means of individual liberty by others, motor vehicles (or, rather, their politically influential owners and the motoring lobby's powerful pressure groups) were soon responsible for road improvement and, in more recent years, as the number of motor vehicles has grown greater, for massive road construction programmes. Studies carried out by the Direction Générale des Routes du Ministère de l'Urbanisme et du Logement (the Roads Department of the Ministry for Housing and Town and Country Planning) make it possible to outline the main features and problems of road-building in the Motor Age: and my own research in the archives of the Fédération Nationale des Travaux Publics (French Federation of Civil Engineering Contractors) and of some of the major road-construction firms, notably the Société Chimique et Routière d'Entreprise Générale and Société Routière Colas, have enabled me to supplement these official studies.

Various major themes emerge from the French experience. Public investment, determined by general economic and political considerations, was very irregular and made it difficult for the construction companies to predict and plan. The system of tendering for public contracts created intense competition within the road-building industry from which a few large and well-connected conglomerates emerged. They were those which, well-supported financially, kept ahead technically, ploughing back much of their profit into research and development.

THE RAPIDLY GROWING DEMAND

In 1914 France was second in Europe only to the United Kingdom in terms of vehicle registrations. (For French vehicle statistics, see App. 15.A.1.) As Chapter 1 showed, this was due to the French motor industry which emerged as European leader in the early

291

years.[1] French engineers already had a reputation for inventiveness. As early as 1862 Alphonse Eugène de Rochas had discovered the principle of the four-stroke engine. The first motor vehicle to be produced in any numbers in France, however, was steam-powered. Designed by Amédée Bollée, known as La Mancelle, it was first built as a short production run in 1876–7. But steam was soon to find a strong competitor in the petrol engine. It was in 1884 that, shortly before Gottlieb Daimler's trial run mentioned elsewhere by Dr Nübel (Chapter 2), Édouard Delamare-Boutteville designed and built a prototype motor vehicle using this novel type of engine. It was, however, never developed.

The growth of the French motor industry was supported from the outset by vigorous home demand and a brisk export trade. (France was the main European exporting country down to 1913.) The early vehicles were unreliable, did not last long and were soon being replaced by improved models. Sales were further boosted over time by reductions in ex-factory selling prices. The French industry did well by creating dealer networks early on. Its publicity was good too, it had its own newspaper (*L' Auto*), and, through the Automobile Club de France, established in 1895, arranged motor trials and races which attracted world-wide attention. The French Motor Show soon became an important international event.[2]

This growth was interrupted by the First World War; but production soon regained pre-war levels and the interwar years saw a sharp rise in registrations, especially of private vehicles. Progress was uneven, however. There was a boom in the 1920s and some of the ground lost to the United States was actually regained; but the Depression of the 1930s brought fresh losses in the face of foreign competition. Growth was resumed after 1949 and the years to 1974 saw the motor car overtake all other means of private transport in France.[3] In spite of the growing number of accidents,[4] the disadvantages of motorisation did not make themselves felt until relatively recently, with congestion in towns bringing down average speeds and parking becoming increasingly troublesome, to motorist and public alike. These urban problems involved local government in higher costs. Environmental pollution, too, (as many believe) caused increasing concern. On the other hand, the enormous advantages of motor vehicles became even more evident. They were the most flexible of all forms of transport and were to play a particularly important role in the social and economic integration of outlying areas, especially if one takes into account that in the 1960s road haulage took over from rail.

BELATED EFFORTS TO MATCH ROAD SUPPLY TO GROWING MOTOR-VEHICLE DEMAND

On the eve of the First World War France was still making the most of its unrivalled road network. A legacy of the eighteenth century, it had benefited from a consistent effort during the French Restoration to bring it back to standard and to ensure adequate maintenance. The Corps des Cantonners, responsible for this and for road surveying, was formed in 1816. The July Monarchy of 1830–48 went further: Acts of 1832 and 1833 provided for the construction of strategic military roads, and a Charter of 1836 made provision for *chemin vicinaux*, the country roads.

These efforts were sustained throughout the Second Empire and into the first years of the Third Republic, despite the much greater popular attraction of the new railways. After the end of the Freycinet programme, however, French roads and road works received much less attention because of the loss of medium- and long-distance traffic to the trains. After an Act passed in 1880, the money spent on roads no longer increased. Indeed, at constant prices it was reduced, especially, if the increased (mainly local) road traffic mileage is taken into account.[5] Thus it came about that, according to a report of 1913, France had by then become 'the country where, for highways, yearly appropriation per kilometre for maintenance work, major repairs and improvements [were] lowest'.[6]

The standards of the road network as a whole were gradually deteriorating, and the practice was spreading of using 'verges for all manner of telegraph lines or ducts and pipes'. The construction of local railways was held to warrant 'the restriction of roadway width to 5 metres'. They also caused 'the proliferation of level crossings and meandering detours'. At the same time road-surfacing practice, being only required to withstand the wear and tear of steel-tyred vehicles, remained unchanged. Sand/water – bound crushed stone aggregate was almost universally used, owing to the high cost of paving-stones.[7] Although road expenditure was increased in the 1912 and 1913 Budgets, the First World War was to prevent further progress: state funding of road works remained at consistent level, but this money was mostly used to maintain the existing network.

From 1918 on, government measures for roads took a new scope and scale in three main successive cycles. The first (1918–27) was concerned with the reconstruction of roads and permanent structures (bridges, tunnels, etc.) destroyed during the war. The second (1927–33) saw the doubling of the national highway network mileage, after the 1930 Finance Bill transferred responsibility for some 40 000 km. of regional highways (*routes départementales*) and local country roads (*routes vicinales*) from local to national government agencies. The third (1933–9) was associated with a succession of major public works programmes, in particular the 1934 Marquet and the 1936 Blum plans. Spurred on by the growth in motor traffic, and further prompted by the aim of reducing unemployment, a recognisable road policy first took shape from 1934 onwards. Its objections were to relieve the burden on local government finances; to improve the existing network; to devise and enforce traffic regulations; and to promote research programmes at the laboratory of the École Nationale des Ponts et Chaussées.

Such initiatives, while they represented a real effort, were far from permanent or certain. Government agencies lacked sufficient funds, with the result that road-users had to be taxed more heavily. There was not enough finance available and state agencies had to resort to public tendering in order to save money. All attempts at reform having come to nothing, be it through the setting up the Office des Routes or the device of staged intalment programmes, the financial constraints and the military needs were to delay until 1938 the start of work on the first major French motorway project, the Autoroute de l'Ouest. Once again, war soon intervened.

Until 1942, nevertheless, appropriations remained steady, at a level comparable to those of the 1927–33 period, with new construction accounting for over 25 per cent of total funds due in particular to the public works programme which was put into effect from October 1940. But 1942 saw the collapse of all this: investment ceased overnight, funds available for maintenance, already on the wane, plummeted, and the achievements in reconstruction were often destroyed by bombing and battle, which tripled the number of bridges knocked out and of road mileage lost.

Even so, in 1945, most of the French road network still seemed to be up to standard; but it soon deteriorated, owing to the rapid increase in road traffic. The priority given to urban and industrial reconstruction meant that funds were not available to prevent the roads from deteriorating further. Provisions for new investment, and, most importantly, for maintenance, were to remain, until 1961, quite inadequate despite the setting up of the FSIR (Fonds Spécial d'Investissement Routier, the emergency road investment fund) in 1952, and notwithstanding the decision, taken in 1955, to entrust the construction of trunk motorways to part-public, part-private corporations (Sociétés d'Economie Mixte). The decision, in 1960, to implement a national plan for roads, together with the havoc wreaked by the serious damage due to the severity of the 'great winter' of 1962–3, led to an awakening to the urgent need to press on with further road construction in order to make up for lost time.

The vigorous resumption of investment, however, was not to result in a more coherent and dynamic road policy until 1970, when the motorway concession system was extended to include private firms. Late in 1971, the government announced a massive devolution scheme: some two-thirds of the national highway network was reclassified, falling under the responsibility of local government at regional (*département*) level, this being offset by substantial subsidies and thus favouring a massive switch in funding. This now concentrated on urban motorway projects, and on projects conforming to the policy of planned development. It consisted in allocating priority to a given line of communications and in deciding which route should go through which *département*. Each *département*, under the direction of the central government agencies, was to be responsible for its respective section of the priority route.

TECHNICAL PROGRESS

France was the first country in which motor-vehicle trials drew attention to the need for improved road surfaces. Holding up the high standards of French roads, the Automobile Club de France, as we have mentioned, encouraged motor trials, which developed into motor races. These events proved spectacularly popular. As early as 1894 the first true motor race, from Paris to Rouen, showed the advantages of petrol over steam. The following year an even greater endurance test, from Paris to Bordeaux and back, attracted world attention. The first international race was held in 1898, from Paris to Amsterdam. But the tragedy of the Paris–Madrid race, in 1903, highlighted the glaring deficiencies of conventional road surfaces for pneumatic-tyred motor traffic. There were seven deaths, including that of Marcel Renault.

French government agencies responded by deciding to apply tar coatings on national highways; and the Automobile Club de France, by creating the Circuit d'Auvergne in 1905, the first purpose-built racing circuit. French engineers found that they could meet most of the requirements of tar application to highway surfacing, with existing, indigenous techniques and processes. In the field of machinery France could still claim a definite lead, with the roller – first contrived by Fortin, in 1830, and later fitted with a steam engine by Gaëtan Brun, in 1874, and with the *monte-jus*, a machine designed for spreading hot tar on roads, developed in 1910 by Jules Lassailly. Adoption of the latter only confirmed the decision to favour tar spraying, which had originated in work undertaken in 1900 by D. Guglielminetti. Tar thus became, through the offices of Deutsche de la Meurthe and of the firm of Lassailly et Bichebois, the principal road-surfacing material. A chronological list of these developments will be found in Appendix 15.A.3.

New surfacing techniques were also adopted: cement first used in road-making in Edinburgh in 1873 and already used in Grenoble in 1876; concrete first used in Bellefontaine (Ohio) in 1892; and cold-mix asphalt, developed in Germany in 1900 and in regular use by 1904. It was introduced into France in 1912. The requirements to be met, however, were to become increasingly complex. This caused the French to turn more and more to foreign equipment. The first American steam shovel was introduced in France by Firmin Deschiron in 1911 and the first self-propelled tar-spreader by Gaëtan Brun in 1913.

The appearance of newer, cheaper materials and the mechanisation of construction led from 1925 onwards to a new wave of construction techniques, taking advantages of advances in compacting, and of the development of gravel mixes. At the same time as poured asphalt was winning general acceptance for urban roadway and footpaths, artificial asphalt gained a commanding lead over rival materials. Coal tar, while still the most widely used surfacing agent in 1939, had finally lost its cost advantage with the British miners' strike of 1926. While tarmacadam, an English invention dating from 1900, was proving highly successful, furnace slag was not employed until relatively late in time; and cement pavements were prohibitively expensive. From 1923, with the benefit of the successful development of anionic emulsions, artificial asphalt proved the most economical material and was superior to all other processes as it could be laid even in rainy weather. To ensure the neutralisation of the negative ions, however (this being indispensable to break up the emulsion, by allowing the bitumen to coalesce and form a stable coating), it was necessary to lay the emulsion on a line gravel base: but this material tended to get compacted too rapidly, and to become destabilised too soon, as normal wear and tear rounded and smoothed the gravel texture. Production processes were mechanised gradually with the massive influx, from 1923, of British, Swiss and, in particular, American equipment. This allowed rapid laying of new types of surface; but the most spectacular advances were achieved in the field of earth-moving operations and soil stabilisation. The tracked mechanical shovel was brought in from the United States in 1922, the bulldozer in 1925 and the tractor-drawn scraper in 1932.

Construction and maintenance practice benefited from research programmes carried out by the Administration des Ponts et Chaussées (government Department of Road and Bridges). After the hard winter of 1962–63, road maintenance became increasingly mechanised and streamlined. At the same time the range of available materials became wider, with the increasing use of 'white products' (natural cement or furnace slag). 'Black products', based on tar or, even more extensively, on asphalt, remained, nevertheless, the basic road-surfacing materials.[8] Owing to the poor performance of bituminous tar and of poured asphalt[9] for road-surfacing in the open country, however, asphalt consolidated its lead, with hot-spread asphalt taking over from cationic emulsion, whose introduction made it possible to build stronger pavements in silica-aggregates. The success of the two processes led to the increasing use of asphalt-aggregate thick-course spreading (*tapis d'enrobé*). These asphalt-aggregate mixes, laid in much thicker courses than previously and therefore stronger and more hard wearing, tended to oust surface coatings of the type favoured in the interwar period. Equipment operated by the road-construction industry steadily improved in performance, bringing about a dramatic fall in earthwork costs. Cost reductions were equally, if less dramatically, effective for surfacing work, as a result of the successful adoption of the slip form (used in laying concrete pavements), of the finisher (for flexible pavements)[10] and of asphalt-aggregate mixing plants of the Barber-Greene type, of U.S. provenance.

ADVANCES IN TECHNOLOGY: THE KEY TO SUCCESS IN ROAD CONSTRUCTION

In the interwar period French road-construction firms enjoyed a high rate of growth, due to the rapid increase in road-maintenance contracts. Depending as they did so directly on public-sector orders, however, they were bound to be adversely affected by the fits and starts in government action. Most importantly, the inflexible guidelines set up for the awarding of public-sector contracts were hardly conducive to the making of high and steady profit. As we have seen, public tenders constituted the regular procedure for contract allocation during that period for both state and local government agencies.

Trade organisations were to function during the 1930s as a means of cushioning the worst effects of such a situation. Some companies, in particular some of the larger ones such as the Société Routière Colas, opted at times to curb growth, in order to conserve balanced financial resources. But only through advances in technology could cost reductions be achieved to offset poor profit expectations in the public works market.

Drive and vulnerability were to remain the salient features of this branch of industry in the period following the Second World War. Full of vigour, the road-construction industry was further able to capitalise on its higher rates of concentration, of specialisation and of productivity per man-hour, compared to civil engineering as a whole; and it was able to turn to good account its extensive vertical integration, made easier by its close links with the oil consortia. None the less, a higher capital–output ratio than that obtaining in other branches of civil engineering, lower working surpluses, coupled to its heavy dependence on the oil industry resulted in the road-construction industry being highly sensitive to government 'stop–go' or to sudden jumps in oil prices, as the Suez crisis had already shown.

Competition was very real and was encouraged by the state. Bolstered by a relative mobility in production factors, and a fairly open access to the market, a large number of businesses were facing a small member of clients. Government agencies dutifully observed the golden rule for the award of public contracts: to keep competitors assured that they would all be treated on an equal and impartial footing when tenders were scrutinised. In this way competition operated to the full. Maintained by the constant influx of newcomers to the field, it always remained a price competition. Moreover it became increasingly wide-ranging with the growing diversification of products offered by the various tendering companies. Most importantly, the intensity of the competition, which drove out the small and weak, resulted in the survivors becoming less dependent on state moneys, having been consolidated into large firms, integrated into powerful consortia, oil-based as a rule, but occasionally centred on banking or on another industry. Some firms, like the Société

Chimique et Routière d'Entreprise Générale or the Société Routière Colas, even graduated to the ranks of what came to be known as 'barometric' firms, those used to indicate changes on the economic weather.

The emergence of the giants, however, did not curb competition. There were no sheltered positions, as evidenced by the fact that even the most strongly entrenched firms were having to adapt or go under. The firm Lassailly et Bichebois, which had been bought up by Compagnie Française des Pétroles in 1966, went out of business in 1975. Other firms, such as Cochery (in 1942), Société Chimique de la Route (in 1966) and Société Anonyme pour la Construction et l'Entretien des Routes (in 1974) were taken over by oil interests.[11] The vulnerability of the older-established firms was due, to a large extent, to the successful challenges of new rivals, latecomers to the field, originating in the chemical industry (Gerland S.A.), colonial operations (Bourdin et Chaussé), or linked to the rise of an entrepreneurial genius (Jean Lefebvre).

Société Chimique et Routière d'Entreprise Générale[12] gradually emerged as the leader in the field because of well-managed adjustment to changing conditions and, most crucially, to an exceptional emphasis on research. Advantages of scale were increasingly proving to be decisive for this dynamic group, strengthened by a well-balanced market strategy allied to a prudently forward-looking investment policy.

A SUCCESS STORY BASED ON THE INVENTION OF A NEW PRODUCT: SOCIÉTÉ ROUTIÈRE COLAS

This firm achieved a remarkable breakthrough in road construction, during the interwar period,[13] on the strength of the particular qualities of the cold-mix asphalt – or 'Colas' – process for emulsifying asphalt. The Colas process brought about substantial cost reduction in road-surfacing and also provided remarkable anti-skid qualities. Patent rights had been acquired by a large civil engineering firm, Société Générale d'Entreprises. The swift success of its asphalt emulsion led the firm to enter into partnership with Shell in 1929, when it was decided to set up a joint subsidiary to market the patent: this was the Société Routière Colas, control of which eventually passed to Shell in 1932. Helped by the development of a diversified range of products the company maintained a steady growth, in spite of the Depression, until the outbreak of the Second World War.

After 1945 it grew spectacularly. This growth was fostered by careful husbanding of resources. Capital was increased by ploughing back profits and great emphasis was placed upon building up research and development. With the development of cationic emulsions – suited to a much wider range of uses than the anionic emulsions produced in the interwar period – Colas was able to offer a widening range of products. It was thus increasingly in a position to meet the requirements of its main customer, the Administration des Ponts et Chaussées. The high rate of return shown by the parent company, together with its remarkable technological achievements, laid the foundations of a powerful group. From a strong base at home it was able to extend its activities in colonial markets, to those of Western Europe and, most importantly, to North America.

CONCLUSION

Whereas motorisation developed rapidly and consistently, French road-construction programmes turn out to have been at best fitful, and often inadequate. Both state and local government agencies were to delude themselves for a long time by their claims of a 'superior' French road network. Overall, government tended to raise funds for maintenance rather than for new construction, until 1934 at least. The slump of the 1950s, which may be attributed to the tailing-off of post-war reconstruction, was also the result of a cutback in maintenance programmes, with no new investment to offset such retrenchment. It was not before the 1960s – not to say the 1970s – that France was able to make up for previous neglect, especially in motorway building.

Road construction had undergone momentous technological change. This began in the 1920s, though it was not until the 1960s that it reached its full potential. The objectives of public-sector clients, to obtain best quality at least cost, tied in with those of industrialists to find outlets for their byproducts, and of road-construction contractors to ensure a steady flow of profitable business. The superiority of hydrocarbon binders, which rapidly became apparent, promoted the formation of large firms whose efforts to develop new products were assisted by the contribution of the laboratories of the Administration des Ponts et Chaussées.

The inflexible guidelines set up for the allocation of public-sector contracts failed to provide regular work for contractors. Their profits margins were low. Public tenders remained one case until 1976.

While most advantageous as a means of saving public money, this intensified competition, especially at times of market contraction or when new tenderers came on the scene. The result was that the better-managed (and often technologically better-suited) firms often shied away from government tenders. Under these conditions top priority was given to innovation, which cut costs and allowed growth. Initially, innovation concentrated on machinery: at first, rollers; later, spreading equipment; and, finally, earth-moving machinery. But the most decisive advances turned out to be in the field of production processes and, above all, of the development of new materials, with the success of asphalt over coal tar, and with the rival claims, since the 1950s, of emulsions and of hot-spread asphalt. Those companies which were prepared to plough back profits into research and development emerged as the small and select band which now dominates the whole business.

Notes

1. C. W. Bishop, *La France et l'automobile – Contribution française au développement économique et technique de l'automobilisme des origines à la deuxième guerre mondiale* (Paris: Librairie Technique, 1971).
   P. Fridenson, *Histoire des usines Renault*, vol I (Paris: Le Seuil, 1972).
   J. M. Laux, *In First Gear. The French Automobile Industry to 1914* (Liverpool: Liverpool University Press, 1976).
2. P. Fridenson, 'Une Industrie nouvelle: l'automobile en France jusqu'en 1914', *Revue d'Histoire moderne*, vol. 19, no. 1012 (1972) pp. 557–78.
   ____ 'Les Premiers inventeurs de l'automobile', *L' Histoire* (Dec 1984).
   P. Gerbold, 'L'Irruption automobile en France (1895–1914)', *L'Information Historique* (Sep–Oct 1983) pp. 189–95.
3. J. Jones, *The Politics of Transport in Twentieth-Century France* (Kingston–Montreal: McGill–Queen's University Press, 1984.)
4. J. C. Chesnais, *Les Morts violentes en France depuis 1926 – Comparaisons internationales* (Paris: Presses Universitaires de France, 1976).
5. N. Spinga, 'L'Introduction de l'automobile dans la société française entre 1900 et 1914' (Étude de presse, M.A. thesis, University Paris X – Nanterre, 1973).
   ____ 'Comment la France a adopté l'automobile', pp. 38–43, in *De Renault Frères, constructeurs d'automobiles à Renault, Régie nationale* (Paris: Privately printed by Renault, June 1985).
6. Report to the Third International Road Congress' (London, 1913). Quoted in *Annales des Ponts et Chaussées* (Dec 1913).
7. P. Le Gavrian, 'La Route à cinquante ans', in *Le Génie civil* (Special 50th Anniversary Issue, 1930) pp. 238–40.

8. This distinction between 'white products' (calcium-based: slag and cements) and 'black products' (tar- or asphalt-based) is quite usual in the road-construction industry. 'White products' refer to the more volatile fractions (paraffin, kerosene, etc.). 'Black products' include fuel oil and lubricants, as well as asphalts and pitches.

9. These techniques were being increasingly employed: the former, in English-speaking countries; the latter, in Federal Germany.

10. Flexible pavements are pavements in 'black products' (tar- or asphalt-based). See note 8.

11. Entreprise Lassailly et Bichebois had been formed in 1892. Société Anonyme pour l'Entretien et la Construction des Routes, constituted in 1920, was the heir to the Brun operation, established in 1870. Cochery had been established in 1926, Société Chimique de la Route in 1928.

12. Société Chimique et Routière d'Entreprise Générale had been formed in 1926.

13. D. Barjot, 'L'Innovation, moteur de la croissance: le Procédé Colas (1920–1944)', in *Histoire, Économie, Sociétés*, vol. 2, no. 1 (1st Quarter, 1983).

Appendix 15.A.1 *Motorisation in France 1895–1975*

Total number of motor vehicles

1. *Estimates at 31 December*

	Cars	Number Motor cycles
1895	300	. . .
1896	500	. . .
1897	1 200	. . .
1898	1 500	. . .
1899	1 700	7 177
1900	2 897	11 252
1901	6 386	12 482
1902	9 207	15 188
1903	12 984	19 816
1904	17 107	27 435
1905	21 523	29 954
1906	26 262	31 863
1907	31 286	27 473
1908	37 586	27 215
1909	44 769	26 840
1910	53 669	27 061
1911	64 209	28 641
1912	76 711	30 418
1913	90 959	35 141
1914	107 535	37 761
1915	102 286	11 284
1916	100 989	13 095
1917	98 534	12 389
1918	94 884	8 394
1919	93 338	28 538

2. *Estimates at 31 December*
(000)

	Cars	Motor lorries and vans	Total	Motor cycles
1920 (1)	157 2	79 4	236 6	50 8
1921 (1)	196 8	92 9	289 7	42 9

(Appendix 15.A.1 continued)

	Cars	Motor lorries and vans	Total	Motor cycles
1922	242 6	120 6	363 2	51 3
1923	293 5	154 7	448 2	70 9
1924	374 0	200 9	574 9	96 4
1925	476 4	244 9	721 3	117 4
1926	541 4	267 4	808 8	157 6
1927	642 7	306 5	949 2	236 8
1928	757 7	331 6	1 089 3	303 6
1929	930 2	366 0	1 296 2	381 9
1930	1 109 0	411 5	1 520 5	443 5
1931	1 251 5	437 9	1 689 4	489 0
1932	1 279 1	433 8	1 712 9	501 0
1933	1 397 1	458 1	1 855 2	541 6
1934	1 479 6	458 9	1 938 5	570 0
1935	1 547 1	457 8	2 004 9	550 0
1936	1 638 5	456 8	2 095 3	527 0
1937	1 720 5	451 2	2 171 7	512 0
1938	1 817 6	451 4	2 269 0	500 0
1939	1 900 0	500 0	2 400 0	400 0
1940	1 800 0	500 0	2 300 0	. . .
1941	. . .	. . .	. . .	. . .
1942	. . .	. . .	. . .	. . .
1943	. . .	. . .	. . .	. . .
1944	680 0	230 0	910 0	. . .
1945	975 0	600 0	1 575 0	. . .
1946	. . .	. . .	1 700 0	. . .
1947	. . .	. . .	1 700 0	. . .

(1) 87 *départements*.

3. *Estimates at 31 December*
(*000*)

	Cars	Motor buses	Motor lorries (1)	Tractors (2)	Total	Motor cycles
1948	. . .	. . .	. . .	. . .	1 850	621
1949	. . .	. . .	. . .	. . .	1 950	699

1950	. . .	. . .	. . .	. . .	2 150	817 5
1951	1 700 0	24 6	715 8	. . .	2 440	1 011
1952	1 800 6	27 1	848 7	168 6	2 844 4	1 446
1953	3 020 0	29 1	1 008 4	208 7	3 266 2	1 982
1954	2 677 0	29 5	1 095 0	242 5	4 044 0	2 563

(1) Vans and road tractors included.
(2) Road tractors excluded.

4. *Estimates at 1 January*
    (000)

	Cars	Motor buses	Motor lorries (1)	Tractors (2)	Total	Motor cycles
1955	2 677	29 5	1 095	242 5	4 044 0	
1956	3 113	30 6	1 194 4	302 8	4 640 8	
1957	3 476	31 2	1 246 7	376 2	5 130 1	4 650
1958	3 972	32 9	1 338 4	474 2	5 818 1	5 170
1959	4 512	34 4	1 429 3	556 9	6 532 6	5 450
1960	5 018	35 6	1 507 0	636 4	7 197 0	4 650
1961	5 546	36 9	1 597 4	705 1	7 885 4	4 950
1962	6 158	38 4	1 684 3	776 9	8 657 6	5 020
1963	7 010	40 2	1 782 3	843 9	9 676 4	5 140
1964	7 800	42 5	1 893 1	914 9	10 650 5	4 900
1965	8 700	45 1	2 023 5	992 0	11 760 6	5 180
1966	9 600	47	2 134 0	1 058	12 839	5 500
1967	10 400	50	2 252 0	1 133	13 385	6 020
1968	11 200	52	2 360 0	1 205	14 817	5 780
1969	11 800	55	2 352 0	1 277	15 484	5 550
1970	12 400	51	2 509 0	1 349	16 319	5 330
1971	12 900	67	2 678 0	1 408	17 053	5 020
1972	13 400	73	2 848 0	. . .	16 321[3]	5 200
1973	13 900	79	3 013 0	. . .	16 992	5 250
1974	14 500	83	3 247 0	. . .	17 830	5 510
1975	15 500	89	3 476 0	. . .	18 565	5 670

(1) Vans and road tractors included.
(2) Road tractors excluded.
(3) Road tractors included, but other tractors excluded.

306

Sources:
1. Ministère des Finances. Chambre syndicale de construction d'automobiles. Chambre syndicale des Constructeurs de motocycles in *Annuaire Statistique de la France*, no. *Retrospectif de 1966*, table v, p. 316 (Paris: INSEE, 1966).
2. INSEE, idem, p. 316.
3. Idem.
4. INSEE, table vi in idem.
5. INSEE, in *Annuire Statistique de la France . . . de 1976*, table 20, p. 464 (Paris: INSEE, 1976).

Appendix 15.A.2 Public expenditure on national highways (1900–59) and turnover of road construction industry (1922–73)

(m. FF. at 1913 francs)

	Total expenditures for national highways	Expenditures for maintenance only	Turnover of road construction sector
1900	39 53	35 08	. . .
1901	40 00	35 53	. . .
1902	40 68	36 13	. . .
1903	40 59	37 18	. . .
1904	40 53	36 19	. . .
1905	39 45	35 36	. . .
1906	38 55	34 55	. . .
1907	39 11	35 10	. . .
1908	41 31	34 90	. . .
1909	39 25	33 54	. . .
1910	38 99	33 26	. . .
1911	36 69	33 09	. . .
1912	38 74	34 05	. . .
1913	42 20	37 30	. . .
1914	43 04	38 28	. . .
1915	53 21	25 56	. . .
1916	33 65	14 31	. . .
1917	51 80	14 57	. . .
1918	15 74	15 23	. . .
1919	78 09	19 53	. . .
1920	60 58	21 59	. . .
1921	132 86	30 93	. . .
1922	118 37	30 34	20 03
1923	114 96	42 86	18 30
1924	96 66	39 43	16 40
1925	84 98	49 50	14 76
1926	60 89	51 36	10 52
1927	83 31	72 48	12 17
1928	86 32	79 96	18 91
1929	95 56	86 34	22 79
1930	112 68	97 06	26 82
1931	162 26	144 17	36 95
1932	124 24	115 72	29 52
1933	157 07	147 61	27 73
1934	116 36	111 55	27 73
1935	202 05	147 42	48 74

(Appendix 15.A.2 continued)

	Total expenditures for national highways	Expenditures for maintenance only	Turnover of road construction sector
1936	248 59	169 90	47 40
1937	188 47	112 78	62 30
1938	172 12	112 00	52 59
1939	143 85	96 43	42 41
1940	120 28	76 42	. . .
1941	189 32	68 59	. . .
1942	107 19	45 26	. . .
1943	61 12	37 81	. . .
1944	62 67	38 08	. . .
1945	141 30	67 37	. . .
1946	233 34	133 15	. . .
1947	247 16	159 06	. . .
1948	186 97	119 73	. . .
1949	208 98	122 70	. . .
1950	162 05	103 34	. . .
1951	157 30	86 48	. . .
1952	152 52	69 61	. . .
1953	166 08	64 07	. . .
1954	186 50	66 49	. . .
1955	236 14	70 41	. . .
1956	233 09	62 34	. . .
1957	193 13	56 64	. . .
1958	167 07	48 32	. . .
1959	222 26	55 37	325 37
1960	. . .	. . .	286 99
1961	. . .	. . .	367 43
1962	. . .	. . .	399 19
1963	. . .	. . .	471 48
1964	. . .	. . .	1108 31
1965	. . .	. . .	1112 92
1966	. . .	. . .	1281 08
1967	. . .	. . .	1367 26
1968	. . .	. . .	1404 95
1969	. . .	. . .	1420 83
1970	. . .	. . .	1885 02
1971	. . .	. . .	1898 33
1972	. . .	. . .	2146 88
1973	. . .	. . .	2239 48

Sources:

1. Sources:
   1900–50. Voted expenditures for national highways.
   1951–9. Actual expenditure.
2. Sources idem:
   Total expenditure – maintenance = investment.
3. Sources:
   1922–39. Census of public sector contracts (concluded by members of Syndicat Professionel des Travaux Publics).
   1959–73. Turnover of road-construction sector.

Appendix 15.A.3 *Landmarks in advances in technology in the field of road construction (1900–74)*

Year	France	Other countries
1820		Development of macadam pavement.
1829	Lyons: first poured asphalt surfacing.	
1830	Compacting roller invented (Fortin).	
1873		First cement pavement (Edinburgh).
1874	Steam roller (G. Brun).	
1892		First concrete pavement (U.S.A.).
1900	Dr Guglielminetti shows the importance of tar applications for road-dust abatement.	Development of tarmacadam (U.K.). and of Westrumite: the first asphalt emulsion (Germany).
1902	First tar application (Nice).	
1904		First application of an oil emulsion, for the surfacing of the Ardennes circuit (Germany).
1905?		First use of tarmacadam on English roads.
1906		First bulldozer (U.S.A.).
1909	First application of bituminous asphalt in France.	
1910	Lassailly's *monte-jus*.	First Warren asphalt spreader (U.S.A.).
1912	First application of asphalt emulsion in France.	
1913	G. Brun introduces from the U.S.A. the first self-propelled tar spreader.	
1920	G. Brun introduces the grader.	H. A. MacKay develops the 'Colas' anionic emulsion (U.K.).
1922	Introduction of the first *tracked* mechanical shovel, from the U.S.A.	First Italian motorway: Milan–the Lake resorts.
1924	First concrete pavements laid in the Nord *département*.	

1925	Introduction of the bulldozer and of the rotary scraper.	First diesel roller (U.K. and Germany) and early development of hydraulic binder techniques of soil stabilisation (U.S.A.).
1926	A. Cochery introduces tarmacadam into France.	
1927	First French diesel roller.	
1928	S.A.C.E.R. develops first asphaltic concrete.	
1929	First artificial asphalt produced in France.	
1931		First travel plant (U.S.A.).
1933		Hitler launches a 7000-km. motorway construction programme.
1934	Generalisation of anti-skid coatings.	
1937	Introduction of travel plant (finisher for 'black' surfacings).	
1938	Work starts on Autoroute de l'Ouest.	First pushed scraper (U.S.A.).
1939	First cationic asphalt emulsion.	
1940		First U.S. motorway.
1945	Introduction of vibrating cement finisher.	First self-propelled scraper (U.S.A.).
1946	Adoption of American asphalt–aggregate mixing plants.	
1949	The Laboratoire Central des Ponts et Chaussées opened.	
1950	Appearance of hot-spread asphalt.	
1951	First Poclain hydraulic crane. First French asphalt–aggregate mixing plant.	First German asphalt–aggregate mixing plant. First asphalt motorway pavement (Fed. Germany).
1954	Cationic emulsions become operational.	
1957	Asphalt–rubber is launched in France and in Federal Germany.	
1958		Adoption of gravel–cement mix for motorway sub-base (U.S.A.).
1959	Adoption of first U.S. slip-form.	
1962–3	Hot-spread asphalts and thick-course heavy aggregates gain ascendancy.	
1965		Development of tar–bitumen.

(Appendix 15.A.3 continued)

Year	France	Other countries
		(Fed. Germany, U.K. and Japan).
1966	Adoption of asphaltic penetration mixes.	
1968	Development of gravel–slag aggregate.	

Index*

accidents, traffic 71, 264–84
  American South 119
  France 142–3
  Japan 231, 233, 234
  Saskatchewan 183
Adams, A. (fuel-cost study) 101
adoption rates of motor vehicles by
  countries 3 fig
advertising 60–1, 135–6, 167
African countries
  car owners 250–1
  motor vehicle penetration 251–2
Age-Herald (Birmingham, Al.) 123
aggression and driving 269
American Automobile
  Association 270
American Automobile
  Manufacturers' Association
  270
American Motor Transportation
  Company 86
American South 115–17
Anderson, A. G. 83
Anglo-American Oil Company 45
Apperson, E. and E. 17
Arnold, W. 18–19, 30
Argentina 165–6
ASAP see SKODA
Ashkhabad 256
asphalt 296–7, 299, 301
Atlanta 117–19, 122–4, 126
Atlanta Constitution 122
Austin, H. 35, 39
Austin Seven 151
Australia 165–6, 266, 269
Austria 148, 202–6
Auto, L' 292
autobahnen 72, 153, 271
Autocar, The 6, 30, 31, 40, 42
  Brighton Run 33
  French roads 23

Automobile Club of America 25,
  27
Automobile Club de France 21,
  23, 292, 295, 296
Automobile Club of Great Britain
  and Ireland 36–9, 44, 45
Automobile Manufacturing
  Business Law 1936
  (Japan) 217
Automotive Safety
  Foundation 270
Automobile Topics 25, 29
automotive sector,
  Saskatchewan 179
autostrada, first 271
Aveling, T. 8
Ayrton, W. B. 10

barges in Central Asia 256
Barjot, D. 23
Bartholomew, H. 124
Bateman, A. H. 9
battery-driven vehicles 10–11
'Be Careful Drive' 126
Beaumont, W. W. 37, 40
Belgium 202–6
Belloc, H. 72
Benz, K. 12–13, 14, 16 fig, 55,
  59–64
Benz and Company 55, 98
Benz Patent Motor Car 59–61
Berliet 132
'bicycle middleman' 137
Bicycling Times 9
Bicyclists' Touring Club 6
'Big Red Cars' 70
Birmingham, USA 119, 125, 126
black communities and
  transport 125–6
Blackburn, A. B. 9
Blaich, F. 4, 265

* Compiled by Mrs J. Butterley

313

Bloomfield, G. 2, 3, *fig.*
Blum plan 294
Board of Highway
  Commissioners 180
Bohemia 198–9, 201–2
Bollée, A. 9, 33, 292
Bollée, L. 21, 33
Bordeaux Race, 20–1, 34
Bosch, R. 31
Bosch Company 98
Bouton, G. 9
breakdown, possibility of 44
Brighton Run 33–4
British Daimler Motor
  Company 31–2, 38–9
British Motor Syndicate 31, 32, 34
Brun, G. 296
B. S. A. 36
Buda 102
Burnham, D. H. 124
buses
  diesel engines 98, 103–4, 105,
    108, 110, 112
  Japan 215, 217–18, 224–7
  road deaths 265
  Soviet Union 258–9
  United States 81–93
  Zaïre 242 *fig*, 243 *fig*, 249, 251
business firms 174
business practices 85
Butler, E. 17

cab companies 138, 141
  see also taxicabs
Cadillac 176
Caesar, O. 89
California
  bus pioneers 82–3
  diesel engines 101, 102, 108,
    110
  speed limits 278
California Taxicab Company 82–3
California Transit 85, 86
camel transport 256
Canada 165–87, 266
Canada Highway Act 1919 180
Canadian Consolidated Rubber
  Company 178
Cannstatt 55–8

Carless Capel and Leonard 44–5
Caterpillar Company 99, 107–8
Census of Merchandising 179
certification (bus companies) 91
Chadwick, E. 266
Chemins de Fer Vicineaux du
  Congo (VICICONGO) 244
Cherdantsev, G. N. 257
Chevrolet 176
Chicago World's Fair 17, 89
Chicago-Waukegan race 24
Christian Advocate 120
Chrysler 104, 177, 217
Circuit d' Auvergne 296
cities, motor vehicles and *see*
  Atlanta
Citröen 131, 132, 135 *fig*
Civil Code 1806 279
Clément, G.–A. 6
clubs
  cycling 6
  motoring 136, 175, 270
  see also under names of
    individual clubs
Cochery 299
Code de la Route 1922 278–9
Colas process 299
Colbert, C. 89
cold-mix asphalt 296, 299
Colonial Sand and Stone
  Company 101
'Columbia' cycle 7
Commerce Colonial, Le 248, 249
Commercial Motor, The 41
commuting 115
Compagnie Francais des Pétroles
  299
compensation, accident 268, 281–2
competition
  France: motor vehicles 139–40;
    road-building 298–9, 301
  USA (diesel engines) 107–8,
    110–12, 113
  Zaïre 250
Complete Motorist, The (Filson
  Young) 44
Concord coaches 67
Conestoga wagons 67
congestion 71, 122–3

Counseil de Gouvernement 245
Consolidated Freight Lines 101
consumption in France 136–8
Container on Flat Car (COFC) 76
controls on bus industry 91
co-ordination of transport
  France 139–40
  Japan 232, 234
Corps des Cantonners 293
costs
  of operating motor
    vehicles 173–4
  of road accidents 269, 278
  social 142–3, 228, 231–2
  transport in Zaïre 240–2, 246–7
courtship 121
crash helmets 282
credit sales 137–8
crime 120
Crown Coach 108
Cummins, C. L. 100, 101
Cummins Engine Company 100,
  102, 105–7
cycles
  motor: accidents 265; decline
    in 280–1, 282; in
    Europe 148, 150–1, 197; in
    Great Britain 42–3; in
    Japan 215; in USSR 259
    (*see also* tricycle, Benz's)
  pedal 5–7, 214–15, 216–17, 218
Cyclists' Touring Club (CTC) 6,
  36
Czech Lands 201–2
Czechoslovakia 194–210

Daily Graphic 34
Daimler, G. 13, 34, 55, 292
  engines 11–12, 14–15, 16
    fig, 17, 56–9, 62
Daimler Motor Syndicate 17, 30,
  31, 36
Daimler Motoren Gesellschaft 62
Darracq, A. 6
Davidson, D. 119–20
de Dion-Bouton 18–19, 21, 23
dealerships 117–18, 133–6, 178
Dearing, C. 26
decentralisation, American

cities 76–7, 122–4
Delamere-Boutteville, E. 292
dependence on motor vehicles
  national comparison 143
Derby, F. A. 16th Earl 41
Deschiron, F. 296
Detroit 27, 28–9
Detroit Automobile Association
  27
Detroit Automobile Company 27
Deutsche 23
Deutsche de la Merthe 296
developing areas 236–61
Development Plan (Zaïre) 247–8
Diesel, R. 97
diesel engines 97–113, 140
diesel fuel 101, 102, 104 *fig*, 106,
  111
Dion, A. de 9, 21, 132
  see also de Dion-Bouton
Direction Générale des Routes du
  Ministère de l'Urbanisme et du
  Logement 291
Dixi 3/15 151
DkW-Reichsklasse 152, 155, 159
doctors, early users of motor
  vehicles 173
Dominion Department of Railways
  and Canals 180, 182
driving licence 141, 143
Du Pont, T. C. 72
Du Pont Road 72
Duff, W. 173
durable goods 141, 228
Dürkopp 6
Duryea, C. E. 7, 17, 68
Duryea, F. 33, 68

earnings, average in Germany 154
  fig
earth highways 181, 182–3, 186 *fig*
École Nationale des Ponts et
  Chaussées 294
Economic Planning Agency 227
Economist, The 36
Edinburgh 272–3
education campaigns 272–3
Edward VII, king of Great
  Britain 38

EEC (European Economic
  Community) 142
'efficient accident rate' 268
electricity 10–11
Eliot, T. R. B. 30
Ellis, E. 30, 31, 37
employment and motor vehicles
  130
endurance tests 18–21, 27–8
engine governor, Daimler's 56
Engineer, The 7, 19
Engineering 34
Europ-Assistance 137
expenditure
  on motoring, national
    comparison 143
  on roads: France 293–5;
    Japan 219, 229, 232, *fig*;
    Saskatchewan 180, 182,
    184–5; Zaïre 238–9, 248

Fageol Brothers 84, 85
Fahrzeugfabrik Eisenach 151
farm-implement manufacturers
  174
farmers 46, 115, 134, 173
fatalities, motor accidents 71
  fig, 264–84
Faulkner, W. 121
Federal Highway Act 1921 68
Federal Highway
  Administration 72
Federal Motor Carriers Act 1937
  71
Fédération Nationale des Travaux
  Publics 291
feed for horses 4
Fiat Company 111
filling stations 178
Filson Young, A. B. 44
Finance Bill 1930 (France) 294
five-year plans (USSR) 261
Fonds Spécial d'Investissement
  Routier (FSIR) 295
Ford, H. 25, 27, 28, 68, 282
Ford Company 107, 108, 176–8
  lorries 215, 217, 237, 249
Foster, M. S. 70

fragmentation of cities 122–5
France
  accident liability 278–9
  developing motor vehicles
    13–24, 58–9
  dieselisation 104
  effects of motor vehicles 130–43
  petrol 44
  publicity 18 seg., 292
  railways 139–40, 153– *fig*, 158 *fig*
  road deaths 265, 266, 267 *fig*
  roads 228 *fig*, 229 *fig*, 230 *fig*,
    231 *fig*, 271, 291–301
  usage of motor vehicles 47–8,
    148, 149, 202–6
franchises, sales *see* dealerships
Franck, L. 238
'freeways' 72
freight transport
  diesel lorries 106, 113
  Germany 156–8
  Japan 218–22
  Soviet Central Asia 258
  USA 75–6, 78 *fig*, 79 *fig*
  Zaïre 236–7, 250–1
Freightliner 111, 113
French Restoration 293
friendly societies 137
From Street Car to Superhighway
  (foster) 70
frontier motorisation 165–87
fuel efficiency, diesel engines 98,
  99, 101, 102
Fuller, M. 142

Gable, C. 89
Garages 44, 130, 133–5, 178–9,
  260
Gas Power Age 167
gasoline *see* petrol
General Motors
  advertising 135 *fig*
  buses 85, 89
  diesel engines 99, 103–4, 106,
    108
  Japan 217
  Saskatchewan 176
Generalanzeiger 59–60

German Federal Republic
  roads 228, *fig*, 229, *fig*, 230 *fig*,
    231 *fig*
  speed restrictions 275, 278
Germany
  autobahnen 271
  Benz and Daimler 11–13, 55–65
  dieselisation 98, 104
  road deaths 265, 266, 267 *fig*
  usage of motor vehicles 148–60,
    202–6
Giffard, P. 18
Good Roads Movement 26, 115,
    175, 179
government policy
  accidents 267–70, 273–84
  in France 139–43, 293–5
  in Japan 218–20, 232, 234
  in Soviet Central Asia 257–8,
    259–60
  see also legislation
Grady, H. W. 117
'grandfather clock' (Daimler and
    Maybach engine) 56–8
gravel highways 182–3, 186 *fig*
Great Britain
  dieselisation 98, 104
  early motor vehicles 29–43
  railways 153 *fig*, 158 *fig*
  road deaths 265–7, 269
  road safety 268, 270, 271–2
  roads 228 *fig*, 230 *fig*, 231 *fig*
  speed limits 274–5
  usage of motor vehicles 47–8,
    148, 149–50, 151, 202–6
Great Depression 72, 74, 81, 89,
    167, 175
Great Horseless Carriage
    Company 31, 36
Greyhound Corporation 87,
    88–90, 92–3, 103
growth
  economic: Saskatchewan 166;
    Soviet Central Asia 257;
    Zaïre 248–9
  motor vehicle usage in USA
    74–8
guarantees for used cars 135

Guglielminetti, D. 296
Habans, P. B. 119
haulage in Zaïre 244
  see also freight
Hayes, A. L. 83
Haynes, E. 17
health 261
Heed, A. 83
Hele-Shaw, H. S. 40–1
Hercules 102, 103
Herodotus 67
Hewetson, H. 18–19
Highways, Department of 180
highways *see* roads
Hildebrand and Wolfmüller 42
Hino 113
hire-purchase 137–8
Hitler, A. 155
Holbrook, D. 264, 265
Holt, A. 41
Homer 55
Hooley, E. T. 31
Hoover, H. C. 69
Horseless Age 25
horses 4–5, 33, 46, 139, 174
hot tube ignition system 56
Houston, Texas 127
Hoyt, H. L. 117
human carriage in Zaïre 236, 237,
    241–2, 249
Humber 7
Hurt, J. 118
hyperinflation 151, 152

ignition system, hot tube 56
Iliad (Homer) 55
Imperial Oil 178–9
imported technology, France and
    130–2
imports, Germany and 158–9
In First Gear (Laux) 13
incentives, economic in USSR
    260–1
income and car purchase 169
Independent (Atlanta) 126
Indiana Toll Road Commission 73
industrialisation and motorism in
    Czechoslovakia 201–2

industry
  French motor   292
  out of cities   76–7
infrastructure, poor in Japan   215
  see also roads
inhabitants per motor car ratio,
  Europe   202–6
innovation, France and   130–3,
  295–7, 298, 300–1
instalments system of payment
  137–8
instruction, driving and
  maintenance   169
insurance   65, 136–7, 268, 279–82
intercity bus industry   86–93
intercity travel (USA)
  freight   75–6, 78 *fig*, 79 *fig*
  passenger   74–5, 76 *fig*, 77 *fig*
'intermédiaire velocipédique, l'
  137
internal combustion   11–13
International Exhibition, Paris
  1869   5
  1889 (Eiffel Tower)   14–15, 58,
  61
International Harvester   99–100,
  104, 106
Interstate and Defence Highway
  Law 1956   107
Interstate Commerce Act (Part
  II)   91
Interstate Commerce Commission
  90, 91, 92–3
Interstate Highway Act   73
Interstate Highway System   74
Isuzu   113
It Happened One Night (film)   89
Italy
  cars imported by Germany   159
  fuel prices   104 *fig*.
  road deaths   266, 267 *fig*.
  toll roads   271
  usage of motor vehicles   148,
  149, 202–6
Iveco   111

Jackson, H. N.   27–8
Jadot, O.   245
Japan   214–34

Jefferson Highway   85
jitney transport   82, 83, 124–6
Johnson, C.   37
Johnson, Junior   116
Jones, A.   40
July Monarchy   293

Kazakhstan   256, 257, 258, 259–60
'KdF'-Wagen   *see* Volkswagen
Kettering, C. F.   99, 108
Kirghizia   260
Kleyer, H.   6
Kohlsaat, H. H.   24
Knight, J. H.   8–9, 30
Koosen, J. A.   30
Koprivnice factory   194
'Kraft durch Freude'   155

L'Hollier, L.   31
L'Orange, P.   97
Labor Advocate (Birmingham,
  Al.)   121, 125
Lacre lorries   237
Lanchester, F. W.   35–6, 39
land, changing use in US   122–4
Lassailly, J.   296
Lassailly et Bichebois   296, 299
Latta, E. D.   118
Laurin et Klement (L&K)   194
Laux, J. M.   13, 20
Lawson, H. J.   31–5, 36, 38
Lea-Francis   7, 36
legislation and motor vehicles   84,
  90, 270, 273–4, 276–83
  see also government policy
Leplae, E.   238
Levassor, E.   16, 20, 23–4, 34, 58
  Panhard and   13, 15
Leviation, The (Davidson)   119–20
Leyland Motors   9
Leyland Steam Motor   41, 42
liability, accident   279–80
liberation, black communities and
  motor vehicles   126–7
licences
  driving   141, 143
  for motor vehicles   *see*
  registration

Liebig, Justus *Baron* von 61
'light locomotives' 8–9
Liverpool 39–41
Lloydminster Times 182
local bus services 82–4
location, motor vehicles in
    Saskatchewan 170–3
Locomobile 25
Locomotion Automobile, La 21
lorries
    adoption rates 3
    diesel engines 98–113
    Japan 215, 217–18, 219 *fig*,
        220–2, 234
    Soviet Central Asia 257, 258
    Zaïre 242 *fig*, 243–4, 245, 249
Lynd, R. S. and H. M. 116

McAdam, J. L. 67
machine-tools 132–3
machinery, road-building 260,
    296–7
Mack Motor Truck Company 85,
    103, 104, 105–6, 111, 113
McKechnie, J. 97
McLaughlin 176–8
maintenance, motor vehicles *see*
    servicing
make of motor vehicle,
    Saskatchewan 176, 177
MAN Company 98
Mannheim, Benz at 59–62
marketing motor vehicles 133–6
Marquet Plan 294
Mancelle, La 292
Maschinenfabrik Esslingen 57
mass transit, US
    encouragement 124–6
Massachusetts Compulsory
    Automobile Liability Security
    Act 1927 279–80, 282
Mavor, J. 179
Maxim, H. P. 17
Maxim, Sir H. 17, 35, 37
Maybach, W. 11–12, 15, 55–8, 62
mechanisation
    road construction 260, 296–7
    transport 7–13
Menier, H. 19, 21

Mercedes 26–7
Mercedes-Benz 110–11, 113
Mesaba Transportation
    Company 83, 85
Messageries Automobile de la
    Province Orientale (MAPO)
    244
Messageries Automobile de l'Ituri
    Oriental (MIO) 244
Messageries Automobile de
    Sankuru (MAS) 244, 248
Michaux, P. 5
Michelin, A. and E. 20
*Middletown in Transition: A Study
    in Cultural Conflicts* (Lynd and
    Lynd) 116
Minnesota 83
Model T Ford 43, 166, 176, 178
Montagu, J. 37–8, 39
monte-jus 296
Moose Jaw 169
moral decline, motor vehicles and
    121
Moravia 201–2
Morris, W. R. 7
Morrison, W. 11
Motor Age 25, 28, 29
Motor Bus Industry Code 90
Motor Car Act (British) 1903 273
Motor Car Club, The 31, 32–3
Motor Car Illustrated, The 38
Motor Carrier Act 91
Motor Carrier Bureau 91
Motor Cycle The, 42
Motor Road Building and
    Maintenance Ministry of 260
Motor Transit Company 86
motor trials 18–21, 27–8, 39
motoring clubs 30–1, 36, 136, 175,
    270
    *see also under names of
        individual clubs*
motoring publications 136
    *see also under names of
        individual publications*
motorists, early 16, 46
Mueller, H. 24
Mulliner, carriage-maker 35
Mumford, L. 116

Münchener Neueste Nachrichten
61
Münchener Tageblatt 60
Muncie, Indiana 116
Munich 60–1
municipalities 180, 183, 184–5

Nader, Ralph 283
National Association of
Automobile
Manufacturers 117
National Association of Motor Bus
Operators 91
National Association of Traction
Engine Owners and Users 8
National Cyclists' Union 6
National Industrial Recovery
Act 90, 91
National Road Safety
Committee 270
National Safety Council 273
National Safety First Association
270
National-Socialist government in
Germany 153–4, 155, 157–8
National Traffic and Motor Vehicle
Safety Act 1966 (US) 283
National Trailways 92
Neue Badische Landeszeitung 59
New, A. G. 42
New Jersey Public Service
Company 103
New Orleans Safety Council 119
'New South Creed' 117
New York 273
New Zealand 165
Nippon Automobile Company 215
Nübel, O. 12, 13, 148
Nuffield, *Lord* 7

occupations, motor vehicle owners
Prague 196 fig, 197–200
October Revolution 1917 257
Office des Routes 294
Olds, R. E. 25, 28
Oldsmobiles 28
OPEC (Organisation of Petroleum-
Exporting Countries) 110

Opel, A. 6, 151
Opel P 4 153, 155
opposition to motorisation in
Zaïre 239–41
Oregon 68–9
Otto, N. A. 55
Otto and Langen 11, 55
Ownbey, *Rev*. R. L. 121
ownership of motor vehicles
Czechoslovakia 195–200
Japan 222–8
Saskatchewan 169–70, 173–6
USSR 259

P and D lorries 107–8
Pacific Electric Company 70
Packard 176
Paller, R. von 61
Panhard, Hippolyte, 21
Panhard, R. 13
Panhard et Levassor 13–16, 21,
58–9, 61
Paris–Amsterdam race 295
Paris–Bordeaux race 19–20, 295
Paris–Brest-Paris cycle race 15, 18
Paris–Madrid race 23, 295
Paris–Rouen race 18–19, 295
parking space 123
Parkyns, *Sir* T. 9
passenger travel
in Germany 153 fig
in Japan 222–3, 226–8
in USA 74–5, 76 fig, 77 fig
in Zaïre 250–1
passengers, motor vehicles and
road deaths 265
Paulis, Col. 240
Peachtree Street, Atlanta 117, 122
pedestrians, road deaths 266
Pedestrians' Association 270
Peltzman, S. 283
Pennington, E. J. 31
Pennsylvania 272
Pennsylvania Turnpike 72–3
People's Car 155–6, 160
Perkins Company 107, 108
Perry, J. 10
petrol
availability 23, 44–5

cost against diesel fuel 98, 104
  fig, 110, 111, 113
sales, Saskatchewan 178–9
taxation 68–9, 111, 175, 280
Petty, R. 116
Peugeot, A. 6, 9, 14
Peugeot Company 15, 16 *fig*, 135
  fig
Pickwick Stages 83, 85, 86
Pierce-Arrow 85
'piggy-backing' (road and rail
  transport) 76, 78
Pitts, T. H. 123–4
planners and fragmentation 124–5
Plowden, W. 274–5
Poland 202, 204–6
pollution 43, 119–20, 142, 214,
  229–30
Ponts et Chaussêes,
  Administration, des 297, 300
Pope, A. A. 7, 25
'Post Houses' 92
PRAGA 194, 195
Prague 196 *fig*, 197–9
Prague Automobile Factory 194
Pravda 260
Preece, W. H. 37
premiums, loaded 281–2
President's Research Committee on
  Recent Social Trends 165
press and advertising 135–6
'Prévention Routière, La' 137
Prewitt, P. J. 119
prices, motor vehicles in
  Saskatchewan 169
Prince Albert Daily Herald 182
promiscuity, sexual 121
Provincial Secretary's
  Department 167
PSA, advertising 135 *fig*
public opinion, against motor
  vehicles in Germany 62–3
Public Vehicles Act 1928 175
Puch, J. 6
purchasing power in Germany
  154–6

Race Meet, Detroit 27

races *see* endurance tests; motor
  trials
radio (countering effect of cars)
  126
Rae, J. 2
railroads *see* railways
railways
  accidents 266–7
  in France 139–40
  in Germany 152, 153 *figs*, 154,
    156–8
  in Japan 220–2, 223, 226–7, 234
  in Saskatchewan 180
  in USA 67–8, 70–1, 75–6, 81,
    85–6
  in Soviet Central Asia 259–60
  in Zaïre 245, 247, 250
Raper, A. 126
*Rapport de la Commission
  Permanente pour la Protection
  des Indigènes* 239
'rational theory of accidents'
  268–9, 272
rationing in Germany 159–60
Red Flag Act 8
Regina 178
registration
  motor vehicles:
    Czechoslovakia 200–2;
    Japan 215, 217;
    Saskatchewan 166–73
  pedal cycles, Japan 215–17
regulations
  buses 84, 90
  France 141–3,
  road safety 273–8
  Saskatchewan 174–5
  see also legislation
Reichsautobahnen enterprise 153
Reichsbahn *see* railways;
  Germany
Reichskraftwagentarif 157
Renault, M. 23, 295
Renault Company 111, 113, 131,
  132, 135 *figs*, 136
retailing, France 133–6
Rheinische Gasmotorenfabrik 12,
  55, 61–2
Ricardo, H. 98

Rigoulet, L. 15
Riley 7, 36
Road Carrying Company Ltd, 41
road haulage *see* freight
Road Improvement Association 6
road–rail co-ordination (France)
    139–40
    see also co-ordination
Road Research Board 271–2
Road Research Laboratory 275,
    278
road-roller 296
road safety 63–45, 229, 268, 270–8
road-surfacing techniques 295–7
Road Traffic Act
    1930 280–1, 282
    1934 271
roads
    accidents on 267, 270–3
    in Canada 179–83, 184–5, 186
        fig
    in France 23, 291–301
    in Germany 152, 153–4
    in Japan 218–19, 228–30, 231
        fig, 232 *fig*
    in Soviet Central Asia 256–7,
        259–60
    in USA 26, 67–79
    in Zaïre 237, 238–40, 241 *fig*,
        245–6, 247–9
Rochas, A. E. de 292
Roger, E. 14, 18, 61
Rolls, C. S. 37, 39
Rolls and Company 37
Rolls-Royce 46
Rome, Ancient 71
Roosevelt, F. D. 90
Roots, J. D. 17
Rothschild, Baron H. de 21–3
Rover 7, 36
Rowe, D. 281
Royal Automobile Club 36
Royal Commission on Motor
    Cars 47
Royal Exchange Assurance 281
Royal Society for the Prevention of
    Accidents 270
rural life, motor vehicles and 115,
    261

Rural Municipality Act 1909 180
Russian Empire 256
Russian Federation 257

safety coach 84, 85
safety legislation (motor
    vehicles) 282–3
sales
    diesel lorries 106, 107, 108, 109,
        110–11, 112
    in France 16 *fig*, 21–4
    in Germany 16 *fig*, 62–5
Salford 273
Salomons, Sir D. 30, 23, 38, 41
Salvation Army (Nashville) 121
Samuel, M. 45
Sarazin, E. 13
Sarazin, L. 14, 15
Saskatchewan 166–87
Scandinavia 270
    see also Sweden
Schacht, H. 159
school buses 105, 108, 110, 112
scooters, motor 259
Seaborn, W. E. 173
seat-belts 141, 282–3
second-hand cars 134–5, 137–8
Second World War 104–5
Self-Propelled Traffic Association
    (SPTA) 30–1, 40–1
self-regulation (buses) 90–1
Serpollet, L. 9
servicing motor vehicles 176,
    178–9, 260–1
settlement, Saskatchewan 179–80
Shell petrol company 45, 299
shipping in Japan 220–1
showrooms, motor 133–4
Shrapnell-Smith, E. 40, 41, 42
Shrewsbury and Talbot, Earl
    of 37
Silesia 201–2
Simms, F. R. 17, 30, 31, 36
Singer 7, 36
SKODA 194, 195
Slovakia 199–200, 201
smog 71
Société Anonyme pour la
    Construction et l'Entretien des

Routes 299
Société Chimique de la Route 299
Sociétés Chimique et Routière d'
Enterprise Générale 291, 299
Sociétés d'Economie Mixte 295
Société des Transports en Commun
du Congo, Le (TCC) 251
Société Générale Enterprises 299
Société Routière Colas 291, 298,
299–300
Society of Motor Manufacturers
and Traders (SMMT) 270
Southern Labor Review 125
Southwestern Greyhound 89
spare-part trade 135
speed restrictions 8, 29–30, 64,
273–8
spirit–petrol mixture
(Czechoslovakia) 195
Sprague, F. J. 10
Spurrier, H. 41
Standard Oil 45
Stanley Cycle Club Show 9
Stanley White steamers 25
state intervention *see* government
policy
state legislation (buses) 84, 90
stationary gas engines 11–12
Statistical Bureau, German 151–2,
155
steam vehicles 7–9, 39–42, 237, 292
Steinway, W. 17
Stephenson, G. and R. 5
Stiwer 6
stock-car racing 115–16
Stock Market Crash 1929 88
Stone, R. 281
streetcars 10, 124, 125
Sturmey, H. 5–6
submarines (diesel engines in) 97
suburbs, American South 122–4
Sunbeam 36
Sunner, J. 9
superhighways 72–4
see also autobahnen; autostrada
support systems for motor
vehicles 176, 178–9, 260–1
Sweden 202–6, 278
Swift 36

Switzerland 202, 204–6, 275

Tang, A. M. 115
tar on roads 295–6
Tate, A. 121
TATRA 194, 195
taxation
motor vehicles 65, 151–2,
156–7, 195
petrol 68–9, 111, 175, 280
taxicabs 82–3, 138, 141, 259
technological advances
buses 84–5
France and 130–3, 295–7, 298,
300–1
Telford, T. 67
Tennessean (Nashville) 118–19,
120, 126
Thompson, F. M. L. 4
Thornycroft of London 41
Times, The 33
Tokyo 215, 230
toll roads 73–4, 106–7, 271
tourism (USSR) 261
townships 180, 181
traction engines 7–8
tractors
first diesel 99–100
in Zaïre 242 *fig*, 243 *fig*
trade, Germany and 159
Traer, G. 89
Trailer on Flat Car (TOFC) 76
tramcars 10, 124, 125
transport policy *see* government
policy
Travis, W. E. 82–3, 86
Trépardoux, C. –A. 9
Trésauget, road-builder 67
tricycle, Benz's 13, 14, 59–62
Triplex safety glass 282
trucks *see* lorries
Tshund'olela, E. 3
Tunbridge Wells 30
Turkestan 256
Turkey, Sultan of 11
tyres 17–18, 20, 159, 178

Unholy Marriage (Holbrook) 264
United Kingdom *see* Great Britain

*United Nations Statistical Yearbook
for 1955* 1
Unsafe at any Speed (Nader)   283
urban life, American South
116–17
USA (United States of America)
buses   81–93
diesel engines   98–113
early motor vehicles   24–9
insurance   279
motorisation   2–3, 67–79, 148
railways   153 fig, 158 figs
regulation, motor vehicles
273–4, 276–7
road deaths   264, 266, 269
roads   228 fig, 229 fig, 230 fig,
231 fig, 272
safety legislation   283
speed limits   278
usage of motor vehicles   47–8
usage of motor vehicles
Czechoslovakia   195–6, 206–10
Europe   202–6
Japan   218, 219, 224–5
penetration   1–4
Saskatchewan   167, 173–6
USA   29, 47–8, 69, 118, 127
Zaïre   242–4, 249
US Commercial Agent in Japan
215
used cars   134–5
USSR (Union of Soviet Socialist
Republics)   257–61
Uzbekistan   258–9, 260–1

Van Zuylen   19, 21
Vanderbilt, W. K. Jr   27
Velo   13, 62
Vickers, Son and Maxim   35
victims of road and accidents   *see*
compensation

Viktoria   13
Volkswagen   155–6
Rabbit   283
Volvo Company   111, 113

war preparations, Germany
and   159–60
water transport
Germany   158 fig
Zaïre   250
Waukesha   102
Werner, M. and E.   42
West Germany
roads   228 fig, 229 fig, 230 fig,
231 fig
speed limits   275, 278
Western Tire and Rubber
Company   178
wheat prospects and motor
vehicles   169
White Company   84
Wickman, E.   83, 89
Wilby, T.   179
Willys   7
Winton, A.   25, 27
Winton Company   7, 28
wire-wheeled car   58
Wolfe, T.   121
Wolseley Tool and Motor Car
Company   35
Worby Beaumont, W.   37, 40
World Health Organisation   264
Wren, C.   83

year of statehood   274
Yokohama   215
Young, A. B. F.   44

Zaïre   236–52
Zschopauer Motorenwerke   159–60